# WADI
# HYDROLOGY

# WADI HYDROLOGY

## Zekai Şen

Istanbul Technical University

Turkey

**CRC Press**
Taylor & Francis Group
Boca Raton London New York

CRC Press is an imprint of the
Taylor & Francis Group, an **informa** business

CRC Press
Taylor & Francis Group
6000 Broken Sound Parkway NW, Suite 300
Boca Raton, FL 33487-2742

First issued in paperback 2019

ISBN-13: 978-1-4200-6154-3 (hbk)
ISBN-13: 978-0-367-38760-0 (pbk)

### Library of Congress Cataloging-in-Publication Data

Sen, Zekâi
    Wadi hydrology / Zekai Sen.
        p. cm.
    Includes bibliographical references and index.
    ISBN 978-1-4200-6154-3 (alk. paper)
    1. Wadis. 2. Hydrology--Arid regions. 3. Rain-water (Water-supply)--Arid regions. 4. Groundwater recharge--Arid regions. 5. Water resources development--Arid regions. I. Title.

GB561.S46 2008
551.480915'4--dc22                                     2008000783

**Visit the Taylor & Francis Web site at**
**http://www.taylorandfrancis.com**

**and the CRC Press Web site at**
**http://www.crcpress.com**

# Dedication

*Bismillahirrahmanirrahim*

*In the name of Allah the most Merciful
and the most Beneficial*

# Contents

# Preface

*Wadi* is the Arabic word for ephemeral water courses in the arid regions, and they are a vital source of water in most arid and semi-arid countries. Catastrophic flash floods occurring in wadis are, on one hand, a threat to many communities and, on the other, major groundwater recharge sources after storms. Population growth, land use, and human settlements in and around wadis increase social and environmental problems because properly planned exploitation is missing, hydrologic calculations are not always based on the best and latest methods, and water resource management programs in these areas may cause inconvenience. This is partially due to the fact that scientific understanding and knowledge of the hydrological processes of wadis are rather poor in most countries. In response to these concerns, many countries have adopted policies for the sustainable management, development, and efficient utilization of their water resources. However, due to increasing water scarcity, climate change, desertification, and drought effects, many countries have been alerted to recognize the urgent need to secure and utilize existing water resources and supplies in order to sustain a minimum resource base.

For many countries, the development and efficient utilization of the renewable sources of water in wadis is the only optimal solution for addressing water shortage problems. The efficiency of harnessing water from wadis will depend on the understanding and knowledge regarding the qualitative and quantitative hydrology and water resource potential of these wadi systems.

In order to properly plan, design, and operate water resources projects, it is necessary to model and measure basic water variables such as rainfall, runoff, infiltration, floods, evaporation, and other relevant hydrologic variables in the field. The assessment and interpretation of these data occur by establishing efficient surface hydrological, hydrometeorological, hydroclimatological, and water resources management to correctly estimate relevant hydrological parameters. This constitutes very important steps in solving water resources problems in arid and semi-arid regions. These regions reveal different hydrological characteristics from humid regions, but the literature of hydrology is full of textbooks that deal with the hydrological and water resources problems of humid areas. Resources for arid region technology are not collectively available in any textbook. Similarly, the term *wadi,* as accepted by UNESCO for describing the water resources and hydrological aspects in arid and semi-arid regions, is rarely encountered.

Readers will find in this book the basic philosophical, logical, and scientific methodologies needed to assess, evaluate, and manage both surface and groundwater resources in arid and semi-arid regions. The problem of sedimentation is discussed with practical formulations for arid and semi-arid regions. The book does not include detailed mathematical derivations but, instead, logical and empirical derivations of the basic concepts, definitions, formulations, procedures, and methods. It is meant for those whose work involves the interdisciplinary topics of water resources evaluation in arid and semi-arid regions with emphasis on wadi hydrology.

Arid regions are characterized by strong climatic contrasts with dry environments suffering seasonal drought, occasional torrential rains, and the consequent floods. These regions produce a characteristic balance of hill slope and wadi channel processes, which are unique from the hydrological point of view. The most characteristic and important features of these areas are that surface flow is ephemeral. Hence, the relative importance of many fluvial processes differs considerably from more humid regions. Some of these processes are not applicable to arid lands and wadi channels. Consequently, arid and semi-arid lands share many attributes that place them outside the domain of humid region hydrology domain, and therefore, they merit separate study.

The purpose of this book is to provide a comprehensive presentation of up-to-date models for utilization with arid and semi-arid region water resources and their applications. In addition to an overview of fundamentals in the field, readers will find a useful summary of current developments as reported in the world's leading hydrology journals. However, a significant part of the book consists of original techniques developed and presented in open literature by the author. Additionally, many unique physical approaches, field cases, and sample interpretations will be presented prior to the application of different models.

I could not complete this work without the love, patience, support, and assistance of my wife Fatma Şen. I also extend my appreciation to General Director Dr. Mohammad Esad Tawfiq of the Saudi Geological Survey (SGS), who provided every opportunity for scientific achievement within the SGS and also introduced me to the different aspects of arid region water resources. I also thank all my colleagues at the hydrogeology department of SGS, and I am also indebted to The Technical University of Istanbul for giving me the opportunity to work in such a variety of arid regions.

# The Author

**Zekai Şen, Ph.D.,** earned his B.Sc. and M.Sc. degrees in civil engineering from the Technical University of Istanbul in 1971 and continued his postgraduate studies at the Imperial College of Science and Technology, University of London. He was granted a diploma from the Imperial College (DIC) in 1972, an M.Sc. in engineering hydrology in 1972, and a Ph.D. in hydrology in 1974. He has worked in England, Norway, Saudi Arabia, and Turkey, and held positions in earth sciences, hydrogeology, astronautics and aeronautics, meteorology, civil engineering, and water resources on various university faculties. Dr. Şen's main interests are renewable energy and especially solar energy, hydrology, water resources, hydrogeology, hydrometeorology, hydraulics, the philosophy of science, and science history. He worked as a chief consultant at the Saudi Geological Survey for almost 5 years. He has published more than 250 papers in over 50 international scientific journals in addition to contributing to many international conferences, symposiums, and numerous technical reports. He has edited proceedings and books, and is the author of *Applied Hydrogeology for Scientists and Engineers* (CRC–Lewis Publishers, 1995). Under his supervision, about 20 students from different countries (Turkey, Saudi Arabia, Yemen, Jordan, Libya, and Pakistan) have obtained Ph.D. degrees in the energy and water sciences fields. Dr. Şen is a member of the American Hydrology Institution and has been recently appointed a member of the United Nations Intergovernmental Panel on Climate Change (IPCC).

# 1 Introduction

## 1.1 GENERAL OVERVIEW

One third of the world's land surface is classified as arid or semi-arid. In most of the world easily developed land is already exploited, and therefore increasing attention is turning to arid areas to relieve population pressures and to provide more food. Soil and water natural resources of arid and semi-arid regions are often in a delicate environmental balance. Arid regions cannot be developed along lines that have been successful in humid areas. These regions are under severe and increasing water stress due to expanding populations, increasing per capita water use, and limited water resources. The world's total population has increased four times in the past 150 years and may double again in the next 30 years. Future projections indicate that by 2025, five billion people will live in countries experiencing moderate or severe water stress (WHO, 1997; Arnell, 1999). Evidently, conditions will be most severe for the driest regions of the world. Effective management is essential, and this requires appropriate understanding of the hydrological processes in arid and semi-arid areas (Wheather and Al-Weshah, 2002).

Climatologic, morphologic, and geologic conditions that affect the hydrological phenomena are different from each other. For instance, in humid and tropical regions abundance of rainfall and consequent runoff and snowmelt provides time and spatial distributions that are more convenient than arid or semi-arid regions, where they are rather haphazard and sporadic. With more pressure on water resources, the natural occurrence, distribution, and movement of water become more sensitive and, accordingly, hydrologic principles should be followed more carefully for better planning, operation, maintenance, and management of water resources, especially in arid and semi-arid regions.

As stated by Pilgrim et al. (1988) and according to UNESCO (1977) classification, nearly half of the countries in the world face aridity problems. There is an obvious need for an improved understanding of arid and semi-arid region hydrology, and for the development of appropriate techniques in hydrologic modeling (Chapter 4). Although some aspects of arid zone hydrology are more amenable to simplified modeling, it is highly probable that greater errors and uncertainty will continue to characterize results for arid zones. Recognition of these problems is fundamental to realistic approaches in arid zone modeling, rational interpretation, and applications.

The majority of countries in arid and semi-arid regions depend either on groundwater or on desalinization for their water supply, both of which enable them to use water in amounts far exceeding the estimated renewable fresh water in the country (IPCC, 1997). Wetlands can retain water, especially during dry (non-rainy) periods, and thus can enhance recharge to major aquifers. However, in arid and semi-arid areas, where groundwater recharge occurs after flood events, changes in the frequency and magnitude of rainfall events will alter the number of recharge events.

Recently, in order to distinguish between humid and arid region methodologies, the term *wadi hydrology* was coined for arid region water resources exploitation and management. It is the sole purpose of this book to present arid region assessment methodologies from different aspects of water resources. More weight is given to the subjects, which fall within the wadi hydrology discipline with innovative and distinctive concepts.

## 1.2 ARID REGION CLASSIFICATION

Many human activities are dependent on climate and, therefore, it is possible to classify arid and humid regions according to the context of interest. Meteorologically, they are classified on the basis of temperature and precipitation as in Figure 1.1.

Almost one third of the world lies in the arid zone and, contrary to common opinion, it can be very warm or cold. Arid regions are defined as having a dry climate with little or no rainfall, very low humidity, and a high annual evaporation rate that exceeds the annual rainfall, with a high deficit in soil humidity that does not give way to even dry farming practices. On the other hand, the scarce and sporadic rainfall is turbulent, brief, and torrential. The nature of this rainfall causes an increase in runoff, erosion, and flooding, which furnishes conditions that must be taken into consideration by water resources planners and operators in selecting a site for human activities or configuration for a settlement center.

Geographically, most arid and semi-arid regions occur between 10° and 35° latitude (e.g., Sahara desert, Kalahari desert), in the interior parts of continents (Australia, Gobi desert) and in rain shadow areas in fold belts (Peru, Nepal). Large parts of the arctic tundra receive less than 250 mm precipitation per annum and qualify as "semi-arid regions," too. Figure 1.2 presents a sketch map of the arid and semi-arid areas of the world based on temperature distribution.

Another classification can be based on whether the area lies within a *desert* or greenery region. In arid regions, the geological outcrops are very distinctive on the surface of the earth. Figure 1.3 shows the Arabian Peninsula arid region features.

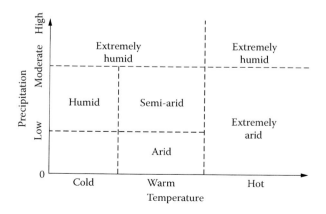

**FIGURE 1.1**  Aridity and humidity.

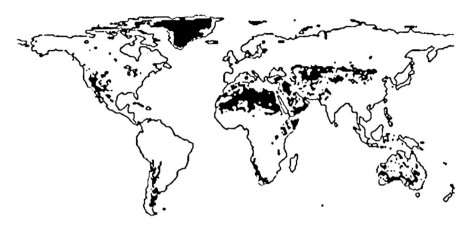

**FIGURE 1.2** Aridity distribution.

In practice, each person or different careers have different connotations for aridity and humidity. Arid and semi-arid regions are also distinguished on the basis of their annual rainfall sums and the following categories can be identified.

1. Deserts where the annual total rainfall sum is less than 50 mm/year and devoid of vegetation. For instance, two-thirds of the Middle East region can be classified as desert. In the northern part of the region, a steppe climate prevails, with cold winters and hot summers. A narrow zone contiguous to the Mediterranean Sea is classified as a Mediterranean zone with wet and moderately warm winters and dry summers.
2. Arid regions where the annual total rainfall sum is between 50 mm/year and 250 mm/year with sparse vegetation.
3. Semi-arid regions where the annual total rainfall sum varies between 250 mm/year to 500 mm/year coupled with steppe vegetation.

Deserts are primarily hot because of their lack of water. When the sun shines, all of the absorbed sunlight goes into raising the ground temperature. If there is moisture in the soil, much of the heat would go into evaporation of the water, transforming water to vapor in the air, and keeping the soil cooler than it would otherwise be. This cooling is from the "latent" heat of evaporation that is required to change liquid water into water vapor. The vegetation itself does not cool the desert, but the water is lost by the vegetation. Deserts are cold at night because of water lacking in the ground, and there is little water vapor in the air.

## 1.3 DESERTIFICATION

Desertification is a slowly creeping phenomenon, which for various reasons takes place in any area over a long period. In general, *desertification* implies decrease in some significant meteorological and agricultural quantities such as rainfall, vegetation coverage, surface water extensions, groundwater level drops, and crop yields.

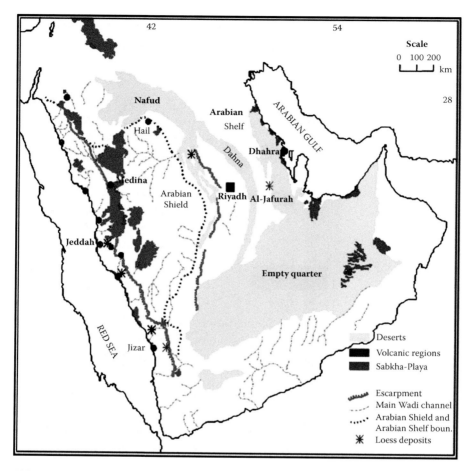

**FIGURE 1.3**    Arabian Peninsula features (Courtesy of the Saudi Geological Survey).

On the other hand, increases in climate temperature, sand coverage, drought coverage, urban area expansion, and sedimentation amounts also imply desertification.

A common misapprehension about desertification is that it spreads from an arid region such as desert core, like a ripple on a pond. Land degradation can and does occur far from any climatic arid region. The presence or absence of a nearby arid region has no direct relation to desertification, which begins usually as a spot on the land, where land abuse has become excessive. From that spot, which might be around a watering point or in a cultivated field, land degradation spreads outward if the abuse continues. Ultimately, the spots may merge into a homogeneous area, but that is unusual on a large scale.

A second misconception is that droughts are responsible for desertification. Droughts cause increase in the likelihood that the rate of degradation will increase on nonirrigated land if the carrying capacity is exceeded. However, well-managed land will recover from droughts with minimal adverse effects when the rainfalls return. The deadly combination is land abuse during good periods and its continuation during periods of deficient rainfall.

Global warming, greenhouse effect, climate change, environmental effects, water resources depletion, and increased demand for more agricultural products put extra pressure on many regions in the world, and it is necessary to measure the scale of desertification in such regions. This is required to plan development in the region in such a way as to preserve at least the present situation and, even better, to improve local conditions with the purpose of combating droughts and desertification.

A complete study of desertification in an area should follow more or less the sequence of meteorology, hydrology, geology, hydrogeology, soil cover, environmental issues, satellite images, geographic information systems (GIS), regional modeling, and simple and readily applicable advice, recommendations, and rules. The following actions should be taken in the initial phases of confirming the initiation of or existence of desertification in any part of the world.

1. The temperature records (monthly, seasonal, or annual) should be processed for any significant increasing trend in the area.
2. The rainfall records should be examined for any obvious or hidden decreasing trends and cyclic phases.
3. The wind direction and speed variations in the study area must be evaluated again with the possibility of increasing or decreasing trends.
4. The dry (nonrainy) and wet (rainy) period durations can be identified from the monthly rainfall records with the possible identification of maximum drought duration (see Chapter 3). It is necessary to make a distinction in an area between extensive drought durations and desertification initiation.
5. Regional (maps) and directional (change with distance) in rainfall, temperature, soil moisture, sedimentation, etc., variables should be sought. Especially, variations from coastal areas toward inland regions, or from the high elevations in different directions are important.
6. Identification of the effects of urbanization on water supply and groundwater level depletions, coupled with the desertification effects.
7. Consideration of global warming, the greenhouse effect, and climate change effects on the region.
8. Crop yield amounts from season (year) to season (year), vegetation coverage, changes in agricultural product types.
9. Sand dune area boundaries (satellite), depths at a set of irregular points (ground) should be measured and mapped.
10. Correlations between the aforementioned variables should be established for the general determination of desertification initiation.

Only after the completion of these points and toward the final stages of the study should the satellite imagery and GIS techniques with pattern recognition be used for the final assessment and delimitation of the desertification effects with the index development and classification.

## 1.4 DROUGHTS

Droughts are manifestations of climatic fluctuations associated with large scale anomalies in the planetary circulation of the atmosphere. They imply rainfall absence or weak rainfall occurrence during a long time over large areas. It is very difficult to identify and clearly determine the onset as well as termination times. It is a creeping phenomenon and its effects accumulate slowly and tend to persist over longer periods of time. The local and regional climate features are important in *drought* generation. Drought depends on the persistence of dryness over months or years. Drought is an extreme event of a time and area process. In driest zones, the variability of rainfall is the highest.

Drought is more than a simple lack of rainfall, which implies *meteorological drought*. It is a persistent moisture deficiency below long-term average conditions that, on average, balance precipitation and evapotranspiration in a given area. Not all droughts are created equally. Similar moisture deficits may have very different consequences, depending on the time of year at which they occur, preexisting soil moisture content, and other climatic factors such as temperature, wind, and relative humidity. Drought can be defined in terms that go beyond the meteorologist's rainfall measurements (Şen, 1980; Şen and Boken, 2005).

*Hydrologic drought* occurs when surface water supplies steadily diminish during a dry (nonrainy) spell. If nonrainy conditions continue, groundwater levels could begin to drop. Various drought characteristics such as duration, magnitude, and intensity are related to each other (Sirdas and Şen, 2003).

*Agricultural drought* occurs when a moisture shortage lasts long and hits hard enough to negatively impact cultivated crops. Soil conditions, groundwater levels, and specific characteristics of plants also come into play in this functional definition of drought. *Ecologic drought* is detrimental to native plants that do not have the benefit of irrigation.

Economic consequences of droughts are more important for humid regions because of the unpreparedness of people for recurrent drought events and the large investments in agriculture that may be impacted with large losses from droughts. *Social drought* is effective when people start to feel that there is not enough water. When drought comes, everybody is concerned; if it lasts, everybody tries to do his/her best to deal with it, but when it passes away, everybody forgets its impact except those who have been significantly hurt (Yevjevich, 1967).

One of the dramatic long-term impacts of droughts, combined with the alteration of human activities, is the degeneration of productive ecosystems into desert in the process called *desertification*, as explained in the previous section. Desertification is not exclusively a consequence of drought, but it may be accelerated by droughts.

Droughts have socio-political, economic, and environmental impacts that are interrelated. Their effects are intricately related to the environmental economics and social fabric of a given region or an entire nation. In order to better understand the drought concept, the following four different events must be considered.

1. *Aridity*: It is a permanent natural condition and a stable climatic feature of a region.
2. *Drought*: It refers to a temporary feature of the climate or to regular but unpredictable climatic changes.
3. *Water shortage*: It is understood mostly as a man-made phenomenon reflecting the concern with temporary and small area water deficiencies.
4. *Desertification*: It is a part of an alteration process in the ecological regime often associated with aridity and/or drought but principally brought about by man-made activities which change the surrounding environment to a significant degree.

## 1.5   CLIMATE CHANGE IMPACTS

Water resources assessments in the arid regions require special techniques due to data scarcity or infrequent measurements and occurrences of rainfall events. Appropriate technical guidance is required for the practical scientific assessments of, especially, the rainfall-runoff process and flood prediction design (see Chapter 4) and, if possible, early warning and groundwater recharge possibilities in order to augment the storage capacities (see Chapter 5). The distinctive hydro-climatic features in arid regions include high levels of incident radiation, generally high diurnal and seasonal temperature variations, low humidity over a short distance inland from the sea, strong winds with frequent dust and sand storms, sporadic rainfall of high temporal and spatial variability, even greater variability of short-duration runoff events in ephemeral drainage systems, generally high infiltration rates in channel alluvium, high sediment transport rates, and relatively large groundwater and soil moisture storage changes (FAO, 1981).

In general, *climate change* is expected to lead to more rainfall coupled with more evaporation, but the important question is: how much of this rainfall will end up in water deficit areas? If it is not adequate, then regional management of water engineering infrastructures comes into view with sustainable water distribution programs. On the other hand, probable rainfall increase in some areas (decline in others) is another indication for regional water resources distribution to needy areas through efficient management programs (see Chapter 6). The main solution for reducing the local and regional vulnerability to climate change requires improved water resources systems management prior to any capacity increase with additional hydraulic structure design and construction. In this manner, the existing supplies will be used efficiently. Long-term management studies will also indicate the necessity of engineering structures, if any for, the region. Efficient management strategies should include rules and regulations for directly controlling land and water use, incentives and taxes for indirectly affecting behavior, the construction of new dams and pipelines to boost supplies, and improvements in water management operations and institutions. Other adaptation measures can include removing levees to maintain *flood plains*, protecting waterside vegetation, restoring river channels to their natural form, and reducing water pollution. Vast and complex infrastructure of *dams*, (especially *subsurface dams*) and pipes can be planned and built to provide justified fresh water resources distribution based on an effective management program.

Climate change is among the triggering agents of unusual *floods.* It causes changes in timing, regional patterns and intensity of precipitation events, and in particular in the number of days with heavy and intense precipitation occurrences. Floods are now being experienced in areas where there were no floods in the past. The recent floods seem to have some effects of global climate change although they cannot be taken as proof that it is already taking place in other parts also.

On the other hand, the potential for increased flooding following climate change would be exacerbated by erosion associated with deforestation and overgrazing both of which are now very widespread in many parts of the Middle East. Such environmental degradations also increase surface runoff and the severity of flooding and contribute to landslides. Hence, in order to effectively assess the future flood occurrence possibilities and loss consequences, erosion areas and rates should also be taken into consideration. Erosion rates are increasing steadily in many parts of the world and are more concentrated in arid regions (Chapter 7).

The vulnerability and sensitivity of water systems and management rules, the strengths and weaknesses of technologies and policies might help to cope with adverse impacts and take advantage of possible beneficial effects. Certain aspects of water resources and infrastructure are very sensitive both to climate and how to manage complex water systems. It is, therefore, necessary to have mediators in the form of engineering structures in order to offset or diminish the sensitivity to various expected and unexpected changes in the future. Changes in management of the engineering systems require understanding what would be most effective factors. Water managers and policymakers must start considering climate change as a factor in all decisions about water investments and the operation of existing facilities and systems.

Records of past climate and hydrological conditions are no longer considered to be reliable guides for the future water resources system design, operation and management. The design and management of both structural and non-structural water-resource systems should allow for the possible effects of climate change, but little professional guidance is available in this area. Further research by hydrologists, water planners, and water managers is needed along the following points.

1. More work is needed to improve the ability of global climate models to provide information on water-resources availability, to evaluate overall hydrologic impacts, and to identify regional impacts.
2. Substantial improvements in methods to downscale climate information are needed to improve our understanding of regional and small-scale processes that affect water resources and water systems.
3. Information about how storm frequency and intensity has changed and will change is vitally important for determining impacts on water and water systems, yet such information is not reliably available. More research on how the severity of storms and other extreme hydrologic events might change is necessary.
4. Increased and widespread hydrologic monitoring systems are needed. The current trend in the reduction of monitoring networks is disturbing.
5. There should be a systematic reexamination of engineering design criteria and operating rules of existing surface and subsurface dams and reservoirs under conditions of climate change.

6. Information on economic sectors most susceptible to climate change is extremely weak, as is information on the socio-economic costs of both impacts and responses in the water sector.
7. More work is needed to evaluate the relative costs and benefits of non-structural management options, such as demand management and water-use efficiency, or prohibition on new flood plain development, in the context of a changing climate.
8. Research is needed on the implications of climate change for international water law and international trade in water. Can "privatization" affect vulnerability of water systems to climate change?
9. Little information is available on how climate changes might affect groundwater aquifers, including quality, recharge rates, and flow dynamics. New studies on these issues are needed.
10. The legal allocation of water rights should be reviewed, even in the absence of climate change, to address inequities, environmental justice concerns, and inefficient use of water. The risks of climate change make such a review even more urgent.

Humans are influencing the global climate and, thereby, altering the hydrological cycle, however inadvertently. Greenhouse warming will have the following effects on water supplies.

1. The timing and regional patterns of precipitation will change, and in many areas more intense precipitation days are likely.
2. General circulation models (GCMs) used to predict climate change and suggest that a 1.5 to 4.5°C rise in global mean temperature would increase global mean precipitation about 3 to 15 percent toward the end of this century.
3. Although the regional distribution is uncertain, precipitation is expected to increase in higher latitudes, particularly in winter. This conclusion extends to the mid-latitudes in most GCM results.
4. Potential evapotranspiration rises with air temperature. Consequently, even in areas with increased precipitation, higher evapotranspiration rates may lead to reduced runoff, implying a possible reduction in renewable water supplies.
5. More annual runoff caused by increased precipitation is likely in the high latitudes. In contrast, some lower latitude basins may experience large reductions in runoff and increased water shortages as a result of a combination of increased evaporation and decreased precipitation.
6. Flood frequencies are likely to increase in many areas, although the amount of increase for any given climate scenario is uncertain and impacts will vary among basins. Floods may become less frequent in some areas.
7. The frequency and severity of droughts could increase in some areas as a result of a decrease in total rainfall, more frequent dry spells, and higher evapotranspiration.
8. The hydrology of arid and semi-arid areas is particularly sensitive to climate variations. Relatively small changes in temperature and precipitation

in some areas could result in large percentage changes in runoff, increasing the likelihood and severity of droughts and floods.

9. Seasonal disruptions might occur in the water supplies of mountainous areas if more precipitation falls as rain than snow and if the length of the snow storage season is reduced.

10. Water quality problems may increase where there is less flow to dilute contaminants introduced from natural and human sources.

## 1.6  WADI HYDROLOGY

The pieces of land that have the known hydrological and geomorphologic properties are referred to as catchments in humid regions, but as wadis in arid and semi-arid regions (see Chapter 2). The major difference appears in the rainfall regime (Chapter 3) and its consequent surface flow pattern (Chapter 4).

Wadis have emerged in arid region hydrology context as distinct scientific areas within the past decade. The *wadi hydrology* is very different from that of humid regions and raises its important scientific, technical and linguistic challenges and hence a special methodological basis is essential to meet current and future needs of water management. In arid and semi-arid environments, rainfall is short-lived and often very intense. As a result of thin soils much of the rainfall runs directly off the surface and infiltration occurs in deep soils down-slope or along wadi courses. The *alluvial fills* (Chapter 2) in arid regions at the depressions generally along faults are known as "wadis," which are natural watercourses with dry conditions most of the year. In some seasons, they become conveyors of runoff, flood or flash flood that carry away large amounts of sediment, leaving a marked imprint on the landscape. Permanently flowing water often exists in the gravels below the surface of a large wadi subsurface terrain. A wadi is a stream that runs full for only a short time (intermittent), mostly during and after a rainstorm. Not every rainstorm, however, necessarily produces surface runoff. It is seldom that wadi flow at a certain section can be described as perennial. If so, the flow is then extremely variable from one year to another. It is important that wadi flow should be viewed as a precious surface water resource in arid regions. Efficient use of wadi flow requires certain conservation measures to be undertaken in order to reduce the evaporation losses.

By definition, wadis differ from humid watersheds in their relative amount of rainfall as compared to the evaporative flux. Following the classification of Potter (1992), arid and semi-arid regions are those subject to precipitation, P, to evaporation, E, ratio (P/E). If this ratio is smaller than 0.5 then aridity prevails, and if it is between 0.5 and 1.0, then the semi-arid conditions appear. The humid regions have this ratio greater than 1.0. In addition to a lower net precipitation rate, rainfall events occur as infrequent, short duration, high intensity storms that bring a major portion of the annual rainfall to the surface during a very short period of time. *Flash flood* events may be a direct result of this type of storm over an arid or semi-arid watershed. Under these conditions, the surface layer is unable to respond to the incident rainfall for infiltration. Hence, excess rainfall results in surface flows that propagate rapidly through the wadi watershed. Even for low intensity rainfall events, the surface crust that develops on arid watersheds can lead to significant surface runoff (Abu-Awwad and Shatanawi, 1997). Once on the surface, water in an arid region is subject to a high evaporative demand from the low humidity and high temperature environment.

In arid regions, such as the Arabian Peninsula or the African deserts, the rainfall amounts throughout any year are not sufficient to have ephemeral streams but rather intermitted surface flows. This is one of the main distinctive features that make a catchment area called a *wadi* in arid regions. It is possible to present some of the major properties of a wadi as follows.

1. Over the whole wadi area the rainfall is not sufficient to cause *ephemeral runoff* occurrences. Instead, occasional rainfalls lead to temporary runoffs that are mostly evaporated due to high temperatures and the remaining infiltrates into subsurface recharging the groundwater reservoirs, i.e., aquifers (Chapter 5).
2. Surface of the wadi is almost bare without vegetation and the geologic outcrops are observable everywhere (Chapter 2).
3. Occasional intensive rainfall occurrences give rise to flash floods especially in the upper or medium wadi portions leading to property and life casualties, (Chapter 4).
4. The main channel is filled with recent Quaternary alluvium due to erosion from the surrounding rocks and consequent sedimentation (Chapter 7).

Episodic heavy downpours in low-relief areas are often followed by overland flash floods and debris flows that follow existing depressions in the landscape. Such arid-region fluvial valleys are also called "wadis." Many wadis that are now found in desert regions are formed during a more humid climatic episode between 13,000 and 8,000 years before present time, at the transition from the Last Glacial to the Early Holocene in the historical geologic scale. Wadis in desert regions carry water only after torrential rainstorms that normally occur once in a few years. At the onset of the rains, water can still infiltrate into the soil. As the downpours continue, the supply of water soon exceeds the infiltration capacity of the soil and excess water is discharged as surface runoff. Consequently, a "flash flood" is set in motion. Slaking and caking of the soil surface enhance surface run-off toward the wadis that become torrential braided streams with high sediment loads. These streams have only one channel, but multiple tributaries. After the downpour, the river will completely dry up again until the next event. Many wadis connect with dry and salty basins where individual floodplains merge into extensive *playas* (*sabkhas* in Chapter 2). The following geomorphologic processes in wadis of arid regions differ from the drainage basins in humid environments.

1. Streams are intermittent or ephemeral (and have very irregular discharges).
2. Mass-wasting processes and unconfined sheet floods are prominent.
3. Many rivers do not reach into the sea but end in inland depressions without outlets.
4. Salt lakes are a common landscape feature.
5. Aeolian processes play an important role, particularly in areas below the 150 mm/year isohyets.

6. Physical weathering processes are prominent whereas hydrolysis of minerals is subdued.

Wadi channels in arid regions are parts of a hydrologic cycle, where surface water transport occurs after the evaporation and infiltration losses from the upstream to downstream parts. Connection of different wadi branches toward the downstream with surface water accumulation may cause flood inundation at downstream lower areas. If the region is drained by a single wadi or wadi system, it is then called as a drainage basin or watershed. The surface area of the drainage basin collects the rainfall water and carries it to the low-lying points within the drainage basin, referred to as a *stream* or *river* in humid regions but as *wadi* in arid regions.

Wadis are the basic transportation system involved with water and rock cycles, leading to erosion, sedimentation, and deposition (see Chapter 7). These two cycles provide useful services for humanity, but the water cycle can cause occasional disasters due to floods and flash floods. However, the rock cycle is the basic agent in the formation of groundwater storage (aquifer) within the wadi courses (Chapter 5). The alluvium deposit surface bounded by the two banks is the transportation medium of surface water in normal circumstances, but in the case of extreme runoff occurrences, the transportation area lies between the high elevations on both sides of the wadi cross-section (hilly or mountainous sides).

The UNESCO Cairo office has initiated a regional program named Wadi Hydrology with the following development objectives (Al-Weshah, 2002).

1. To improve the understanding and knowledge of the hydrological processes in arid and semi-arid zones with emphasis on wadi hydrology.
2. To develop the concept of integrated and sustainable development and management of wadi systems, and to improve methodologies to cope with water scarcity in dry regions.

Wadi hydrology has emerged as a distinct scientific area within the past decade, mainly due to the initiative of a small number of individuals within and outside the Arab region, and the active support of UNESCO, assisted by Arab Centre for the Studies of Arid Zones and Dry Lands (ACSAD) and Arab League Educational, Cultural and Scientific Organization (ALECSO). Hydrology of arid and semi-arid areas is very different from that of humid areas and raises important scientific, technical and logistical challenges, and that an improved scientific base is essential to meet current and future needs of water management (Wheather and Al-Weshah, 2002). The following main results are expected from any study concerning wadi hydrology.

1. Improvement and consolidation of the knowledge on the physical processes of wadi hydrology through research and development.
2. Enhancement of the water resources development and management capabilities with respect to wadi systems.
3. Strengthening of human resources and institutions in arid and semi-arid regions through training and capacity building activities at various levels.

Although *wadi* is the name given to a seasonal water course in the Arab region, increasingly this has become a term recognized internationally and used in most hydrological publications all over the world in research on arid and semi-arid regions. In spite of its great role as a vital source of water supply, a natural drainage system, as well as a source of threat of catastrophic floods in many arid-region countries, the scientific understanding and knowledge-base of the wadi's hydrological processes are rather poorly understood in most of these countries.

Achieving an adequate scientific understanding of wadi systems in arid and semi-arid regions is a challenging task. For this purpose, it is necessary to consider the coordination of meteorological, hydrologic, hydrogeological, and hydrochemical data collection and interpretation activities often conducted by many government agencies.

## 1.7  METHODOLOGICAL DISTINCTIONS

Although the same hydrological laws apply in arid (as well as semi-arid) and humid regions, the physical characteristics of the two are often so different that caution is needed in employing commonly accepted methods for hydrological problem solving. The distinctive characteristic of hydrological processes in arid and semi-arid regions is variability in time and space. For instance, in arid regions, long-term cycles or secular swings in climate are important. Likewise, the marked variation, in semi-arid climates, among the seasons of the year may require segregation of basic data by seasons. Hydrological factors common in one season may not exist for the next season. In arid and semi-arid regions a very significant percentage (say, 90%) of the runoff may originate in very small percentage (say, 10%) of its drainage area. Vegetation cover may vary radically across the wadi.

The main limitations in the development of arid region hydrologic assessments are due to the paucity of both data and high quality measurements. Data collection is sparse for many countries in any arid region for a range of social and physical reasons. This lack of data will have to be addressed in the near future for better assessments, interpretations, and conclusions. Most often, field observations are taken as the sole basic data for water resources evaluation. Additionally, the literature on arid region hydrology and water resources development is not only scarce but insufficient. One of the reasons for data paucity is the recent nature—principally within three to four decades—of interest in arid region settlements. Here, the usual economic patterns of urbanization do not exist; the population is sporadically distributed, economic resources are limited, and extremely dry and harsh environmental circumstances prevail. The recent prosperity of these arid regions originated from the rich oil fields or closeness to such areas. This led to accumulation of refined, reliable, and good quality basic data such as rainfall records, occasional runoff measurements, infiltration tests, groundwater level records, aquifer test results, etc.

Due to the aforementioned difficulties, the hydrology of arid regions has not received as much attention as other climatic regions in the past. Almost all hydrological methods are developed for humid regions, where the human activities are the most concentrated for social, economical, and infrastructural developments. Research is either occasional or missing in the water-stressed regions of the world, where the importance of sustainable water resource management is of great interest.

Only recently hydrological studies have been attempted to understand the variability in fluxes and processes that occur within arid and semi-arid wadis. Such studies require extensive field campaigns at the reconnaissance and later stages in an adaptive manner for problem solving. Recently, among the arid region hydrological research interests are field trips and experiments, empirical and rational approaches, theoretical developments, and numerical computer modeling. A recent review by Scanlon et al. (1997) highlights some of the major issues in flow and transport within arid surface and unsaturated subsurface systems. A common trend in each new discovery has been the realization that hydrologic processes in arid or semi-arid regions are distinctly different from their humid climate counterparts. Although much remains to be done, arid region features and water resource developments appear in the literature rather meagerly, and specific approaches are needed more often than before for understanding the nature of dominant hydrological features.

There is a presumption that arid region streams are essentially much the same as humid region rivers, but this is not correct. It is now clear that dryland hydrology cannot be predicted by simple extrapolation of humid region hydrology (McMahon, 1979; McMahon et al., 1987). In arid regions variations in flood magnitudes are larger than for humid regions, and the discharge of the largest flood is very much greater than the mean flood discharge. Another fundamental difference is that significant amounts of water are lost in the downstream locations in arid and semi-arid regions. These are mainly due to evaporation and infiltration losses. In some locations such transmission losses can be up to 100% of the original discharge, which means there is no surface flow in the wadi. In general, in humid regions, discharge increases in the downstream direction, whereas in arid regions just the opposite is valid. In humid regions, groundwater contributes to rivers during dry periods, but in arid regions the surface flow is very much reduced or absent because the groundwater table is usually already depressed.

## REFERENCES

Abu-Awwad, A. M. and Shatanawi, M. R., 1997. Water harvesting and infiltration in arid areas affected by surface crust: examples from Jordan. *J. Arid Environ.*, 37: 443–452.

Al-Weshah, R., 2002. The role of UNESCO in sustainable water resources management in the Arab World. *Desalination*, 152, 1–13.

Arnell, N. W., 1999. Climate change and global water resources. *Global Environ. Change*, 9: S31–S49.

FAO, 1981. Arid zone hydrology for agricultural development. FAO, Rome, *Irrigation Drainage*, Paper 37, 271 pp.

IPCC, 1997. An introduction to simple climate models used in the IPCC Second Assessment Report. In Houghton JT, Filho LGM, Griggs DJ, Maskell K, IPCC Working Group I. Available online at http://www.ippc.ch.

McMahon, T. A., 1979. Hydrological characteristics of arid zones. In *Hydrology: Areas of Low Precipitation*. IAHS Press, International Association of Hydrological Sciences Publication 128, Wallingford, England, 105–123.

McMahon, T. A., Finlayson, B. L., and Srikanthan, R., 1987. Runoff variability: a global perspective. In: Solomon, S. I., Beran, M., and Hogg, W. (Eds.) *The Influence of Climate Change and Climate Variability on the Hydrologic Regime and Water Resources.* IAHS Press, International Association of Hydrological Sciences Publication 168, Wallingford, England, 3–12.

Pilgrim, D. H., Chapman, T. G., and Doran, D. G., 1988. Problems of rainfall-runoff modeling in arid and semi-arid regions. *Hydrol. Sci. J.*, Vol. 33, No. 4: 379–400.

Potter, L. D., 1992. Desert characteristics as related to waste disposal. In *Deserts as Dumps? The Disposal of Hazardous Materials in Arid Ecosystems.* Reith, C. C. and Thomson, B. M. (Eds.) Univ. of N.M. Press., Albuquerque, N.M., 21–56.

Scanlon, B. R., Tyler, S. W., and Wierenga, P.J.,1997. Hydrologic issues in arid, unsaturated systems and implications for contaminant transport. *Rev. Geophys.*, Vol. 35, No. 4: 461–490.

Sirdas, S. and Şen, Z., 2003. Spatio-temporal drought analysis in the Trakya region, Turkey. *Hydrol. Sci. J.*, Vol. 48, No. 5, 809–820.

Şen, Z., 1980. Statistical analysis of hydrologic critical droughts. *J. Hydraul. Eng. ASCE*, 106 (HY1), 99–15.

Şen, Z. and Boken, V. K., 2005. *Techniques To Predict Agricultural Droughts.* Oxford University, 40–54.

UNESCO, 1977. World Distribution of Arid Regions. Map Scale: 1/25,000,000, UNESCO, Paris.

Weather, H. S. and Al-Weshah, R., 2002. Introduction. Hydrology of wadi systems IHP regional network on wadi hydrology in the Arab region. In cooperation with the Arab League Educational, Cultural and Scientific Organization (ALECSO) and the Arab Centre for Studies of Arid Zones and Dry Lands (ACSAD).

World Health Organization, WHO, 1977. *Comprehensive Assessment of the Freshwater Resources of the World.* WHO, Geneva, 34 pp.

Yevjevich, V., 1967. An Objective Approach to Definitions and Investigations of Continental Hydrologic Drought, Hydrology Paper No. 23, Colorado State University, Fort Collins, CO, 43 pp.

# 2 Wadi Characteristics

## 2.1 GENERAL OVERVIEW

Arid region water resources are dependent on surface features, which play a most significant role in the formation and occurrence of rainfall, and also characterize surface water types (runoff, flood, flash flood) and the formation of groundwater reservoirs, which are replenished through infiltration. It is assumed in this chapter that surface vegetation or forest cover is almost nonexistent, and only geological features in terms of rock outcrops and geomorphologic characteristics are important for water resources occurrence, distribution, and movement in any wadi system.

The geomorphology of arid zones is characterized by marked breaks in topographic slope where mountains, hills, or major rock outcrops with little or no true soil cover give way to typically flat and deltaic alluvial land forms. Up-gradient of this break as the dominant geomorphologic process is erosion by runoff from denuded slopes (see Chapter 7). Basins are frequently *endoreic* (without surface outlet), and there is active channel capture by younger water courses (FAO, 1981). Drainage lines, which may start as simply eroded and boulder-filled gullies high up on immature slopes, become successively gravelly and then sandy as bed slopes flatten; valley bottoms tend to have more or less deeply incised drainage and *quaternary alluvial* terraces are common. Reduction of the surface flow due to infiltration and lateral dispersion in wider channels cause abrupt deposition of bed load and much suspended load, with the resultant formation of gravel/boulder areas. Fluvial channel processes are especially apparent in the transition zone at the emergence of well-defined drainage channels from hard-rock mountainous regions.

This chapter presents different surface features in arid regions that play a significant role in the formation of runoff and groundwater storage. Some of the geomorphologic features take part in the estimation of surface runoff discharge that results after each storm rainfall.

## 2.2 GEOMORPHOLOGY

The surface features of a drainage basin determine the characteristics of runoff, flooding, and groundwater, and to some extent rainfall occurrences in arid regions. The runoff calculations in arid and semi-arid regions require empirical relationships between the water amount and different geomorphologic features such as the drainage area, main channel slope, its length, drainage density, etc. These can be calculated in the office, provided that either a detailed topographic map is available or, better, *digital elevation model* (DEM) values stored in computers. Wadis exist in all areas of arid regions all over the world and can be determined as long as one can identify the enclosing water divide line, preferably by using topographic maps at a scale of 1/50,000. A section of such a map is presented in 3-D form in Figure 2.1.

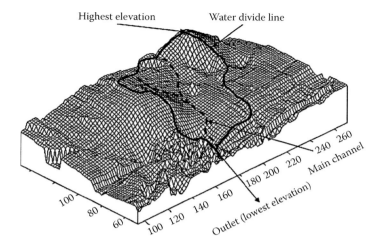

**FIGURE 2.1**   Geomorphology of a wadi.

Within a wadi basin there are sets of local highs and lows. The main channel has its lowest elevation at the *outlet point*. It has tributaries in the forms of subchannels, which have their outlets on the main channel. These channels are also called watercourses, streams, creeks, rivulets, or rivers. The channel system is referred to as the drainage pattern of the watershed, and its surface sheet flow is defined by surface water collectors on both sides. They transport water toward the lower elevation points in the form of stream flow.

The drainage pattern is like veins on a leaf. Among the channels there is one which starts from the catchment outlet point and extends to the upstream foothill area, and even then may proceed onward over a cliff (*escarpment*) up to mountainous terrain. This is referred to as the main channel of the drainage area. As in Figure 2.1 the wadi area is very irregular in shape. It has the following general features.

1. Wadi-delimiting *water divide lines* are irregular in shape, but they have a regular property that at any point on these lines has a decline direction (or incline in the opposite direction).
2. On this delimiting line, there is one point with the highest elevation. It is at the most distant part from the outlet, which constitutes the lowest elevation point on the wadi boundary line. All the surface waters flow toward the outlet (Figure 2.1).
3. The whole wadi might be thought of at least in three major parts, namely, upstream, midstream, and downstream, depending on its distinctive meteorological, hydrological, topographical, and geological features (Figure 2.2). It is obvious that the downstream has the least slope, whereas the upstream has the greatest slope due to foothills and mountains. Additionally, the upstream portion has the maximum rainfall amounts, because, in addition to meteorological frontal systems, these areas receive orographic rainfall, which is due to high elevations. This is the main reason why even during dry periods the upstream parts of the drainage basin feed the main channel,

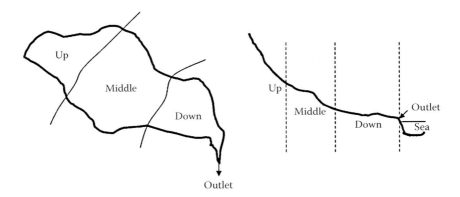

**FIGURE 2.2**   Longitudinal wadi cross-section portions.

which carries the surface water (river water) through the mid- and downstream portions to the outlet. During dry (nonrainy) periods, due to high hydraulic gradients, groundwater storage feeds the main channel, especially upstream. Consequently, the quality of surface and ground waters upstream is better than midstream or downstream. Most of the year, mid- and downstream portions receive relatively less rainfall amounts. In general, there are no dense industrial developments in the upstream parts of any wadi.

4. From upstream to downstream across the wadi, there are sequences of lowest points relative to its banks, with the lowest point being at the outlet, which ends either at the seashore, lake, or desert. A sequential collection of the lowest points will define the *main channel* of the catchment.

5. The upper reach of a wadi has comparatively steeper slope, higher surface water velocity, bigger erosion rates, better groundwater quality, rather small groundwater recharge, and more intense rainfall.

6. Wadi channels and depressions are filled with *quaternary deposits*, but other areas are rock surfaces almost without vegetation. Figure 2.3 indicates typical quaternary deposit and rock surface in a wadi. Wadi deposit has a narrow width at the upstream portions, which widens along the channels

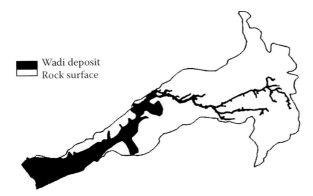

■ Wadi deposit
□ Rock surface

**FIGURE 2.3**   Wadi drainage boundaries.

toward the downstream. In general, wadi alluvium is finely layered and has depths that may reach on the average 30 to 50 m and at some places up to 100 to 120 m. Commonly, the very word *wadi* is the name of the area, which is covered by quaternary deposits at lower elevations than the surrounding areas. The whole unit is a wadi area (drainage basin), which collects the surface water from different points and directs them to a common outlet point. During such a transitional direction, surface water runs over the alluvial fills, which are referred to as *wadi channels* (courses) by local people. Consideration from this figure indicates that in arid regions, natural water courses may be divided roughly into three major groups as follows (FAO, 1981):

a.  Stable, rocky, steep, deeply incised, irregular channels, which almost wholly control the flow characteristics of the water/sediment mixture.
b.  Unstable, disorganized, braided alluvial channels whose slope, depth, shape, and bed form are controlled by the water sediment flow characteristics in a state of dynamic equilibrium with them.
c.  Minor water courses in flat alluvial plains or approaching the terminal reaches of major basins. In these water courses, vegetal growth dominates the water flow. There may be no obvious singly preferred channel.

The first alternative degrades, and the discharge measurements in the channels are subject to large errors due to high velocity, turbulence, and debris load. The last two are of aggrading type and subject to large errors due to the configuration of the water course.

7.  The depth of the wadi deposits also increases toward the downstream. The groundwater quality deteriorates at the downstream, where the groundwater reservoir has relatively thicker alluvial deposit (see Figure 2.4).
8.  Erosion rate is the maximum at high elevations, whereas the deposition rate is more at the lower elevations along the wadi alluvium.

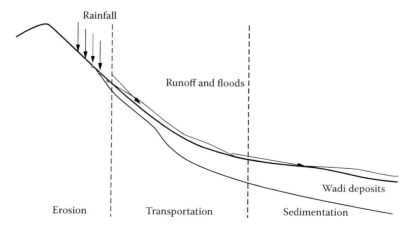

**FIGURE 2.4**   Longitudinal wadi cross-section sedimentation.

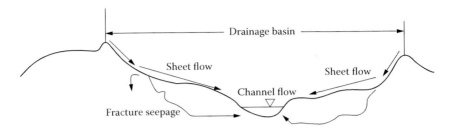

**FIGURE 2.5**   Sheet flow and channel flow.

There are no *perennial streams* in wadis, and after the rainfall occurrence, depending on its intensity, intermittent surface flows appear within the lower portions. Hence, the whole wadi area may be viewed as *sheet flow* and *conveyance* (channel) areas (see Figure 2.5). The former includes all points higher than the plain areas at the lower locations, whereas the latter has the most concentrated surface flow in visible channels.

In the sheet flow area, the rainfall cannot infiltrate directly in significant amounts, but there are flash surface flows leading toward the confluence area from convenient directions, depending on the geomorphologic features of the sheet flow area. The conveyance area is the most important region for the strategic groundwater resources both from the surface and subsurface features points of view. In this area, surface features are important for *infiltration rate* calculations, and the subsurface for the *groundwater storage* volume. In arid regions, the concentration of extensive study must be on the conveyance area by taking into consideration the groundwater storage calculations, the sheet flow area, and subsurface fractured, weathered, and structural geological features. The wider and deeper this area, the more the groundwater storage, and hence there is plenty of domain for strategic groundwater planning where the contribution through the fractured area and subsurface geology must be considered, even by simplified methodologies (see Chapter 5).

## 2.3   SOIL AND LANDFORMS

Arid-zone climatic constraints to a large extent determine the nature of the *soil cover* and *landform*. The dryness, intense radiation, and diurnal temperature fluctuations also result in the absence of vegetation and strong air movement. Additionally, limited sporadic and short-duration storm rainfalls cause intense erosion and transportation of large soil quantities to the lowlands, where the wadi channel courses are. All these phenomena add up to a continual and long-term process of change in the arid regions. The most common soil types in arid zones are the *sabkhah* (*playa*), gravel, *sand dunes,* and *loess.* On the other hand, the most common landforms are *piedmont* and *alluvial plain, alluvial fan,* flood, and coastal plains.

Surface and subsurface geological compositions play direct roles in the occurrence, distribution, and movement of water resources in arid and semi-arid regions, more so than in humid regions because groundwater recharge takes place from the naked (without vegetation) cover of the rock outcrops through fractures, shear zones, and faults and from the alluvial deposit surfaces through the pores (see Figure 2.3).

In regard to groundwater, the geological layers can be categorized into two broad categories as *groundwater reservoirs* (porous, fractured, or karstic rocks) and *non-reservoirs* (intact rock masses) (Şen, 1995). In more practical terms, these are permeable or impermeable rocks, respectively.

Mass wasting and fluvial and aeolian processes are the most important landform-shaping factors in arid and semi-arid regions. The following subsections will solely discuss mass wasting and fluvial and lacustrine landforms in arid environments. Fluvial processes in arid regions produce typical landforms. These are different in high- and low-relief areas.

## 2.3.1 ALLUVIAL FANS

In arid regions, alluvial deposits are rock debris that are eroded into fine sediments and subsequently transported by ephemeral (intermittent) streams to the wadi (valley floor) and deposited there as a result of gradient decrease of the mountains at higher elevations on the drainage basin periphery (except the outlet point). These sediments are then subsequently distributed into fan-shaped landforms called *alluvial fans*, which are gently sloping fan-shaped deposits that are formed at a sharp break in stream gradient, where they flow from a mountain range onto a plain (see Figure 2.6).

These are dry most of the time except during and right after the occasional rainfall events. Alluvial fans are very noticeable and abound especially in arid and semi-arid regions (Şen, 1995). The extent of an alluvial fan depends on wadi slope, size, climate, and the character of the rocks in the source area. Individual forms may have radii up to several kilometers. The gravels tend to be poorly sorted and angular to poorly rounded. Fine-grained debris is deposited farther downstream and may be cross-bedded and massive.

Alluvial fans provide groundwater in coastal arid regions, where they reach larger dimensions than in humid regions and play a more important role as potential aquifers.

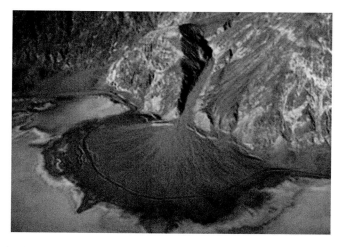

**FIGURE 2.6**    Alluvial fans. Courtesy of the Saudi Geographical Survey.

## 2.3.2 ALLUVIAL FILLS

Alluvium fills are formed as a result of weathering. Water flow takes place in fills or where the relief is favorable if the rainfall is sufficient to provide the driving force for movement. Gravels are the coarsest products of erosion, and they are moved shorter distances from their sources and deposited in more restricted areas than finer deposits such as sand, clay, or mud. In arid regions, fluvial gravels are widespread and fill wadis, surface depressions, or fault zones. The thickness of an *alluvial fill* may reach several hundred meters. They are commonly coarse, interbedded with coarse and fine sands, and show a progressive decrease in size toward downstream, accompanied by a notable increase in roundness. Generally, bedding and sorting are comparatively better than in alluvial fans. Usually, the medium is a coarse-grained type, as the fine grains are either washed away by flash floods or blown away during the long dry spells.

In general, the groundwater is found in the voids of the alluvial fills, which make up potential *groundwater reservoirs* for local uses. In arid regions, wadi alluvial fills are the primary locations for water-well excavation to supply the nearby villages. Figure 2.7 shows numerous well locations all along the wadi alluvial fills in Wadi Fatimah, Kingdom of Saudi Arabia (Al-Sefry et al., 2004).

The convenient features of a wadi are the *floodplain*, terraces, meander scrolls, swamps, and natural drainage toward the main wadi channel, with the result that

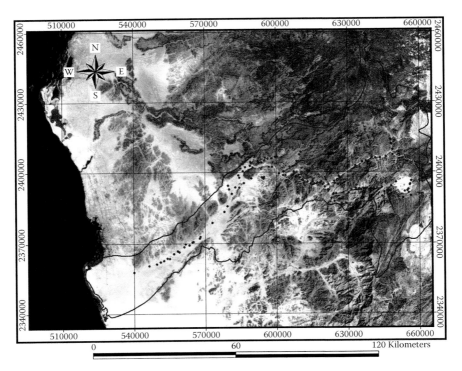

**FIGURE 2.7**  Wells in the alluvial fills of Wadi Fatimah (courtesy of Saudi Geological Survey, SGS).

silt-loaded flood water remains on the flood plain *swamp*. Evaporation of the swamps in wadis causes salt deposition, which creates salinity problems when irrigation is later practiced on the swampy soil. The rainfall imposes a direct *groundwater recharge* to alluvial fills.

Although significant groundwater reservoirs may be found within these deposits, they occur far below the surface. Surface depression in the alluvial fills may have impounded water, which indicates a close proximity to the water table. In arid zones, excess of evaporation over rainfall causes the capillary zone to move downward, leaving behind its dissolved salts or alkaline deposits.

### 2.3.3 SAND DUNES

Sand dunes are accumulations of loose, well-sorted, windblown sand grains (mostly very fine to medium) in wave-like mounds or ridges whose characteristic shapes are maintained by periods of wind-induced, grain-by-grain movement (see Figure 2.8).

*Sand dunes* occur wherever topographic and climatic conditions permit the deposition of wind-borne sand material without vegetation role. They tend to migrate from sand sources, such as wadis, beaches, and *playas*, to topographically controlled accumulation sites. Due to the long dry spells in the desert regimes, the deposits of water are worked over by the wind.

In areas of occasional rainfall, the base of a dune may store enough moisture to nourish grasses in the inter-dune adjacent flat areas, and seepage of moisture may occur around the dune perimeter. At depths of sand dunes, groundwater storages are available. For instance, at about 500 m depths of Rub-Al-Khali ("Empty Quarter") desert of the Arabian Peninsula there are groundwater storages (see Figure 1.3).

**FIGURE 2.8**   Sand dunes.

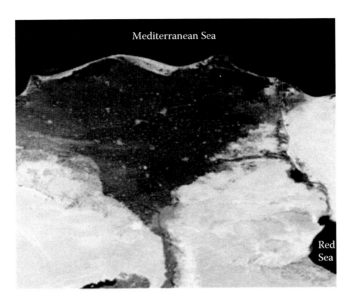

**FIGURE 2.9**   Nile River delta formation.

## 2.3.4   Delta Deposits

Surface water in natural channels usually flows into seas or large lakes. When flowing water meets with standing water, it gives rise to what is known as a *delta*, which is a body of sediment deposited at the mouth of a natural channel when its slope and accordingly water velocity have the lowest possible values. The delta shape also depends on the balance of sediment from the natural channel and the erosive power of waves and sea currents. The channel networks rework themselves into new channel systems or distributaries and flow across older *delta deposits* (Figure 2.9).

A modern definition describes a delta as "the sub-area and submerged contiguous sediment mass deposited in a body of water (ocean or lake) primarily by the action of a river" (Moore and Asquith, 1971). Fundamentally, deltas are of terrestrial deposition, not marine. However, marine sediments may be incorporated in delta fronts intercalating with alluvial deposits, if phases of subsidence alternated with phases of delta make up. Deltas are at the downstream ends of a drainage basin; therefore, both the gradient and the flow velocity decrease and suspended sediments and bed loads consequently settle down. Unconsolidated deposits in deltas grade seaward, starting with fine sand followed by silt, open marine clay, and mud.

Deltas are always associated with water, and because of the flat topography, the water table occurs within a few meters of the ground surface. Generally, deltas contain fairly well-sorted granular deposits and are well drained along the surface as any rainfall infiltrates to the water table close below. The groundwater table elevations in deltas are fairly constant, reflecting the elevation of the nearby water body. Therefore, deltas formed near freshwater lakes are excellent locations for groundwater exploitation. However, there is always salt water intrusion into the fresh groundwater body from the oceans. The extent of intrusion depends on the difference in elevation between the groundwater table in the delta and the ocean surface, as well as the nature of the delta deposits.

### 2.3.5 COASTAL PLAIN DEPOSITS

Coastal plains are found all over the world as unconsolidated sediments, bounded on the continental side by a highland such as a cliff, reef, hill, or escarpment, and separated on the marine side by a shoreline from a surface water body, either a lake or an ocean (Figure 2.10).

The southwestern Arabian Peninsula, most of which is in Yemen, has a narrow *coastal plain* (*tihamah*) along the Red Sea about 30 to 80 km wide, interior highlands, and mountains that descend to a great sandy desert. The coastal plain is hot and virtually rainless with high humidity, alluvium and talus carried down from the highlands, and little vegetation cover. The tihamah coastal plain receives little or no rainfall, but can record high levels of humidity. *Coastal plains* include deposits of both continental and marine origin. Close to the foothills of the highlands continental deposits predominate, gradually giving place to marine deposits seaward. With the regular tidal fluctuations these two types of deposit become intercalated. Their source material is supplied by river, ice, wind, and coastal erosion (Şen, 1995).

Groundwater is provided directly by infiltration from the rainfall and indirectly by inflow of water from the adjacent hills. If there is no adequate replenishment, the fresh groundwater is replaced by saline waters. The fresh *groundwater table* is usually higher than the adjacent surface water bodies and follows the plain surface topography smoothly. Coastal-plain formations can act as a groundwater reservoir by holding the freshwater supplies slightly above sea level and the saltwater table. In such situations overabstraction of the freshwater may allow the saltwater to contaminate the groundwater supply. The greater the distance from the sea, the higher the groundwater table elevation and the greater the freshwater depth.

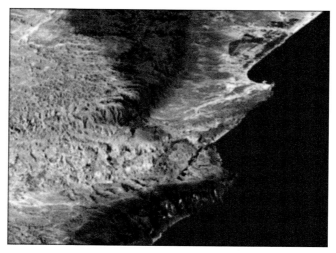

**FIGURE 2.10**   Coastal plain deposits.

**FIGURE 2.11** Sabkhah.

### 2.3.6 SABKHAH (PLAYA)

In arid regions, if a low-lying basin has no outlet, incoming water from *flash floods* evaporates inside the basin where its dissolved salts accumulate in the lowest parts. Such salt lakes indicate that annual evapotranspiration is greater than the sum of incoming flood water and rainfall. These salt-concentrated locations especially along the coastal areas are referred to as a *playa* in English (borrowed from the Spanish) and as a *sabkhah* in Arabic. When it dries out, the muddy lake floor shrinks and cracks. Accumulated salts crystallize and form crusts on top of the playa floor and in cracks in the surface soil (Figure 2.11). Much of the accumulated salts stem from evaporitic marine sediments outside the basin; many Mesozoic (Triassic, Jurassic) and Tertiary sediments are very rich in evaporates (Şen, 1995).

Sabkhahs are formed where wind erosion removes surface materials down to the water table. Water is always associated with sabkhahs in the form of flooding, runoff accumulation, capillary rise, and tidal fluctuation. The sediments that fill sabkhahs consist of sand, silt, clay, and salts in varying combinations. Their flat surfaces mark the elevation to which soil moisture rises above the static water level. Below this surface the materials are damp, wet, or saturated; above, they dry out and blow away.

## 2.4 ROCK ENVIRONMENTS IN ARID REGIONS

In most arid regions, *volcanic rocks* are either associated with young volcanoes or are a part of an ancient volcanic system. The old versions tend to be very poor aquifers except in fault zones. Some of the young volcanic rocks are quite impervious, but they can provide good water supply if rainfall is high enough. In low-rainfall areas such as in arid zones, however, younger volcanoes are very permeable and therefore the groundwater drains away quickly to the toe of the volcano or to the sea. As *aridity* increases, the discovery of spring outlets from the volcanic rocks is easier on land but if recharge is limited, these springs may be dry much of the time due to the rapid discharge of recharge on the water.

In arid regions, a considerable part of the landscape is underlaid by granitic and metamorphic "hard" rocks, which are practically impervious except where fractured and weathered. Tectonic conditions define the fracturing, whereas depth of weathering depends on time and past climates. The storage capacity of fractured rocks is usually very limited, so that in arid regions, little of the rare recharge can be stored. *Weathered rocks* may have good storage capacity but low permeability. Therefore, the ideal locations for well drilling are in *fractured rock* environment overlain by saturated weathered rock. If weathering is extensive, then usually there is fractured rock below. However, in most of the hard rocks, fractures are rare below 50 m and normally not found below 100 m. For instance, in the Arabian Peninsula, these rocks seldom yield water because of the limited rainfall, but they may yield some water when they are located beneath wadi alluvium. In cases of poorly permeable hard rocks, the small amount of water contained is of very poor quality due to extremely slow circulation in connection with the present-day hydrologic cycle. Therefore, existence of fresh water in these environments indicates that there is circulation and therefore some replenishment and some significant permeability values.

The most significant rock features in terms of water availability are orientation, aperture, density, and roughness (Şen, 1996). These characteristics depend on the resistance offered by the rock to the force involved. For instance, in hard rocks fractures are extensive, large, and dense as compared with those in *softer rocks*, which are of limited extent, relatively small and less dense. In either type of rock, however, fractures (unless filled with cementing material) facilitate the flow of water through the rock. A fractured medium offers less resistance to water movement because the water molecules are not subjected to barriers within the fracture. It is logical to say that the longer the fractures and the smoother their walls, the easier is the water transmission. Because fractures dominate the movement of groundwater, knowledge of their geological settings is indispensable in hydrology and hydrogeology, as well as water engineering, especially in arid regions. Şen (1995) provided an extensive literature and explanation on various rock environments in arid regions and on their groundwater potentiality.

## 2.5 DRAINAGE BASIN FEATURES

The drainage pattern and arrangement of the natural channel network determine the efficiency of the drainage system. With other factors being constant, the time required for water to flow a given distance is directly proportional to flow path. A well-defined wadi system reduces the flow distance, thus reducing the travel time, and the resulting outflow hydrograph will usually have a shorter time to peak.

The most significant information for some hydraulic structure design at any site is the background area that conveys the whole water into this site. Although there is a single natural outlet for each drainage area, depending on the water resources development purpose, the outlet can be any point along one of the wadi channels. Such a point corresponds to dam location, as seen in Figure 2.12.

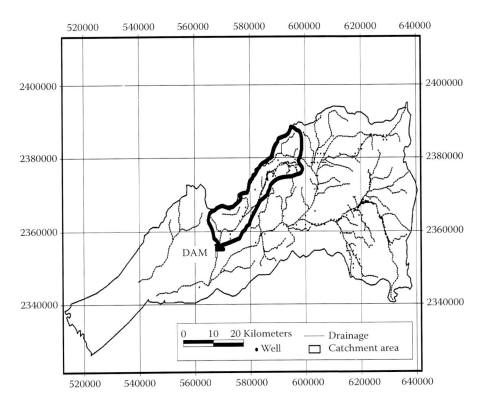

**FIGURE 2.12** Dam location.

Determination of a natural outlet is certain by geological and topographical facts, but water structure site selection is not so easy in practice. In addition to geologic and topographic features as morphology, the selection depends also on hydrological properties, social activities and water demand specifications. Field appreciation of small drainage areas is possible by visual inspection.

If the primary drainage channels of arid zones do not terminate at inland sur-face-water depressions or the sea, they eventually braid into a complex of micro-channels and are lost. It is unusual for the active reach of an ephemeral water course to extend more than 100 m downstream of its point of emergence from the mountain zone onto an outwash fan or alluvial plain.

In any water resources development and assessment not only the time and areal extend of water related phenomena are important, but also the surface features of the drainage basin play a significant role on the time and spatial distribution of rain-fall water volumes. In any water budget study, hydrologic cycle components such as rainfall, evaporation, infiltration, runoff, depression, and detention are significant without knowledge of wadi morphological features, but it is not possible to perform a complete study. Geomorphologic patterns are concerned with the geometrical fea-tures of the landscape. These patterns are controlled by factors such as slope, cli-mate, vegetation, and soil and rock resistance to erosion (Chapter 7). Although there

**TABLE 2.1**

**Drainage Basin Area**

| Area (km²) | Classification |
|---|---|
| A >1000 | Very big |
| 1000 < A < 100 | Big |
| 100 < A < 5 | Middle |
| A < 5 | Small |

are a set of geomorphologic parameters, in the following sequel only those most frequently related to arid region hydrology are explained.

### 2.5.1 DRAINAGE AREA

The term *drainage area* refers to the projected horizontal surface area that lies upstream of a point on a stream or wadi course (main channel). The water from rain or irrigation ultimately reaches this point. Its most important feature is the horizontal area of the drainage basin, which can be easily calculated from a topographic map after the delimitation of the water divide line. Simple logic states that increase in the area will result in increase in the rainfall share on the drainage basin and likewise the surface water and groundwater shares also increase. According to the drainage area, $A$, the drainage basins can be classified into the four categories as shown in Table 2.1.

There are many empirical formulations that relate the drainage basin area to peak discharge, $Q_P$, produced after a long duration of rainfall. The most widely used, simplest empirical formulations appear as

$$Q_P = aA^b \tag{2.1}$$

where $a$ and $b$ are the parameters that can be obtained after statistical regression method applications (see Chapter 4).

### 2.5.2 STREAM SLOPE

The slope of a wadi is its vertical drop per unit of horizontal distance, and it plays one of the dominant roles in runoff and flood velocity determinations. In general, *slope* is steepest at higher elevations in the upstream areas and is much reduced as the wadi approaches its base level at the downstream portions. It is possible to appreciate the slope along the wadi from upstream toward the downstream along the longitudinal profile of the main channel, which has generally a concave shape. In high elevations, the surface water erodes a deep wadi in the hilly and mountainous terrain due to the high runoff velocity. The steeper the slope, the more rapidly flows the surface water. Therefore, the time to peak will be shorter, and the peak water amount will be higher. On the other hand, infiltration capacity, and consequently the groundwater recharge, tend to be smaller as the slope gets steeper.

Another important factor that affects surface water velocity is the slope of the main channel. There are different slope measures, but the *main channel slope, S,* is the most widely used in practical work. It is defined simply as the ratio of difference between the outlet, $E_o$, and the furthest point, $E_f$, elevations to the main stream length, $L$, as,

$$S = \frac{E_f - E_o}{L} \tag{2.2}$$

The steeper the slope, the faster the surface flow arrival to the outlet. Of course, apart from the main channel, there are slopes of other surfaces within the drainage basin and they play an additional role in the distribution and movement of surface flow, streamflow, infiltration, vegetation, sedimentation, etc. Provided that the topographic map of the whole wadi is available, the overall *average slope, $S_a$,* can be defined as,

$$S_a = \frac{\Delta E}{A} W_T \tag{2.3}$$

where $\Delta E$ is the elevation difference between two consecutive contour lines on a topographic map and $W_T = (W_1 + W_1 \ldots, W_n)$ is the total length of contour lines that remain within the drainage area (see Figure 2.13).

### 2.5.3 DRAINAGE DENSITY

The *drainage density, $D_d$,* is a measure of the total stream channel length, $\Sigma L_i$, per unit area, $A$, of drainage basin, which can be expressed as,

$$D_d = \frac{\Sigma L_1}{A} \tag{2.4}$$

The smaller the drainage density of a watershed, the slower the surface flow and runoff move, but in the meantime infiltration increases. This leads to delay in the hydrograph discharge peak and causes a small peak discharge (see Chapter 4). Drainage density cannot explain solely the drainage behavior of the catchment area. Two drainage basins with the same area and drainage density cannot make a distinction

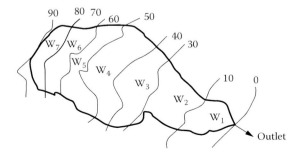

**FIGURE 2.13**   Topographic maps.

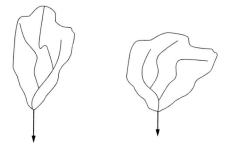

**FIGURE 2.14**  Two watersheds with the same area and drainage density.

even though one of them may have a better drainage system than the other because of difference in shapes (Figure 2.14).

### 2.5.4  CENTROID LENGTH

In synthetic hydrograph analysis (Chapter 4) one of the significant watershed parameters is the distance between the nearest point to the drainage basin centroid on the main channel and the outlet point. The *centroid length,* $L_c$, is measured along the main channel between A and C (see Figure 2.15).

The centroid point can be determined by hanging a wadi template made of cardboard from at least two different points. Vertical straight lines are drawn from each hanging point, and the intersection between these two straight lines is the location of the watershed centroid. Its projection on the main channel gives the nearest point, C, of the channel to the centroid.

### 2.5.5  ELEVATION FEATURES

In order to describe the elevation features of a drainage basin in a refined manner, the subareas between successive equal elevation lines (contour) are calculated (see Figure 2.16). The percentages, $p_i$, of each subarea are calculated out of the total watershed area. This helps to calculate the total percentage of subareas below any given contour line. If the area between two successive contours is shown by $a_i$, and

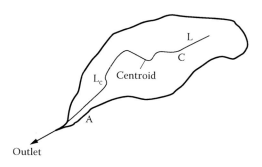

**FIGURE 2.15**  Centroid length.

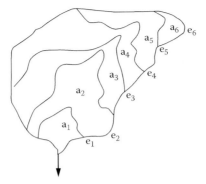

**FIGURE 2.16** Equal elevation lines within a drainage area.

the average elevation of this subarea by $\bar{e}_i$, then the average elevation of the watershed, $\bar{e}$, can be calculated as,

$$\bar{e} = \left(\frac{a_1}{A}\right)\bar{e}_1 + \left(\frac{a_2}{A}\right)\bar{e}_2 + .. \left(\frac{a_n}{A}\right)\bar{e}_n \qquad (2.5)$$

or, succinctly,

$$\bar{e} = p_1\bar{e}_1 + p_2\bar{e}_2 + ... p_n\bar{e}_n \qquad (2.6)$$

Among the useful specifications is also the overall average elevation of the drainage area. For this purpose, a graph is drawn between the elevation and the area above the elevation considered. As the elevation increases, the area below the given level also increases (Figure 2.17).

This is referred to as the elevation area or *hypsographical graph*. Such a graph is useful for rainfall, temperature, and vegetation-type variation assessments with elevation.

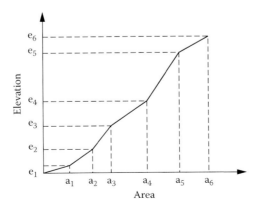

**FIGURE 2.17** Elevation area graph.

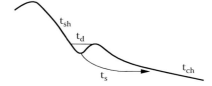

**FIGURE 2.18**  Time of concentration components.

## 2.6  TIME OF CONCENTRATION

One of the most important characteristics of any drainage basin is the time that a raindrop requires from the farthest upstream point to reach the outlet. This is the *time of concentration,* $t_c$. Depending on the geological composition and wadi surface features, it is composed of different durations. The first is the *saturation time* duration, $t_s$, required after the rainfall start until the rainfall intensity becomes equal to an infiltration rate at which the soil is completely saturated. The second duration is required for the filling of depressions on the surface, which is called the *depression time,* $t_d$. Another duration starts after the completion of depression filling. This is the time required for the arrival of drops to surface as a channel network, and is called the *sheet flow time,* $t_{sh}$. Still another duration, $t_{ch}$, is the time that a water drop needs to reach the outlet point along the channel. These are schematically shown in Figure 2.18, and by definition the time of concentration can be expressed as

$$t_c = t_s + t_d + t_{sh} + t_{ch} \tag{2.7}$$

In practice, it is not possible to measure these time durations separately. For this reason, they are calculated through some rational approaches.

### 2.6.1  AVERAGE TIME OF CONCENTRATION DERIVATION

If the channel length at which the water flows from the far distant point to the outlet is $L$ and arrival time is $t_c$, then the average velocity, $\bar{V}$, can be expressed by definition as

$$\bar{V} = \frac{L}{t_c} \tag{2.8}$$

which yields

$$t_c = \frac{L}{\bar{V}} \tag{2.9}$$

The velocity is dependent on many factors such as the average slope, $S$, of the drainage basin; roughness of the basin surface (*Manning coefficient,* n); cross-section characteristics such as the cross-sectional area, $A_c$; and the wetted perimeter, $P$. Average velocity is defined empirically for natural channels by different researches. For instance, the Manning formulation states that

$$\bar{V} = \frac{1}{n} R^{2/3} S^{1/2} \tag{2.10}$$

where $R$ is the hydraulic radius defined as the ratio of flow cross-sectional area to wetted perimeter, which can be approximated simply as the summation of the cross-section width, $W$, and twice of the water depth, $D$, as

$$P = W + 2D$$

or

$$P = W\left(1 + \frac{2D}{W}\right)$$

In many arid regions, and especially in the mid- and downstream portions of wadis, $W \ggg D$, and therefore one can approximately rewrite this as

$$P \approx W \tag{2.11}$$

Hence, the hydraulic radius definition yields

$$R = \frac{A_c}{W}$$

Because $A_c = WD$, this last expression reduces to

$$R = D \tag{2.12}$$

After all these explanations, Equation 2.10 can be rewritten as

$$\bar{V} = \frac{1}{n} D^{2/3} S^{1/2}$$

The substitution of this expression into Equation 2.9 leads to,

$$t_c = \frac{nL}{D^{2/3} S^{1/2}} \tag{2.13}$$

Consideration of unit surface flow depth ($D = 1$) renders Equation 2.13 into the following practically usable forms as

$$t_c = \frac{nL}{\sqrt{S}} \tag{2.14}$$

This expression can be generalized into an empirical form as

$$t_c = aL^b S^{-c} \tag{2.15}$$

where $a$, $b$, and $c$ are parameters that should be determined from a regional study in the area.

For many catchments in the United States, Kirpich (1940) has shown that time of concentration can be given as,

$$t_c = 0.000323L^{0.77}S^{-0.385} \tag{2.16}$$

In this equation, if $L$ is substituted in meters, then $t_c$ is obtained in hours.

## REFERENCES

Al-Sefry, S., Şen, Z., Al-Ghamdi, S. A., Al-Ashi, W., and Al-Baradi, W., 2004. Strategic ground water storage of Wadi Fatimah—Makkah region Saudi Arabia. Saudi Geological Survey, Hydrogeology Project Team, Final Report.

FAO, 1981. Arid zone hydrology for agricultural development. FAO, Rome, Irrig. Drain. Paper 37, 271 pp.

Kirpich, Z. P., 1940. Time of concentration of small agricultural watersheds. *Civ. Eng.*, 10 (6), p. 362.

Moore, G. T. and Asquith, D. O., 1971. Deltas: Term and Concept. *Bull. Geol. Soc. Am.*, 82, pp. 2563–2568.

Şen, Z., 1995. *Applied Hydrogeology for Engineers and Scientists.* Boca Raton, CRC Lewis Publishers, 467 pp.

Şen, Z., 1996. Theoretical RQD-Porosity-Conductivity-Aperture Charts. *Int. J. Rock Mech. Min. Sci. Geomec. Abstr.,* Vol. 33, No. 2, pp. 173–177.

# 3 Rainfall Pattern

## 3.1 GENERAL OVERVIEW

The major agent of the climate and the hydrology in any region is rainfall, and especially in the arid regions, its lack over long periods gives rise to aridity and water stress (see Chapter 1). The availability of rainfall measurements does not mean that their temporal and spatial distribution characteristics can be used for engineering, agriculture, and social development studies. In arid regions, even a passing, short- duration shower is regarded as a useful water resource, and its accumulation in ditches is used for domestic and local irrigation purposes. For within-year developments in arid regions, it is possible to sustain the design of a water resources system in a simple but effective manner. Over inter-annual periods, large-scale water structures (surface and *subsurface dams*, *well fields*, extensive irrigation) require rainfall-sourced surface water, and especially groundwater availability, in arid regions. The volume of rainfall over any wadi is important for the strategic planning of water resources in arid regions (see Chapter 5). Rainfall is the sole source for *groundwater recharge*, which furnishes almost all the demand for domestic and irrigation use in arid zones.

Arid zones have most often the minimum network of instrumentation (daily, totalizing and recording raingages and observation wells for groundwater monitoring). On the other hand, some tens or hundreds of relatively recently established measurement instruments have unrepresentative spatial distribution due to short term and ad hoc water resource surveys or investigations. Reasonably reliable assessments of rather scarce point rainfall temporal and spatial characteristics are necessary for proper estimations of runoff and groundwater recharge levels in arid regions. A limited amount of rainfall data, in addition to inherent variability in the rainfall events, presents practical difficulties in arid zones for water resources planning, development, and management studies. Within-year (monthly and seasonal) variations in rainfall records have utmost significance in arid regions for exploitable water resources development and management studies.

This chapter presents the general occurrence of rainfall in arid zones and the simple calculation methods for practical uses in the design, development, and management of water resources.

## 3.2 RAINFALL FEATURES

The most important features of arid and semi-arid rainfalls can be summarized as follows:

1. Rainfall can be very varied and erratic, spatially as well as temporally.
2. Individual storm-total can be very high where, in many cases, the single storm rainfall far exceeds the mean annual rainfall.

3. Rainfall intensities can be very high, and sediment yield could be greatly increased (see Chapter 7).
4. The amount of runoff is increased by the scaling effects of rainfall impact, which increases runoff transport capacity.
5. Due to the seasonal pattern of the rainfall, erosion and sediment yields follow a similar pattern, where the most valuable period for erosion and sediment yield is the early part of the wet season when the rainfall is high but the vegetation has not grown sufficiently to protect the surface.

Weather patterns in arid regions are most often under the effect of small scale *orographic* and *convective* rainfall occurrences rather than occasional large scale *frontal* rainfalls. Spatial variability is directly related to the local and regional topography. At high elevations orographic rainfall occurs, and this happens especially if surface water and nearby escarpments, cliffs, or high hills exist within 150 to 200 km. For instance, along the Red Sea coastal planes (tihamah), there are sudden escarpments of 3,000 m height at 5 to 150 km distance, where moisture laden air moves inlands, rising and cooling in the meantime (Figure 3.1).

The rainfall initiation and intensity is augmented due to windy *storms* prior to rainfall, which makes the nucleus of condensation. At times, even though the moist air is available, in the absence of winds the rainfall does not occur. This is one of the main reasons for spatial and temporal variation of rainfall in arid regions.

On the other hand, summer monsoon rainfalls, primarily from tropical storms and convectional thunderstorms, represent some 20 to 30% of the total annual rainfall in the coastal plains. Over the highlands (more than 1,000 m), it represents only 80% of the total rainfall. In contrast, winter rainfalls are of cyclonic frontal origin

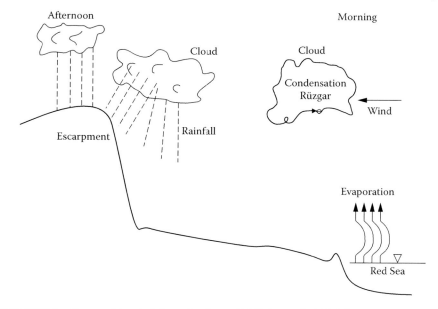

**FIGURE 3.1**    Orographic rainfall mechanisms. (Al-Sefry et al., 2004. With permission.)

and represent the bulk of rainfall on the coastal plain, decreasing gradually to the east from 80% on the coast to 20% in the highlands. Occasional sporadic rainfall occurrences on the *coastal plains* are due to convective cells. These are very local rainfalls that cause extremely spatial rainfall variations in arid regions.

In humid regions, there are recorded rainfall amounts in each month with many rainy days. In arid regions, dry months do not have any rainfall occurrence and some years are without rainfall events. Even the wet periods may have few rainy months with one or two rainy days. Such an unpredictable situation in arid regions is completely due to the haphazard interaction of various meteorological and topographical events.

Significant sources of uncertainty in the rainfall assessments in arid regions arise from the record length and scarce number of raingage stations. The former imposes temporal restriction, whereas the latter implies limitations in spatial (areal or regional) assessments. Detailed storm structure can be identified with rainfall densities approaching at least one measurement station per 5,000 km$^2$.

## 3.3   TEMPORAL DISTRIBUTION OF RAINFALL

The rainfall records at a station have a particular distribution around the averages. The frequency distribution of maximum daily temperature records indicates more or less higher and lower values than the average with a rather symmetric temperature distribution. This is not the case for rainfall frequency distribution, which is skewed to the right as in Figure 3.2. This means that low rainfall amounts are recorded more frequently than the high rainfalls.

In arid regions, most of the days have zero rainfall, and therefore, in the preparation of frequency distribution diagrams as in Figure 3.2, zero rainfall days are treated separately (see Section 3.6). In Figure 3.2, only rainy days are taken into account and, accordingly, three parts are identified as follows:

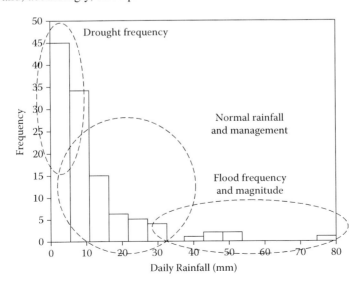

**FIGURE 3.2**   Rainfall frequency distribution.

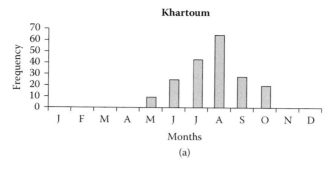

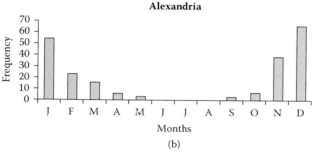

**FIGURE 3.3** Frequency distribution of mean monthly rainfall at (a) Khartoum, and (b) Alexandria.

1. Low rainfall events indicate *drought* conditions and, hence, on the left-hand side there is a vertical ellipse that signifies this situation.
2. High rainfall events indicate the possibility of *floods* and/or *flash floods*, which are represented by a horizontal ellipse on the right-hand side.
3. In-between rainfall events are important for the management of water resources so as to reduce the effects of extremes (drought and flood).

There are many days in arid zones without rainfall (zero rainfall) and *ground-water recharge*. The rainfall in arid zones usually pours down and, consequently, flash floods may pose risks even in the most desiccated desert regions. There are very scant flood records, and people in arid zones have experience in dealing with these dangerous natural water hazards. However, catastrophe may occur when a flash flood is unusually large or when nothing has happened for such a long time that the settlers have been lulled into a false sense of security.

In order to show the spatio-temporal rainfall variation in arid regions, the monthly average rainfalls are presented in Figure 3.3 for Khartoum (Sudan) and Alexandria (Egypt). The distinctive difference is obvious between the inland and coastal rainfall events. In both cases there are months without rainfall, which indicate the implications of semi-arid regions.

Natural or artificial groundwater storage is essential for people to live permanently in places with such a seasonally variable water supply. In arid zones, because the rainy season is relatively short, it is possible for the annual local supply to be concentrated in a few intense local storms.

Relative anomalies (or residual rainfalls), $r$, are deviations of the annual rainfall, $R$, from the long-term temporal mean, $\overline{m}_r$ divided by the standard deviation, $s_r$, which is the measure of the average variability within the record. This is also referred to as the standardization procedure in the statistics literature (Benjamin and Cornell, 1960).

$$r = \frac{R - \overline{r}_r}{s_r} \tag{3.1}$$

The relative anomalies are without unit, and their average is equal to zero with unit standard deviation. In arid regions a distinctive rainfall pattern has a number of below average rainfall years more than the number of years above the rainfall. Otherwise, the number of rainfall occurrences above and below the average level will be almost the same as in Figure 3.4.

Such plots are useful in comparing different rainfall patterns at different locations. If the rainfall relative frequency is symmetric with almost equal number of cases above and below the average, one can expect to find positive (negative) anomalies in excess of one standard deviation in about 16% of all cases. A rainfall anomaly which is larger than two standard deviations would be found in about 2%, and one larger than three times the standard deviation is very rare.

Another temporal feature of arid region rainfall is the internal independence structure, which can be identified graphically. The independence between the dry and wet years can be depicted by plotting previous year residual rainfall value versus the next year residual rainfall as in Figure 3.5, where there is no distinctive trend.

The random scatter of points in this figure implies that the *dry* and *wet* year (*spells*) sequences are independent from each other. The independence is observed

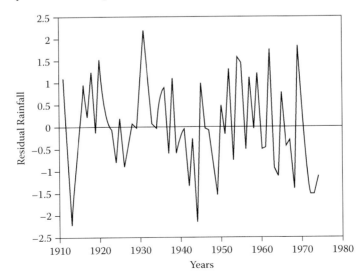

**FIGURE 3.4**   Average annual rainfall variability.

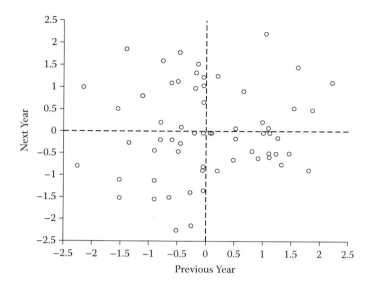

**FIGURE 3.5**   Independence of wet and dry spells.

because there are scatters of almost the same number of points in each quadrant. Furthermore, in arid zones, annual rainfall amounts have a positive skew in their histograms, which is in contrast to temperament zones where the annual rainfall histograms are more or less symmetric with comparatively smaller skewness (see Figure 3.6).

Skewed distribution implies that there are more annual values below the arithmetic average. The mode is smaller than the median, and they are both smaller than the arithmetic average. Hence, the small intensity rainfalls occur more frequently than big intensities.

## 3.4   ARIDITY INDEX

Many *aridity* and *semi-aridity* definitions appeared in the literature but none can be entirely satisfactory, and the terms arid and semi-arid remain somewhat imprecise and vague. Generally, the aridity implies that rainfall does not support regular rainfed farming. According to Walton (1969) this definition encompasses all the seasonally hot arid and semi-arid zones classified by means of rainfall, temperature, and evaporation indices. Among the aridity indices, those based on rainfall only state that any duration without rainfall for 15 consecutive days is considered a dry period. In some regions, 21 days or more with rainfall less than one-third of normal is considered as a dry period. There are also percentage-dependent indices such as in the case of annual rainfall less than 75% and monthly rainfall less than 60% of normal. However, in an individual case, any rainfall less than 85% is also considered a dry case. A simple quantitative *aridity index*, $A_I$, measure of a region is defined as the rainfall amount per temperature degree or

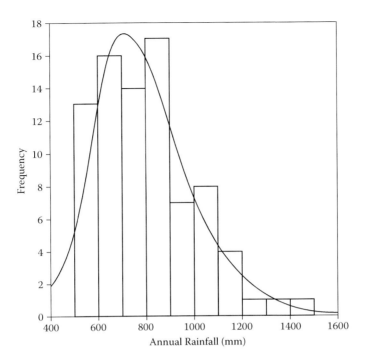

**FIGURE 3.6** Annual rainfall histogram of Istanbul (Turkey).

$$A_I = \frac{\overline{R}}{\overline{T}} \qquad (3.2)$$

where $\overline{R}$ and $\overline{T}$ are the average monthly rainfall and temperature values, respectively. Figure 3.7 represents the monthly aridity indices for the cities of Jeddah on the Red Sea coastal plain and for Taif at the top of an *escarpment* with more than 2,000 m above mean sea level (Al-Sefry et al., 2004).

The aridity index in Equation 3.2 has positive values only with zero as the lowest limit corresponding to no rainfall case. Generally, it takes values around 1. The greater the aridity index, the higher the rainfall amount associated with low temperatures. Hence, big aridity indices correspond to better groundwater recharge possibilities.

## 3.5 ACCUMULATIVE RAINFALL CURVE (ARC)

The amount of accumulative rainwater during a storm is measured either by a totalizing or recording raingage. The former yields a single rainfall measurement after the storm (Figure 3.8), whereas the latter provides a nondecreasing accumulative rainfall curve (ARC) within storm (Figure 9a). The *total rainfall* measurement is useful in engineering designs concerning daily, monthly, annual, or multiannual duration decisions.

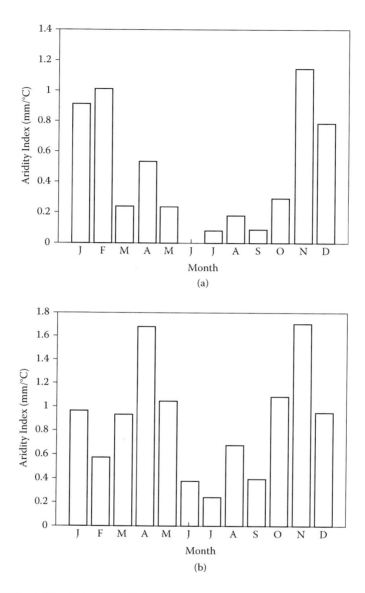

**FIGURE 3.7**    Aridity index (a) Jeddah, (b) Taif.

The slope of the chord (or tangent) during a given time interval, Δt, is referred to as the *rainfall intensity*, i, as

$$i = \frac{dR(t)}{dt} \approx \frac{R(t)}{t} \tag{3.3}$$

In practical studies Δt is fixed as 5, 10, 15, etc. minutes. The change of intensity with time in Figure 3.9b is the *hyetograph*.

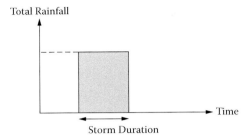

**FIGURE 3.8** Total rainfall heights.

### 3.5.1 RAINFALL INTENSITY–DURATION

In general, the maximum rainfall intensity, $I$ (mm/hr), during a single storm can be expressed (Maidment, 1993) simply as

$$I = \frac{R}{T + C} \qquad (3.4)$$

where $T$ is the duration of the storm, and $R$ and $C$ are constants. This expression can be rewritten as

$$\frac{1}{I} = \frac{1}{R}T + \frac{C}{R} \qquad (3.5)$$

It is represented as a straight line on $(1/I)$-$T$ coordinate system with slope, $1/R$, and the intercept, $C/R$ (Figure 3.10).

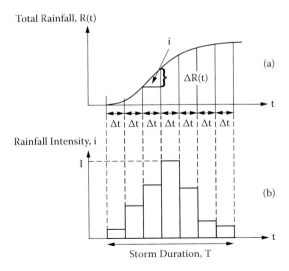

**FIGURE 3.9** (a) Accumulative rainfall curve, (b) hyetograph.

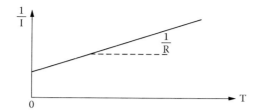

**FIGURE 3.10** Intensity–duration relationship.

**TABLE 3.1**
**Average Value of Constants**

| Wadi Name | R (mm) | C (h) |
|---|---|---|
| Wadi Fatimah | 60.0 | 1.2 |
| Wadi Na'man | 95.0 | 1.1 |

Historical storm rainfall data for wadis Fatimah and Na'man, in the western part of the Arabian Peninsula yield R and C values as shown in Table 3.1 (Sirdaş and Şen, 2007).

The average value of C is equal to 1.15. Unless T is very small, it will not greatly matter whether C is adopted as 1.15 or 1.0 (Richard, 1955).

On the other hand, it is well known regionally that as the area, $a$, from the storm center increases, the areal average rainfall intensity, $\bar{i}$, decreases. Such a relationship can be expressed by areal reduction factor, $0 < f(a) < 1$, which is equal to 1 at $a = 0$, but decreases to an asymptotic value as the storm area increases (Wiesner, 1970). Leclerc and Shaake (1972) have shown that the areal reduction factor is sensitive to the catchment area and rainfall duration.

The temporal maximum, $I$, and areal average, $\bar{i}$, intensities are related to each other through the areal reduction factor as

$$\bar{i} = If(a) = \frac{Rf(a)}{T + C} \tag{3.6}$$

The storm isohyetal maps are representatives of the areal average rainfall intensity distributions, and therefore it is possible to calculate $f(a)$ from a given isohyetal map over the catchment area. The areal average rainfall intensity up to each isohyet line is computed, and the ratios of these averages to maximum intensity are given in Tables 3.2 and 3.3 for three storm events in wadis Fatiman and Na'man (Al-Sefry et al., 2004).

Hence, the change of $f(a)$ by $a$ is presented in Figures 3.11 and 3.12 for wadis Fatimah and Na'man and approximately to constant values of 0.63 and 0.68, respectively.

**TABLE 3.2**
**Average Areal Rainfall Calculations for Wadi Fatimah**

| Storm | Duration (hour) | Rainfall (mm) | Average Rainfall Intensity Over Area (km²) | | | | | |
|---|---|---|---|---|---|---|---|---|
| | | | 1.00 | 60.5 | 470.0 | 920.0 | 1800.0 | 2500.0 |
| I | 2.80 | 67.80 | 1.00 | 0.90 | 0.86 | 0.76 | 0.62 | 0.60 |
| II | 6.70 | 95.20 | 1.00 | 0.92 | 0.86 | 0.80 | 0.70 | 0.67 |
| III | 3.00 | 48.80 | 1.00 | 0.91 | 0.84 | 0.77 | 0.63 | 0.61 |
| Average | 4.20 | 82.60 | 1.00 | 0.91 | 0.85 | 0.78 | 0.66 | 0.63 |

**TABLE 3.3**
**Average Areal Rainfall Calculations for Wadi Na'man**

| Storm | Duration (hour) | Rainfall (mm) | Average Rainfall Intensity Over Area (km²) | | | | |
|---|---|---|---|---|---|---|---|
| | | | 1.00 | 16.00 | 223.0 | 474.0 | 607.0 |
| I | 2.30 | 128.50 | 1.00 | 0.96 | 0.85 | 0.73 | 0.69 |
| II | 2.70 | 108.50 | 1.00 | 0.96 | 0.87 | 0.75 | 0.68 |
| III | 4.30 | 99.20 | 1.00 | 0.99 | 0.87 | 0.74 | 0.68 |
| Average | 3.10 | 112.0 | 1.00 | 0.97 | 0.86 | 0.74 | 0.68 |

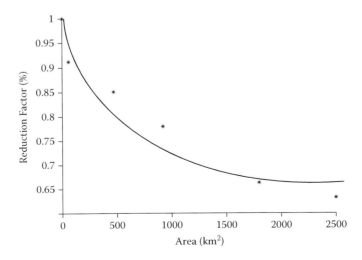

**FIGURE 3.11** Areal reduction ratio change by area for Wadi Fatimah.

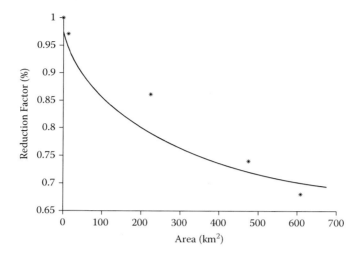

**FIGURE 3.12**  Areal reduction ratio change by area for Wadi Na'man.

## 3.6  INTENSITY–DURATION–FREQUENCY (IDF) RELATIONSHIP

Rainfall *intensity–duration–frequency* (IDF) relationship is of fundamental importance in hydrology and water resources systems design. Many procedures and formulas, mainly empirical, have been proposed for IDF relationship determination in
the literature (Chow, 1964; Bell, 1969; Aron et al., 1987). A mathematical approach
has been proposed by Burlando and Rosso (1996) and Koutsoyiannis et al. (1998).
In practice, division of the rainfall duration into three groups—namely, durations
from 1 m to 1 h as "short," from 1 h to 24 h as "intermediate," and more than 24 h as
"long"—has led to some meaningful interpretations and works on the regionalization of IDF relationships in different geographical areas for several countries (Froehlich, 1995a,b,c; Garcia-Bartual and Schneider, 2001).

In many parts of the world, past and current IDF statistics are computed by
considering the Gumbel distribution (Gumbel, 1958) of Fischer-Tippett type I probability distribution function (PDF) to describe the frequency of extreme rains. This
law of distribution is also used by most of the official meteorological services of the
world, especially in humid regions. The rainfall regime in the arid regions, however,
does not abide by the Gumbel PDF, but an *exponential PDF* is suitable.

### 3.6.1  IDF CURVES AND PDFS

In the derivation of IDF curves, any PDF implies the following basic information,
which is necessary for meaningful interpretations.

1. The form of the PDF shows the rainfall feature in the study site and it is symmetrically distributed or skewed. For instance, in arid regions, it is climatologically meaningful to expect storm rainfalls at high (low) intensities with low (high) frequencies as already mentioned in Section 3.3. This is a suitable representation by the two-parameter gamma distribution and especially by its special case of exponential PDF.
2. The same PDFs give way to calculate the risk values attached with a set of engineering *design periods*, T, such as 2-year, 5-year, 10-year, 25-year, 50-year, 100-year, and 500-year.

At this point it is useful to relate the risk, $R$, to the design period (*recurrence interval*), which are inversely related to each other as

$$R = \frac{1}{T} \tag{3.7}$$

Herein, $R$ is defined as the probability of exceedence once during the design period. Hence, the *design intensity*, $i_D$, which constitutes the basis of IDF curves can be obtained from a suitable PDF, $f(i)$, after integration $a$,

$$R = \int_{i_D}^{+\infty} f(i)di \tag{3.8}$$

Another requirement in the derivation of IDF curves is the relationship between the storm rainfall duration, $T$, and the maximum intensity, which has already been mentioned in Section 3.5.1. The two-parameter relationships are available in the literature in one of the following three mathematical forms:

1. The most widely used two-parameter ($R$ and $C$) relationship was proposed in the 1930s by Sherman (1932), and was already given in Equation 3.4.
2. Bernard (1932) suggested a hyperbolic form as

$$I = \frac{R}{T^C} \tag{3.9}$$

3. The relationship between $I$ and $T$ can also be expressed as a straight-line on a semi-logarithmic paper as

$$I = R - CLnT \tag{3.10}$$

Equations 3.9 and 3.10 are particular cases of the general four-parameter intensity–duration relationship for a specified return period cited recently in Koutsoyiannis et al. (1998). These equations have been widely used in hydrology (Keifer and Chu, 1957; Aron et al., 1987).

Three-parameter functional relationships may be suggested, for instance, by incorporating a third parameter into Equations 3.9 and 3.10 or some other types, as follows:

$$I = \frac{R}{(T+C)^B} \tag{3.11}$$

and

$$I = \frac{R}{T^B + C} \tag{3.12}$$

Additional three-parameter equations, which are used in by Garcia-Bartual and Schneider (2001), are

$$I = R + \frac{B}{T+C} \tag{3.13}$$

and

$$I = \frac{R}{B + CT} \tag{3.14}$$

The equation used and recommended in the U.K. Flood Studies Report (Keers and Wescott, 1977) is given as

$$I = \frac{R}{(1+BT)^C} \tag{3.15}$$

For all used equations, the $R$, $C$, and $B$ parameters (for three-parameter equations) are estimated using a nonlinear estimation procedure.

### 3.6.2 STUDY AREA AND DATA

The study area is Wadi Damad in the southwestern part of the Arabian Peninsula along the Red Sea coast where the rainfall amounts are relatively higher than other regions. The rainfall intensity records due to different storms are provided by the Ministry of Agriculture and Water publications (Zahrani et al., 2007). Although rainfall intensity data are available for many years, herein only several storm records are presented for the sake of argument (Table 3.4). This table presents event-based rainfall intensity data in Wadi Damad from 1971 only.

It is obvious that the rainfall intensities are presented in a cumulative manner for each event at 10, 20, 30, and 60 min in addition to 1, 2, 3, 6, and 12 h. The last two columns indicate the total storm rainfall with the corresponding storm duration.

One of the main features between the storm rainfall duration and the total storm rainfall is that they occur independently, as is obvious from Figure 3.13.

There is no systematic relationship between the storm duration and total rainfall, and therefore they are independent from each other.

**TABLE 3.4**
**Wadi Damad Storm Rainfall Intensity Data**

| No. | Date | 10 (min) | 20 (min) | 30 (min) | 1 h | 2 h | 3 h | 6 h | 12 h | Total Rainfall (mm) | Total Duration (h) |
|-----|------|----------|----------|----------|-----|-----|-----|-----|------|---------------------|--------------------|
| 1 | 3/26/1971 | 1.0 | 1.2 | 1.2 | 1.2 | 1.2 | 1.2 | 1.2 | 1.2 | 1.2 | 0.20 |
| 2 | 3/31/1971 | 1.3 | 1.3 | 1.3 | 1.3 | 1.3 | 1.3 | 1.3 | 1.3 | 1.3 | 0.17 |
| 3 | 4/2/1971 | 2.9 | 2.9 | 4.5 | 5.6 | 5.6 | 5.6 | 5.6 | 5.6 | 5.6 | 0.67 |
| 4 | 4/3/1971 | 1.8 | 2.1 | 2.1 | 2.1 | 2.1 | 2.1 | 2.1 | 2.1 | 2.1 | 0.33 |
| 5 | 4/6/1971 | 1.2 | 1.4 | 1.4 | 2.2 | 4.0 | 4.2 | 4.2 | 4.2 | 4.2 | 2.67 |
| 6 | 4/9/1971 | 10.6 | 11.1 | 11.1 | 11.1 | 11.1 | 11.1 | 11.1 | 11.1 | 11.1 | 0.20 |
| 7 | 4/11/1971 | 3.2 | 3.2 | 3.2 | 3.2 | 3.2 | 3.2 | 3.2 | 3.2 | 3.2 | 0.17 |
| 8 | 4/17/1971 | 4.2 | 7.2 | 9.3 | 9.7 | 9.7 | 9.7 | 9.7 | 9.7 | 9.7 | 0.53 |
| 9 | 4/18/1971 | 4.0 | 4.7 | 4.7 | 4.7 | 4.9 | 5.1 | 5.1 | 5.1 | 5.1 | 2.35 |

### 3.6.3 BASIC DATA DERIVATION

It is necessary to calculate durational rainfall intensities from Table 3.4. For this purpose, 5-min, 10-min, 20-min, 30-min, and 60-min are adopted as basic intensity durations. Although, it is possible to calculate the rainfall intensities for all these durations from the values given in Table 3.4, 5-min duration intensities are not recorded, and they are very much required in the practical applications. For this purpose, the first assumption is considered as the following: "Only 60% of the first 10-min rainfall intensity will be considered as values of 5-min."

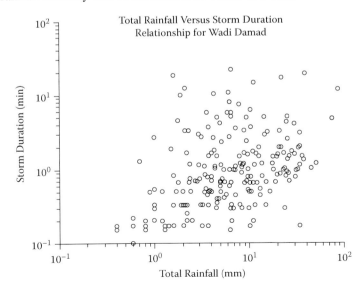

**FIGURE 3.13** Storm duration and total rainfall relationship.

**TABLE 3.5**
**Maximum Rainfall Intensity Values**

| Event No. | 5 min | 10 min | 20 min | 30 min | 60 min |
|---|---|---|---|---|---|
| 1 | 0.60 | 0.40 | 0 | 0 | 0 |
| 2 | 0.78 | 0.52 | 0 | 0 | 0 |
| 3 | 1.74 | 1.16 | 1.6 | 1.1 | 0 |
| 4 | 1.08 | 0.72 | 0 | 0 | 0 |
| 5 | 0.72 | 0.48 | 0 | 0.8 | 0.2 |
| 6 | 6.36 | 4.24 | 0 | 0 | 0 |
| 7 | 1.92 | 1.28 | 0 | 0 | 0 |
| 8 | 2.52 | 1.68 | 2.1 | 0.4 | 0 |
| 9 | 2.40 | 1.60 | 0 | 0 | 0.2 |

Other durations are calculated and presented in Table 3.5, again only for given storms in Table 3.4.

The specification of "maximum" implies the case when there is more than one duration; the maximum of the rainfall intensity during the same durations is taken into consideration. This point can be explained further by considering, say, 10-min storm rainfall durations that can be derived from Table 3.4, where the record durations are 10-min, 20-min, 30-min, 60-min and 1-h, 2-h, 3-h, 6-h, and 12-h. Hence, from this sequence of time durations, it is possible to have three 10-min durations, namely, the first 10-min as in Table 3.4, then the difference as (20 − 10) = 10-min and finally, (30 − 20) = 10-min. There is no other possibility for 10-min duration calculation from the given data in Table 3.4. Hence, the maximum 10-min rainfall amount is adopted as the representative for this duration.

The data in Table 3.5 are for indicated durations but in practice, it is necessary to have 1-h rainfall duration intensity. In this case, Table 3.5 intensities must be converted to uniform intensities such that the unity is [mm/h]. For this purpose, the following procedures are applied to each one of the columns. The first 5-min column is multiplied by 0.5 because the shape of the rainfall is considered as a triangle starting from the storm rainfall initiation. Hence, its area as a triangle should have the multiplication of 1/2 = 0.5. Besides, within 1-h, 5-min is available 12 times, and hence the resultant value must be multiplied by 12 so as to convert to 1-h rainfall. This means that the first value in Table 3.5, which is 0.6 must be multiplied first by 0.5 for the area and then by 12 for the 1-h period. Accordingly, 0.6 × 0.5 × 12 = 3.6 mm results as the value in the first column of the uniform rainfall in Table 3.6. A similar procedure is applied for other columns. The collective results are presented in Table 3.6.

This is the basic table which is used in the following sections for rainfall IDF curve derivations.

**TABLE 3.6**
**Uniform Intensity Conversions to mm/hour**

| 5 min | 10 min | 20 min | 30 min | 60 min |
|-------|--------|--------|--------|--------|
| 3.60  | 1.20   | 0      | 0      | 0      |
| 4.68  | 1.56   | 0      | 0      | 0      |
| 10.44 | 3.48   | 2.4    | 1.1    | 0      |
| 6.48  | 2.16   | 0      | 0      | 0      |
| 4.32  | 1.44   | 0      | 0.8    | 0.2    |
| 38.16 | 12.72  | 0      | 0      | 0      |
| 11.52 | 3.84   | 0      | 0      | 0      |
| 15.12 | 5.04   | 3.15   | 0.4    | 0      |
| 14.40 | 4.80   | 0      | 0      | 0.2    |

### 3.6.4  PROBABILITY DISTRIBUTION FUNCTIONS (PDFS)

The frequency distribution function (*histogram*) of 5-min intensity values in Table 3.6 is shown in Figure 3.14.

A first glance indicates that the rainfall intensity frequency distribution is of arid and semi-arid regions characteristically similar to Figure 3.2, where low rainfall intensities occur with relatively higher frequencies than medium and low intensities. This is the behavioral trend of an exponential PDF model, $f(i)$, given as (Feller, 1967),

$$f(i) = \frac{1}{\mu} e^{-\frac{1}{\mu}i} \tag{3.16}$$

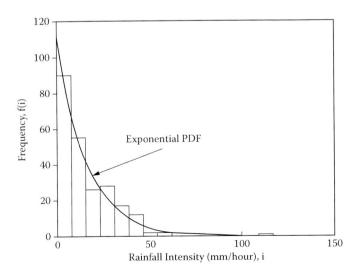

**FIGURE 3.14**   Frequency distribution of 5-min intensities.

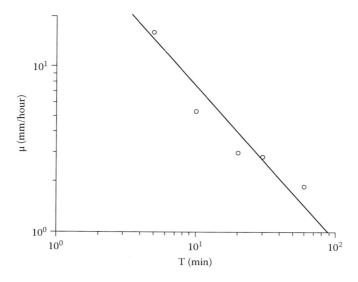

**FIGURE 3.15**   Average intensity–duration relationship.

where $i$ is the rainfall intensity and $\mu$ is the sole model parameter, which is equivalent to the arithmetic average of the available rainfall intensities. Hence, the average intensity for 5-min data is $\mu = 15.92$ mm/h. The corresponding exponential model is given in Figure 3.14.

The same concept is applied to other time duration intensities (10-min, 20-min, 30-min, and 60-min), but as the time duration increases the number of intensities decrease, and, therefore, many zero values enter the frequency distribution calculations (Table 3.6), which are not considered in further calculations. This will mean that the number of storm events will decrease, but still there is enough data even for long time durations. The rainfall intensity frequency distribution curves for 10-min, 20-min, 30-min, and 60-min have arithmetic average rainfall intensities as 5.32 mm/h, 2.96 mm/h, 2.80 mm/h, and 1.87 mm/h, respectively.

### 3.6.5   AVERAGE INTENSITY–DURATION MODEL

It is obvious from the calculations that as the time duration increases the model parameter, $\mu$, decreases. It is possible to identify the relationship between the average intensity and duration as in Figure 3.15.

The scatter of points yields to a straight-line on a double-logarithmic graph paper and hence the mathematical relationship is a power function, which can be obtained after the regression method application as,

$$\mu = 0.0151T^{-0.934} \tag{3.17}$$

This relationship provides the ability to know the average intensity for any desired duration during the intensity calculations. For instance, if the time duration is 15 min then the average rainfall intensity can be estimated from Equation 3.17 as 5.25 mm/h.

### 3.6.6 INTENSITY–DURATION–FREQUENCY CURVES

In practical applications, all over the world, the *design periods* are taken as T = 2, 5, 10, 25, 50, 100, and 500 years. These *return periods* imply physically that there will be only one extreme event, which may exceed the design value during the whole period. This sentence leads to the *probability of occurrence* (frequency) or *risk*, R, of future events as the reciprocal of the design period, T, as in Equation 3.7. Hence, the corresponding frequency values attached to the return period are 0.50, 0.20, 0.10, 0.02, 0.01, and 0.002, respectively. The substitution of Equation 3.16 into Equation 3.8 after some simple algebra leads to

$$R = \int_{i_D}^{+\infty} \frac{1}{\mu} e^{-\frac{1}{\mu}i} \, di = e^{-\frac{1}{\mu}i_D} \tag{3.18}$$

and finally,

$$i_D = -\frac{1}{\mu} \log R \tag{3.19}$$

Hence, substitution of each R and μ values for each duration yields design rainfall intensity values as in Table 3.7.

Finally, the plot of time duration versus intensities for each return period gives IDF curves on a double logarithmic paper as shown in Figure 3.16.

## 3.7 RAINY AND NONRAINY DAYS

In arid regions, any day with rainfall amount not equal to zero is considered as wet (rainy), otherwise, it is dry (nonrainy). Hence, an uninterrupted sequence of rainy (nonrainy) days preceded and succeeded by at least one nonrainy (rainy) day is referred to as rainy, R, (nonrainy, NR) spell. In humid regions, the alternating nature of R and NR periods of any hydrologic sequence can be identified objectively by its

**TABLE 3.7**
**IDF values**

| Return Period | Frequency | Duration (min) | | | | | |
|---|---|---|---|---|---|---|---|
| | | 5 | 10 | 15 | 20 | 30 | 60 |
| 2 | 0.50 | 11.03 | 3.68 | 2.86 | 2.05 | 1.94 | 1.29 |
| 5 | 0.20 | 25.62 | 8.54 | 6.65 | 4.76 | 4.50 | 3.00 |
| 10 | 0.10 | 36.65 | 12.22 | 9.52 | 6.81 | 6.44 | 4.30 |
| 25 | 0.04 | 51.24 | 17.09 | 13.31 | 9.52 | 9.01 | 6.01 |
| 50 | 0.02 | 62.27 | 20.77 | 16.17 | 11.57 | 10.95 | 7.31 |
| 100 | 0.01 | 73.31 | 24.45 | 19.04 | 13.63 | 12.89 | 8.61 |
| 500 | 0.002 | 98.93 | 32.99 | 25.69 | 18.39 | 17.40 | 11.62 |

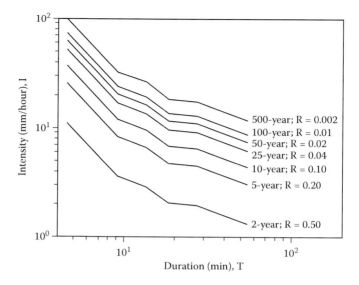

**FIGURE 3.16**   IDF curves.

truncation at a certain level, which may correspond to zero rainfall amounts in arid regions as in Figure 3.17.

The main points in the truncation of a series for the wet and dry duration definitions are the up-crossing and down-crossing points where the rainfall event changes its state from *dry* to *wet* or from wet to dry spell, respectively. The dry (wet) duration is the difference between the up-crossing (down-crossing) and previous down-crossing (up-crossing) occurrence times. The following important points emerge from a truncation procedure.

1. Wet and dry periods follow each other alternatively.
2. The total duration of wet and dry periods is equal to the whole duration of record.
3. The occurrences of up- and down-crossing points along the truncation level are irregularly distributed.

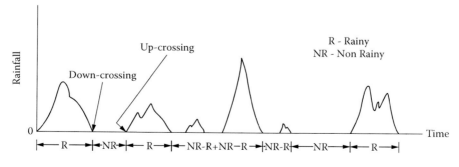

**FIGURE 3.17**   Schematic rainy and nonrainy spells.

4. The wet and dry period durations occur randomly.
5. The total duration of wet periods is invariably smaller than total dry periods at any station in arid regions. In other words, dry periods are more persistent than wet periods.
6. By definition, the duration of any wet (dry) period is equal to the number of uninterrupted sequence of wet (dry) days.
7. In arid regions, the smaller the wet spell duration the more intensive is the rainfall amount. It is tantamount to saying that the overall maximum rainfall amount appears within a short duration wet period.

In general, the significance of wet and dry periods can be summarized as follows on the basis of various hydrological phenomena.

1. Wet spell durations play an important role in the water supply, agricultural planning, and forest management.
2. Dry and wet durations are helpful in quantitative description of drought and flood or flash flood occurrence assessments (predictions), respectively.
3. They guide to actual evaporation calculations on a daily basis. Such an approach is very essential especially in arid and semi-arid regions.
4. Rainfall runoff relationship identification and runoff prediction in order to control surface flow discharges. There are no *perennial* streams in arid regions, and consequently, there are runoff records only for short periods after the occurrence of consecutive rainy days.
5. Sequence of dry and wet durations is significant in the operation of flood protection, water supply, and recreational dam storages.
6. Groundwater management studies need information on consecutive wet and dry spells. During dry spells groundwater abstraction is expected to increase.

All the classical methods applied so far in the dry (wet) duration evaluations are invariably based on the statistical, probabilistic and stochastic techniques for their objective descriptions (Yevjevich, 1967, 1972; Şen, 1976, 1980a, 1984).

### 3.7.1 FRACTAL DIMENSION OF RAINY AND NONRAINY DURATIONS

Let us suppose that there are n dry durations, $d_1$, $d_2$, $d_3$, ..., $d_n$ over a truncation level. The following steps are necessary for *fractal dimension* derivation of dry and similarly wet durations.

1. Select basic dry unit duration, d(1), conveniently as one day, month, or year, depending on the available data. For instance, each month may be conveniently represented by 1 cm.
2. Take a pair of dividers and set it to this basic unit segment.
3. Walk the dividers along the first drought duration and count the number of segments, $n_1$, for the basic unit.
4. Repeat step (3) for each drought duration and the result is a sequence of integer numbers, $n_1$, $n_2$, $n_3$, ..., $n_n$.

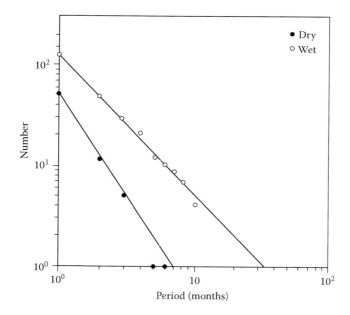

**FIGURE 3.18**  Fractal dimension of wet–dry spells.

5. Add the step numbers so as to find the total steps, $N(1) = n_1 + n_2 + n_3 + ...+ n_n$. The value of $N(1)d(1)$ is an estimate of the total drought duration, $D(1)$, with $d(1)$ unit duration.

6. Increase the length of segment (dividers) length by one and then repeat steps 3 to 5. The result is $N(2)$ step number with $d(2)$ segment length and $N(2)d(2)$ drought duration. Continue to do the same procedure until the dividers have segment length equal almost to the maximum drought duration length in the record.

7. There are now several estimates of total drought duration, $D(j)$, at different segment lengths, $d(j)$ and numbers $N(j)$ where $j = 1, 2, ..., m$, and m is the number of possible segments. It is obvious that an increase in $d(j)$ will cause decrease in $D(j)$.

Figure 3.18 gives the plot $D(j)$ versus $d(j)$ on a log–log paper, which is referred to as Richardson's plot in the literature (Mandelbrot, 1979). The methodology presented herein gives rise to double logarithmic plotting of dry (wet) spell duration versus their number of occurrences, and consequently, straight lines are fitted through the points. If the wet period straight-line is above the dry straight line, then the location is regarded as having a rather *humid* climate; otherwise, *aridity* prevails. If the wet and dry period straight lines are very close to each other, then the area is said to have a semi-arid climate. The straight-line method on a double logarithmic graph paper provides a basis for the identification of a corresponding generation model for wet (dry) periods (Şen, 1984).

The points, however, may not fall exactly on a straight line since there will be scatters due to either the paucity of data and/or sampling variability. The application

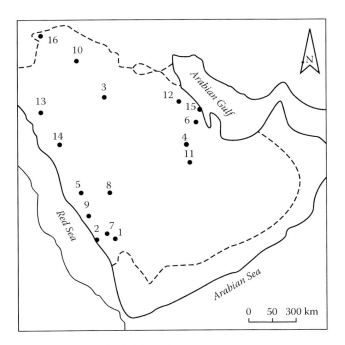

**FIGURE 3.19**   Location map of rainfall stations.

of a simple regression analysis will give the best fitting straight line, which will have the following three important implications:

1. The slope of the straight line on log–log paper is related somehow to the irregular occurrences of dry (wet) durations. Logically, steeper lines will imply occurrence of frequent droughts with relatively smaller variability.
2. The more the segment (dividers) length is reduced, the more accurate will be the estimation of total drought durations. By extending the straight lines in both directions, one can see the dry (wet) period variations for smaller or larger scales than available in the record.
3. The degree of irregularity involved at each step is almost the same such that both the large and small scale details of the drought feature are similar geometrically except their scale. This is the *self-similarity* property of *fractal dimensions* (Mandelbrot, 1979).

The methodology developed is applied to monthly rainfall sequences by Bazuhair et al. (1997) at 16 observation stations representatively distributed within different climatologic regions of the Kingdom of Saudi Arabia (Figure 3.19). Station characteristics are represented in Table 3.8.

Some of the stations are located on coastal plains, which show extreme dry periods. However, the stations on the escarpment receive comparatively higher rainfall amounts more frequently, and therefore, their dry durations are shorter.

**TABLE 3.8**
**Rainfall Record Characteristics in Saudi Arabia**

| Serial Number | Station Name | Record Duration | Annual Average Rainfall (mm) | Wet Period (%) | Dry Period (%) | Elevation (m) |
|---|---|---|---|---|---|---|
| 1 | Al-Amir | 1966–1986 | 33 | 79 | 21 | 2,100 |
| 2 | Qamah | 1965–1986 | 24 | 31 | 69 | 20 |
| 3 | Ba'Qaa | 1967–1986 | 9 | 47 | 53 | 755 |
| 4 | Harad | 1966–1986 | 18 | 49 | 51 | 300 |
| 5 | Turabah | 1965–1986 | 17 | 43 | 57 | 1,126 |
| 6 | Hofuf | 1967–1986 | 15 | 45 | 55 | 160 |
| 7 | Abha | 1965–1986 | 30 | 87 | 13 | 2,200 |
| 8 | Beljurshi | 1965–1986 | 38 | 86 | 14 | 2,400 |
| 9 | Ranyah | 1965–1986 | 21 | 54 | 46 | 810 |
| 10 | Sakakah | 1965–1986 | 11 | 45 | 55 | 574 |
| 11 | Yatrib | 1968–1986 | 11 | 36 | 64 | 119 |
| 12 | Sarar | 1968–1986 | 16 | 49 | 51 | 75 |
| 13 | Al-Ula | 1966–1986 | 17 | 36 | 64 | 650 |
| 14 | Mosajid | 1966–1986 | 18 | 39 | 61 | 471 |
| 15 | Qatif | 1967–1986 | 18 | 40 | 60 | 5 |
| 16 | Qurrayat | 1967–1986 | 9 | 46 | 55 | 549 |

Few stations are at elevations over 2,000 m above m.s.l. and, therefore, their wet durations are expected to be significantly long.

In order to present the application of the methodology in finding the durations of wet and dry spells as well as their numbers, monthly rainfalls at the Al-Amir station from 1966 to 1979 are given in Table 3.9.

The sequence of monthly rainfall amounts are converted into wet- and dry-spell numbers of preset segments along the available record. If the interest lies in the number of wet and dry periods of, say, a 2-month (i.e., two successive months) duration only, then calculations from Table 3.9 show that there are 49 wet and 12 dry cases, respectively. Similarly, for wet or dry periods of six uninterrupted consecutive months 10 and 1 occurrences are found within the Al-Amir monthly records. Hence, for all the possible segments the results are presented in Table 3.10.

As expected, in general, a first glance at Table 3.10 shows that there is an inverse relationship between the duration (period) and number of wet and dry months. Figure 3.18 is for Al-Amir station, and it shows such a straight-line relationship on a log–log graph paper. The slope of the straight line is in reverse relationship with the duration of wet or dry periods. The steeper the slope, the shorter will be the duration change. Due to its elevation and hence the effect of orographic rainfall occurrences, there is a comparatively smaller number of dry spells than wet spells, and longer wet spells occur rather frequently.

It is possible to obtain a preliminary prediction of the possible longest wet (dry) spell duration, on the average, from Figure 3.18 for the Al-Amir station,

**TABLE 3.9**
**Al-Amir Station Monthly Rainfall Amounts**

|      |      |       |       |       |       | Month |      |       |      |      |      |      |
|------|------|-------|-------|-------|-------|-------|------|-------|------|------|------|------|
| Year | J    | F     | M     | A     | M     | J     | J    | A     | S    | O    | N    | D    |
| 1979 | 49.0 | 00.0  | 25.5  | 02.5  | 65.5  | 25.5  | 12.0 | 138.0 | 24.3 | 20.4 | 00.0 | 18.0 |
| 1978 | 29.2 | 30.4  | 07.0  | 27.2  | 61.0  | 04.0  | 72.4 | 59.0  | 10.3 | 00.0 | 00.0 | 01.5 |
| 1977 | 20.8 | 02.5  | 20.0  | 19.5  | 66.9  | 07.3  | 33.9 | 108.0 | 00.0 | 16.2 | 00.0 | 00.0 |
| 1976 | 06.4 | 00.0  | 102.3 | 37.8  | 45.5  | 11.8  | 20.7 | 49.9  | 00.0 | 07.0 | 35.0 | 00.0 |
| 1975 | 13.0 | 11.5  | 09.5  | 157.0 | 08.0  | 21.2  | 29.1 | 90.9  | 07.9 | 00.0 | 03.5 | 00.0 |
| 1974 | 00.0 | 00.0  | 212.0 | 18.5  | 56.2  | 38.0  | 17.8 | 116.0 | 16.5 | 00.0 | 00.0 | 00.0 |
| 1973 | 26.3 | 00.0  | 00.0  | 08.2  | 35.7  | 10.6  | 18.7 | 34.6  | 01.0 | 02.5 | 07.0 | 54.8 |
| 1972 | 08.2 | 05.2  | 103.0 | 141.0 | 50.3  | 14.5  | 69.9 | 38.8  | 00.0 | 06.5 | 03.0 | 09.5 |
| 1971 | 00.0 | 00.0  | 00.0  | 129.0 | 180.0 | 08.7  | 00.0 | 48.6  | 00.0 | 00.0 | 00.0 | 12.2 |
| 1970 | 40.3 | 00.0  | 91.1  | 00.0  | 00.0  | 16.5  | 30.2 | 21.8  | 16.8 | 00.0 | 00.0 | 00.0 |
| 1969 | 77.5 | 98.1  | 63.5  | 34.1  | 74.1  | 10.1  | 30.2 | 18.1  | 10.1 | 00.0 | 00.0 | 00.0 |
| 1968 | 00.0 | 76.4  | 00.0  | 66.6  | 58.6  | 55.7  | 09.7 | 096.3 | 00.0 | 02.1 | 22.4 | 00.0 |
| 1967 | 00.0 | 00.0  | 30.7  | 75.9  | 70.3  | 11.5  | 35.3 | 29.4  | 19.7 | 00.0 | 70.5 | 00.0 |
| 1966 | 00.0 | 119.0 | 00.0  | 65.6  | 14.5  | 16.3  | 11.6 | 27.9  | 00.0 | 00.0 | 29.3 | 00.0 |

**TABLE 3.10**
**Wet and Dry Periods and Their Durations**

|                  | Number |     |
|------------------|--------|-----|
| Periods (months) | Wet    | Dry |
| 1                | 120    | 49  |
| 2                | 49     | 12  |
| 3                | 30     | 5   |
| 4                | 21     | 1   |
| 5                | 12     | 1   |
| 6                | 10     | 1   |
| 7                | 9      | 0   |
| 8                | 7      | 0   |
| 9                | 4      | 0   |

corresponding to one month. Hence, at Al-Amir the expected longest wet (dry) spell durations appear as intercepts of the straight lines on the horizontal axis in Figure 3.18 as 33 months (7 months). Similar calculations are carried out for other stations and Table 3.11 summarizes the straight-line intercept and slope values for monthly wet (dry) spell durations and represents the wet and dry spell duration number charts for all the stations.

**TABLE 3.11**

**Intercept and Slope Values for Various Stations**

| Number | Station Name | Intercept (month) | | Slope (1/month) | |
|:---:|:---|:---:|:---:|:---:|:---:|
| | | Wet | Dry | Wet | Dry |
| 1 | Al-Amir | 65 | 7 | 1.25 | 2.00 |
| 2 | Qamah | 9 | 38 | 2.20 | 1.53 |
| 3 | Ba'Qaa | 24 | 35 | 1.55 | 1.39 |
| 4 | Harad | 16 | 52 | 1.44 | 1.26 |
| 5 | Turabah | 17 | 27 | 1.65 | 1.53 |
| 6 | Hofuf | 30 | 40 | 1.39 | 1.32 |
| 7 | Abha | 90 | 5 | 1,19 | 2.29 |
| 8 | Beljurshi | 100 | 6 | 1.20 | 2.12 |
| 9 | Ranyah | 21 | 17 | 1.60 | 1.70 |
| 10 | Sakakah | 32 | 55 | 1.40 | 1.25 |
| 11 | Yatrib | 18 | 55 | 1.58 | 1.25 |
| 12 | Sarar | 32 | 42 | 1.39 | 1.26 |
| 13 | Al-Ula | 6 | 35 | 2.38 | 1.49 |
| 14 | Mosajid | 12 | 26 | 1.80 | 1.50 |
| 15 | Qatif | 27 | 34 | 1.44 | 1.39 |
| 16 | Qurrayat | 26 | 34 | 1.44 | 1.39 |

The following significant interpretations can be drawn from Table 3.11 and the corresponding graphs in Figure 3.20.

1. The station location is referred to as humid if the wet period straight line is significantly above the dry line (Al-Amir in Figure 3.18; Abha and Beljurshi in Figure 3.20). Otherwise, the location is under arid conditions (Harad, Yatrib, Al-Ula, Mosajid, and Qamah in Figure 3.21). *Semi-aridity* prevails if these two lines are close to each other (Ba'Qaa, Turabah, Hofuf, Ranyah, Sakakah, Sarar, Qatif and Qarrayat in Figure 3.20). Hence, out of the 16 rainfall stations three, eight, and five are *humid*, *semi-humid*, and *arid* locations, respectively. Furthermore, it is important to notice that humid locations are situated in the southwestern part of the Arabian Peninsula, where the Asir Mountains reach elevations up to 3,000 m above mean sea level. This area experiences high rainfall amounts due to orographic and monsoon effects.
2. There are no crossings of wet and dry straight lines. This is tantamount to saying that wet and dry periods are always shorter or longer than each other.
3. The slope of straight lines indicates the intensity of wet and dry periods. The smaller the slope, the shorter and more persistent is the wet or dry period. For instance, among the plots in Figure 3.21, Beljurshi station has the least wet-period slope, which implies that this station has the longest and most persistent durations of wet spells. Similarly, the shortest dry period occurs

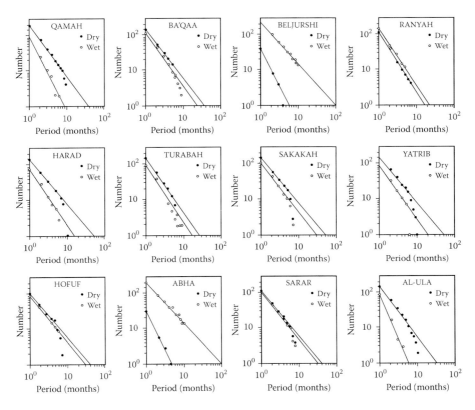

**FIGURE 3.20**  Wet (dry) spell period number of occurrences.

at Abha station because it has the minimum slope (1.19 per month) among other stations (see Table 3.11).

## 3.8  RAINFALL DISTRIBUTION FUNCTIONS AND PREDICTION

The random behavior of rainfall patterns in arid or semi-arid regions such as in Libya makes their prediction rather difficult. However, probability and statistical methodologies provide the necessary ways for such a goal. It is, therefore, a primary prerequisite to derive the frequency distributions of rainfall records at each available meteorological station with a theoretically convenient PDF. Such studies have been considered by many researchers (Taylor and Lawes, 1971; Fiddes, 1977; Martin, 1988; Abouammoh, 1991; Wallis and Hosking, 1993).

The study area lies within Libya at approximately between 34°N to 33°N latitude and 10°E to 23°E longitude. This area stretches about 950 km away from the Mediterranean coast toward the south and about 1,450 km from west to east along the Mediterranean Sea (Figure 3.21). There is a fair number of meteorological stations in the study area; however, the period of rainfall records are frequently different. Most of the meteorological stations are concentrated on the northwestern part.

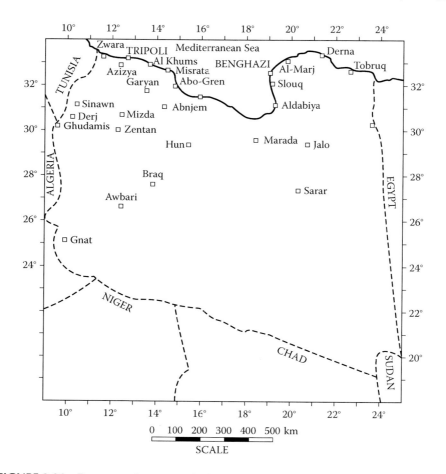

**FIGURE 3.21** Representative meteorological stations.

Herein, monthly rainfall data are used rather than daily because most daily rainfall records have zeros as in any arid region.

Prior to processing the rainfall data, it is first necessary to assess the reliability and the quality of the observations. It is observed during the study that interstation correlation of monthly rainfalls was very low as expected in arid regions. Missing data for many stations are filled from observations at least three nearby stations by using the *normal ratio* method. The rainfall amounts at the index station are weighted by the ratio of normal annual rainfall values (Summer, 1988). Kohler (1949) and Berndtsson and Niemczynowicz (1986) noted that, *double mass curve* may be derived to test short and long periods of rainfall data for homogeneity over time. The mass curve for monthly and annual rainfall will normally approach a straight line, assuming that there is no consistent trend in month to month or year to year rainfall totals. Over a long data period therefore, the normal expectation is that, while there may be irregularities in the curve, the overall trend will be clear and consistent. If, however, consistent changes in the rainfall regime are present, then a clear inflexion will appear on the mass curve near the point when the change occurs. Here, the mass curves are examined in order to know whether the available data are homogeneous or not. Mass curve results show that, the rainfall data at 18 stations became a straight line which implies that the data are homogeneous. However, data at 11 stations consist of a series of straight lines, and they are corrected accordingly.

After the necessary reliability checks, 29 stations are considered for calculation, analysis, and prediction purposes (Eljadid, 1997). The rainfall records run at least 23 years in duration, which is enough to assess the temporal and special characteristics in arid regions.

### 3.8.1  FREQUENCY DISTRIBUTION

Frequency distributions (histograms) and their theoretical counterparts in the form of PDFs are basic tools in any climatological analysis. They help to find relevant probabilities that may provide a basis for climatological predictions. Several fundamental PDFs are available, including the normal, gamma, extreme value, binomial, Poisson, and negative binomial distributions (Walpole, 1982). It is possible to observe quickly from a histogram and its PDF that if most observations are concentrated near zero, then *gamma PDF* or as its special case *exponential PDF* is suitable. The histograms and corresponding exponential PDFs are drawn for some stations in Libya as shown in Figure 3.22.

All the histograms appear as skewed to the right (positively skewed), because they have long right tails with almost zero frequencies at the higher data values. The large values in the right tail are not offset by correspondingly low values near zero, which means that the arithmetic mean, $\mu$, is greater than the median, $m$, and mode (the most frequently occurring value in the record). The difference between the mean and median divided by the standard deviation, $\sigma$, can be used as a definition of the skewness, $\gamma$, measure as

$$\gamma = \frac{3(\mu - m)}{\sigma} \qquad (3.20)$$

In general, the values of $\gamma$ falls between $-3$ and $+3$ (Walpole, 1982). For a perfectly symmetrical distribution, the mean and median are identical and the value of $\gamma$ is zero. Libyan rainfall dataset skewness is always positive as a reflection of arid zones. A glance at Figure 3.22 indicates the following distinctive characteristics (Şen and Eljadid, 1999).

1. All the histograms have their maximum frequencies at the lowest monthly rainfall amounts. This means that Libya is an arid country (Section 3.2), which is to say that very low intensity rainfalls occur rather frequently all over Libya.
2. Very intensive rainfall occurrences are rather rare but, when they happen, the amount of rainfall is quite high.

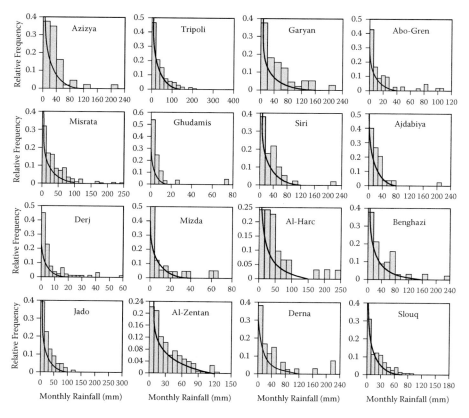

**FIGURE 3.22**    Histogram and PDF for some stations in Libya.

3. In some stations the transition from very low to high intensity rainfall amounts is smooth in a continuously decreasing form such as at stations Tripoli, Misrata, and Jado.

4. In some of the stations there are moderately high frequency occurrences between small and large intensities—for example, at Zwara and Ghudamis.

5. In some of the histograms there are continuous rainfall amounts without "zero" frequency over the range of the data. If all the rainfall amounts have continuous nonzero frequencies, this corresponds to a rather *humid* location.

6. The distribution of frequencies within each class interval over the data has different variability ranges.

7. It is possible to make interpretations as for the change of frequencies with the data value. Hence, one can decide about the most frequently occurring data values.

8. The frequency of extreme value occurrences is easily observed.

9. It is also possible to identify the skewness of the data toward to lower or higher values.

Although histograms give the impression that they are of *exponential* type but the statistical tests on the basis of chi-square ($\chi^2$) criterion have shown that they all come from a theoretical gamma PDF family. The most convenient theoretical gamma PDFs are also shown as continuous curves in Figure 3.22. Especially, the rainfall frequency distributions at Tripoli, Zwara, Azizya, Misrata, Jado, Ghudamis, Mizda, Al-Zentan, Abo-Gren, Benghazi, Al-Marj, Derna, and Slouq comply almost perfectly with the gamma PDF.

The choice of suitable PDF is often dictated by convenience or procedural policy, primarily owing to the lack of a basis for selecting the distribution of best fit. According to histograms in Figure 3.22, the most suitable procedure is based on fitting gamma PDFs. $\chi^2$ test has been used to examine the type of the PDF, which is suitable for such analysis. In general, the goodness of fit test between observed (histogram), $O_i$, and expected (PDF), $E_i$, frequencies is based on the calculation of $\chi^2$ quantity as,

$$\chi^2 = \sum_{i=1}^{k} \frac{\left(O_i - E_i\right)^2}{E_i} \tag{3.21}$$

where k is the number of classes in each histogram and the sampling distribution of $\chi^2$ is approximated very closely by the chi-square distribution. If $O_i$'s are close to $E_i$'s, then the $\chi^2$ value is small, which indicates a good fit.

### 3.8.2 GAMMA PDF PREDICTIONS

The application of gamma PDF is suggested by Barger and Thom (1949), and since then it has been adapted to rainfall dataset by various researchers (Mooley and Crutcher, 1968; Wihl and Nobilis, 1975; Mooley, 1973). The gamma PDF gives good fit to rainfall climatological series (WMO, 1966). It is one of the most useful continuous PDFs available for many natural event occurrences, especially in arid regions. It is defined by shape, $\alpha$, and scale, $\beta$, parameters as

$$P_X(X) = \frac{1}{\Gamma(\beta)} \alpha^\beta X^{\beta-1} e^{-\alpha X} \tag{3.22}$$

where $X$ is the rainfall amount, and $\Gamma(\beta)$ is called the gamma function with the properties,

$$\Gamma(\beta) = (\beta - 1) \qquad \text{for } \beta = 1, 2, 3, \ldots \tag{3.23}$$

$$\Gamma(\beta) = \int_0^\infty t^{\beta-1} e^{-t} dt \qquad \text{for } \beta > 0 \tag{3.24}$$

and

$$\Gamma(1) = \Gamma(2) = 1, \qquad \Gamma(1/2) = \sqrt{\pi} \tag{3.25}$$

On the other hand, the cumulative gamma PDF, $p(X)$, is given as

$$p(X) = \frac{\int\limits_{0}^{X} \alpha^{\beta} t^{\beta-1} e^{-\alpha t} dt}{\Gamma(\beta)} \tag{3.26}$$

which can be evaluated using a table of the incomplete gamma function given in any handbook of mathematics (Pearson, 1957).

The moments in this instance give poor estimates of the parameters, and thus the maximum likelihood fit is recommended. Sufficient estimates are, however, available and these are closely approximated by the following formula (WMO, 1966)

$$\beta = \frac{1}{4A}\left(1 + \sqrt{1 + \frac{4A}{3}}\right) \tag{3.27}$$

$A$ is given by,

$$A = Ln\overline{X} - \frac{\sum\limits_{i=1}^{n} LnX_i}{n} \tag{3.28}$$

where $X_i$ (i = 1, 2, 3, …, n) is the rainfall data and $\overline{X}$ is the arithmetic average. The parameter $\alpha$ is then estimated by

$$\alpha = \frac{\beta}{\overline{X}} \tag{3.29}$$

As summarized by Stevens and Smulders (1979), the values of $\alpha$ and $\beta$ can be obtained from the method of moments, the method of energy pattern factor, the maximum likelihood method, the Weibull PDF paper method, or by the use of percentile estimators.

### 3.8.3 Application

All 29 meteorological stations across Libya are selected to develop the future probability predictions of the rainfall amounts. Two maps, one for the shape and the other for scale parameters, are developed as in Figures 3.23 and 3.24. Especially at sites where the stations are not available, i.e., given the latitude and longitude, it is possible to obtain the values of $\alpha$ and $\beta$ from these maps.

In order to estimate the probability of rainfall exceeding 10 mm, 25 mm, 50 mm, and 100 mm over the study area, it is necessary to consider the standard forms of the prediction. For instance, prediction of rainfall exceedence of 10 mm is given explicitly as $P_X(\text{rain} > 10 \text{ mm}) = 1 - P_X(\text{rain} < 10 \text{ mm})$, which can be calculated from Equation 3.22 as

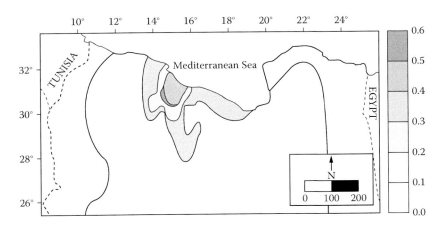

**FIGURE 3.23**   Shape parameter distribution ($\alpha$).

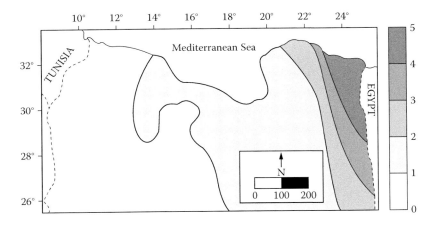

**FIGURE 3.24**   Scale parameter distribution ($\beta$).

$$P_X(rain < 10) = \frac{1}{\Gamma(\beta)} \int_0^{10} \alpha^\beta X^{\beta-1} e^{-\alpha X} dX \qquad (3.30)$$

The use of this equation together with the incomplete gamma function tables lead to calculations for rainfall exceedence predictions as presented in Table 3.12.

According to the general pattern of the Libyan actual rainfall dataset, it is clear that Libya receives less rainfall within its peak periods. Any future prediction based on the actual dataset should follow the same pattern. Results show that in most of the stations the probability of rainfall amounts with 10 mm has a strong chance to occur rather than the other rainfall estimated probability. It is clear from Table 3.12 that there is an inverse relationship between the probability of occurrences and the

**TABLE 3.12**
**Probabilities of Rainfall Amounts Based on Gamma Function**

| Number | Station Name | 10 mm | 25 mm | 50 mm | 100 mm |
|---|---|---|---|---|---|
| 1 | Zwara | 0.37109 | 0.15730 | 0.04550 | 0.00468 |
| 2 | Beniwalid | 0.57241 | 0.91889 | 0.01857 | 0.00017 |
| 3 | Azizya | 0.67032 | 0.36788 | 0.13534 | 0.01832 |
| 4 | Hun | 0.54881 | 0.22313 | 0.04979 | 0.00248 |
| 5 | Tripoli | 0.67032 | 0.36778 | 0.13534 | 0.01832 |
| 6 | Jaghbub | 0.63763 | 0.27253 | 0.08209 | 0.01111 |
| 7 | Al-Khums | 0.54881 | 0.22313 | 0.04979 | 0.00248 |
| 8 | Awbari | 0.63776 | 0.36788 | 0.13534 | 0.01111 |
| 9 | Misrata | 0.63776 | 0.27253 | 0.08209 | 0.01111 |
| 10 | Brag | 0.49363 | 0.11161 | 0.00738 | 0.00008 |
| 11 | Ghudamis | 0.08327 | 0.00584 | 0.00011 | 0.00000 |
| 12 | Derj | 0.22067 | 0.04550 | 0.00468 | 0.00011 |
| 13 | Sinawn | 0.05778 | 0.00336 | 0.00004 | 0.00000 |
| 14 | Mizda | 0.19229 | 0.03197 | 0.00270 | 0.00002 |
| 15 | Tobruk | 0.29427 | 0.08327 | 0.01431 | 0.00091 |
| 16 | Marada | 0.23672 | 0.05778 | 0.00815 | 0.00018 |
| 17 | Jado | 0.47950 | 0.31731 | 0.15730 | 0.02535 |
| 18 | Al-Zentan | 0.37109 | 0.15730 | 0.04550 | 0.00468 |
| 19 | Sarar | 0.90484 | 0.91889 | 0.80125 | 0.57241 |
| 20 | Garyan | 0.75300 | 0.39163 | 0.11161 | 0.00738 |
| 21 | Gnat | 0.50653 | 0.27253 | 0.08209 | 0.00674 |
| 22 | Abo-Gren | 0.34278 | 0.15730 | 0.04550 | 0.00270 |
| 23 | Abnjem | 0.09426 | 0.00815 | 0.00018 | 0.00000 |
| 24 | Sirt | 0.61493 | 0.20354 | 0.02929 | 0.00044 |
| 25 | Ajdabiya | 0.40657 | 0.10026 | 0.01111 | 0.00012 |
| 26 | Benghazi | 0.80879 | 0.40601 | 0.09580 | 0.00302 |
| 27 | Al-Marj | 0.77707 | 0.39163 | 0.11161 | 0.01173 |
| 28 | Derna | 0.95592 | 0.47626 | 0.04026 | 0.00006 |
| 29 | Slouq | 0.68227 | 0.30802 | 0.07190 | 0.03600 |

assumed rainfall amounts over all regions in the study area. That is, the probability of occurrences decreases as the rainfall amounts increase. In addition, results reflect some special distribution of the rainfall over the coastal and noncoastal stations. Coastal stations especially at the western and eastern zones such as at Zwara, Sarman, Tripoli, Al-Khums, Benghazi, Derna, Al-Marj, and Slouq have rainfall amounts more than other stations and in general, rainfall amounts decrease southward as shown in Figure 3.25.

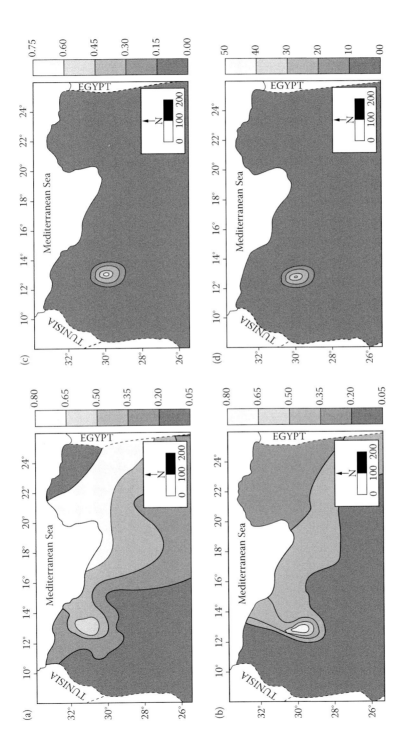

**FIGURE 3.25** Probability distribution of rainfall amounts: (a) 10 mm, (b) 25 mm, (c) 50 mm, and (d) 100 mm.

**TABLE 3.13**
**Stations with 50% or More**
**Rainfall Occurrences at 10 mm**

| Number | Station Name | % |
|--------|--------------|-----|
| 1 | Sarman | 57.2 |
| 2 | Al-Zawiya | 67.0 |
| 3 | Azizya | 54.8 |
| 4 | Tripoli | 67.0 |
| 5 | Al-Qrabolli | 63.7 |
| 6 | Al-Khums | 54.8 |
| 7 | Zlitan | 63.7 |
| 8 | Misrata | 63.7 |
| 9 | Yefrn | 90.0 |
| 10 | Garyan | 75.3 |
| 11 | Trhona | 60.0 |
| 12 | Sirt | 61.4 |
| 13 | Benghazi | 80.0 |
| 14 | Al-Marj | 77.7 |
| 15 | Derna | 95.5 |
| 16 | Slouq | 68.2 |

A total of 16 meteorological stations have more than 50% probability of occurrence with 10 mm rainfall amounts (see Table 3.13). On the other hand, there are some percentages for rainfall to occur with the values of 25 mm and 50 mm, especially over the coastal stations, but the percentages of occurrences with 100 mm are quite insignificant within that region.

## 3.9   SPATIAL DISTRIBUTION OF RAINFALL

Spatial rainfall interpolation (SRI) procedures are available in the literature in such a way that the whole drainage basin is subdivided into a set of mutually exclusive and exhaustive subareas in the forms of triangular, polygon, or contour lines. Each one of these methods provides a uniform rainfall depth over subareas, which cannot reflect the real situation. In fact, the rainfall records are expected to change spatially in such a manner that each point has different but spatially related rainfall amount.

A single point rainfall measurement is quite often not representative of the volume of rainfall falling over a given catchment area. A dense network of point measurements and radar estimates can provide a better representation of the true volume over a given area. A network of rainfall measurements can be converted to areal estimates using any of a number of techniques, which include the classical techniques such as arithmetic mean, *Thiessen polygon*, isohyetal analysis, distance weighting, and percentage weighting (Şen, 1998). Tabios and Salas (1985) compared several SRI methods and concluded that a geostatistical method (ordinary and universal Kriging) with spatial correlation structure is superior to Thiessen polygons, polynomial interpretation, and inverse-distance weighting. Hevesi et al. (1992a,b) suggested the use of multivariate geostatistical techniques for SRI in mountainous terrain. Reliable estimates by these techniques are particularly difficult when the regional coverage of stations is sparse or when rainfall characteristics vary greatly with locations. Such situations frequently occur in arid regions due to sporadic and haphazard rainfall occurrences. Kedem et al. (1990) have shown by considering satellite images and simple probability models that the higher the rainfall intensity, the smaller the affected area over large regions. They are not as practical as conventional procedures such as the arithmetic mean, Thiessen polygons, or isohyetal map techniques, which do not require much data and computation time (Fiedler, 2003).

In practice, most often average areal rainfall (AAR) is estimated for any required period in a number of different ways. The method to be used depends on the rainfall measurement station configuration, altitude, aspect and locality, and the availability of topographic maps, in addition to the skill of the hydrologist. Whichever method is applied, the first requirement is the reliability of the data.

### 3.9.1 CLASSICAL GEOMETRIC WEIGHTING FUNCTIONS

In any SRI analysis technique the main idea is that interpolation at any point is considered as a weighted average of the measured values at a set of irregular sites. In general, if there are $n$ measurement sites with rainfall records, $R_i$, $(i = 1, 2, \ldots, n)$, then the spatial interpolation site, $m$, has rainfall, $R_m$, which can be obtained as,

$$R_m = \frac{\sum_{i=1}^{n} W(r_{i,m}) R_i}{\sum_{i=1}^{n} W(r_{i,m})} \tag{3.31}$$

where $r_{i,m}$ and $W(r_{i,m})$ are distance and corresponding weighting function between sites $i$ and $m$, respectively. In the literature, all the weighting functions that are proposed by various researchers appear as functions of the distance between the sites, such as the inverse distance or inverse distance square methods.

Figure 3.26 shows various dimensionless weighting functions used in the meteorology literature so far by different researchers. Unfortunately, none of these functions are rainfall record dependent but suggested on the basis of the logical and geometrical configuration. The weights are expected to represent partial spatial behavior of the rainfall phenomenon. For instance, Thiebaux and Pedder (1987) suggested weightings, $W(r_{i,m})$, in general, as

$$W(r_{i,m}) = \begin{cases} \left( \dfrac{R^2 - r_{i,m}^2}{R^2 + r_{i,m}^2} \right)^{\alpha} & for \quad r_{i,m} \leq R \\[2ex] 0 & for \quad r_{i,m} \geq R \end{cases} \tag{3.32}$$

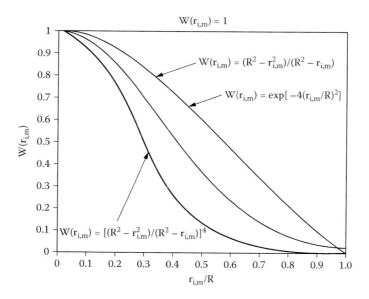

**FIGURE 3.26**  Geometric weighting functions.

where $R$ is the *radius of influence,* and $\alpha$ is a parameter that reflects the curvature of the weighting function. Equation 3.32 is the generalization of the classical Cressman (1959) method where power parameter is equal to one. In general, the two unknowns, $R$ and $\alpha$, in Equation 3.32 are determined subjectively by expert view. The objective determinations of R and $\alpha$ are already presented by Şen (1997) and Şen and Öztopal (2001) through the geostatistical and genetic algorithm procedures. One of the most commonly used alternative weighting functions in Figure 3.26 has $\alpha = 4$.

Again, another form of geometrical weighting function was proposed by Sasaki (1960) and Barnes (1964) as

$$W(r_{i,m}) = \exp\left[-4\left(\frac{r_{i,m}}{R}\right)^{\alpha}\right] \qquad (3.33)$$

The significance of Equations 3.32 and 3.33 lies in the implications that the weightings decrease with increase in distance. The weighting functions must reflect the spatial behavior of the meteorological phenomena, in addition to the station configuration geometry.

### 3.9.2  Arithmetic Average

This method yields plausible AAR values, provided that the rainfall amounts at each station are not different more than 10% relative difference percentage, $r$, which is calculated as

$$r = 100 \frac{R_{max} - R_{min}}{R_{max}} \tag{3.34}$$

where $R_{min}$ and $R_{max}$ are the minimum and maximum rainfall records at a given set of stations, respectively. Provided that $r < 10$, then the AAR, $\overline{R}_A$, can be calculated according to the arithmetic average formula as

$$\overline{R}_A = \frac{1}{n} \sum_{i=1}^{n} R_i \tag{3.35}$$

where $n$ is the number of stations and $R_i$ is the amount of rainfall at station $i$. It is important to notice that this expression is a special case of Equation 3.31 with equal weightings.

### 3.9.3 THIESSEN POLYGON

The arithmetic average method does not consider the subareal contribution of each raingage location. Logically, each rainfall recorded at a raingage should have its areal domain of influence, i.e., a representative subarea. If the relative difference between the stations is more than 10%, then in the calculation of AAR the role of representative subareas must be taken into consideration. It is not always clear how to get the most representative subarea values. Although the advent of GIS has greatly streamlined the problem of determining spatial statistics, sometimes it is not practical to use GIS. Instead, the Thiessen (1911) polygon approach is probably the most common method used in hydrometeorology for determining AAR over a drainage basin when there are several raingage sites. The basic concept is to divide the drainage basin into several polygons, each one around a site, and then take the weighted average of the measurements based on the size of each polygon area. In this way, measurements within large polygons are given more weight than within small polygons.

The construction of subarea polygons can be achieved as follows:

1. Consider the drainage basin boundary as in Figure 3.27. There are three raingages inside the drainage basin, and the fourth is outside. The application of the Thiessen areal weighting method consists of first constructing a series of triangles by joining pairs of adjacent rainfall stations. The nearest stations should always be jointed, and the triangle should be kept near the equiangular shape as much as possible.
2. In Figure 3.27 there are four measurement sites. Connect them by dashed lines.
3. Draw perpendicular lines to bisect each of the "connecting" lines and extend these bisecting lines until either they intersect the watershed boundary or another bisecting line.
4. Now four polygons have been generated. Note that measurements made at point 4 will contribute less to the final average than, say, the measurements at point 2.

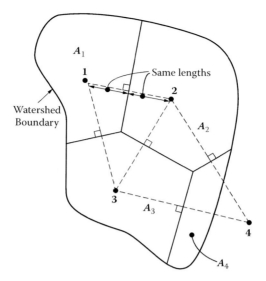

**FIGURE 3.27**   Thiessen polygonal division.

If there are $n$ stations then there are $n$ polygons and n representative subareas, $A_i$, ($i =$ 1, 2, ..., $n$) for each station with the rainfall record, $R_i$. Finally, the AAR calculation can be achieved, according to the weighted average, as

$$\bar{R}_A = \frac{\sum\limits_{i=1}^{n} R_i A_i}{\sum\limits_{i=1}^{n} A_i} = \frac{1}{A}\sum_{i=1}^{n} R_i A_i = \sum_{i=1}^{n} R_i \frac{A_i}{A} = \sum_{i=1}^{n} R_i a_i \qquad (3.36)$$

where $a_i$'s are the percentage representative areas and, therefore, $0 < a_i < 1$ and $a_1 + a_2 + ... a_n = 1$. If the representative polygon subareas are all equal to $A_s$, then $A_i/A = A_s/nA_s = 1/n$ and hence, Equation 3.36 becomes equivalent to Equation 3.35.

### 3.9.4   ISOHYETAL MAPS

The word *isohyet* is a compound Greek word consisting of *iso* and *hyet* meaning "equal" and "rain" in English. Hence, collectively it implies equal rainfall lines (contour lines) similar to the equal elevation lines in a topographic map. The construction of the isohyetal map is exactly the same as drawing any topographic map, the only difference being that instead of triangularization points for the topographic map there are raingage locations and rainfall amounts for a specific duration of time instead of the elevations. In order to calculate the AAR from this map the following steps are necessary.

1. Find average rainfall between isohyets. This is the arithmetic average of the two isohyets.

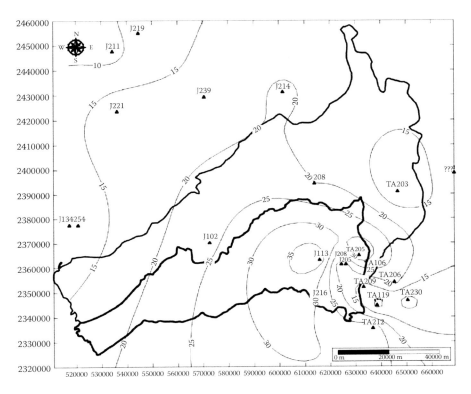

**FIGURE 3.28**    January isohyetal map.

2. Find the area on the map between two successive isohyets.
3. Multiply the area between isohyets by the average rainfall amounts adjacent to these isohyet lines.
4. Find the sum of figures obtained in step 3, and then divide by the total area of the drainage basin to obtain the AAR value.

Because the rainfall records are the sole reliable hydrological measurements in the arid zones, the isohyetal maps are useful for depicting the regional (areal) distribution of the rainfall. Every isohyetal map has a contour interval, which implies a difference between the two successive contour lines. Such a map is shown in Figure 3.28 over wadis Fatimah and Na'man drainage basins near the western part of the Red Sea for January (Al-Sefry et al., 2004).

This is the most accurate among all the previous methods, and it depends heavily on the skill of the person in drawing the isohyets. It is preferable that the mapmaker should know the terrain within the wadi and be able to interpolate the isohyets to truly reflect natural conditions in the field. In mountainous terrain the isohyets would approximate topographic contours, because orographic rainfall usually increases by going upslope toward the mountainous areas. If some measuring points show unusually heavy rainfall, then the isohyets can be clustered around such points. The advantages of the isohyetal maps can be stated as follows:

1. They permit analysts to exercise their own judgment and knowledge of average and specific rainfall distribution within the area.
2. Suspected or missing data can be estimated.
3. No artificial weighting is applied to rainfall records at raingage stations.
4. Topographic controls are taken into consideration.

However, there are also disadvantages as they are more time-consuming than most of the other methods and more subjective, in that the accuracy depends upon the skill of the hydrologist.

### 3.9.5 PERCENTAGE WEIGHTED POLYGON (PWP)

The major drawback in the Thiessen polygon approach is that whatever the rainfall amounts at the raingages, the subarea polygons remain the same because they are entirely dependent on the raingage location configuration. In order to alleviate this drawback, it is suggested that the representative subarea polygons are not dependent on the station configuration only, but in the meantime on the rainfall amounts (Şen, 1998). If the rainfall values at three apices of a triangle are $R_1$, $R_2$, and $R_3$, then their respective percentages are calculated as

$$p_1 = 100 \frac{R_1}{R_1 + R_2 + R_3} \tag{3.37}$$

$$p_2 = 100 \frac{R_2}{R_1 + R_2 + R_3} \tag{3.38}$$

and

$$p_3 = 100 \frac{R_3}{R_1 + R_2 + R_3} \tag{3.39}$$

respectively. A two-dimensional plot of percentages can be shown on a triangular graph paper as in Figure 3.29 (Koch and Link, 1971).

In order to demonstrate the method explicitly, the following step-by-step algorithm is presented.

1. Draw lines between each adjacent pair of rainfall stations. Hence, a set of triangles is obtained, similar to the Thiessen method, that cover the study area.
2. For each triangle calculate the rainfall percentage at its apices according to Equations 3.37 to 3.39. Consider that each apex has the value of 100 percentage with zero percentage starting from the opposite side.
3. Consider bisector, which connects an apex to the midpoint of the opposite side and graduate it into 100 equal pieces.
4. By making use of one rainfall percentage calculated in step 2, mark it along the convenient bisector, starting from the opposite side toward the apex.

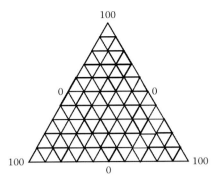

**FIGURE 3.29**   Triangular coordinate paper.

5. Draw a parallel line from this marked point in step 4 to the side opposite to the apex considered with its rainfall percentage.

6. Repeat steps 4 and 5 for the next rainfall percentage and find a similar parallel line this time to another opposite side.

7. The intersection of these lines defines the key point for the triangle considered.

8. Repeat steps 4 and 5 for the third percentage. If the parallel line to the side crosses through the key point, then the procedure of finding the key point for the triangle is complete. Otherwise, there is a mistake either in rainfall percentages or in the location of marked points along the bisectors in steps 3 through 6, inclusive.

9. Return to step 2 and repeat the process for the triangles constructed in step 1. In this manner each triangle will have its key point. The greater the rainfall percentage for an apex, the closer the point will lie to this apex.

10. Key points at adjacent triangles are connected with each other to form polygons, each enclosing a single rainfall station.

11. The boundaries of polygons around the basin perimeter are defined by drawing a perpendicular to the sides of triangles from the key points. Now, the division of the whole basin into subareas is complete.

Finally, for each rainfall amount, $R_i$, at station $i$, there is a corresponding subarea, $Ai$, which helps to calculate the AAR, or $\overline{R}_A$ by the PWP method, according to Equation 3.36.

The application of the PWP method is presented by Bayraktar et al. (2005) for the southeastern provinces of Turkey where a semi-arid climate prevails (Figure 3.30).

In order to determine the AAR from 10 different meteorological stations (Table 3.14), the PWP method is applied together with the other conventional methods (arithmetic mean, Thiessen polygon, and isohyetal map techniques). For each method, AAR values are calculated with the help of Equation 3.36.

Areal values of Thiessen polygons are merely given in Figure 3.31 because the key points remain the same as long as the meteorological stations do not change.

Although all monthly isohyetal maps are prepared, herein for the sake of argument only January and July isohyetal maps are given in Figure 3.32.

**FIGURE 3.30**  Study area location.

---

**TABLE 3.14**
**Total Monthly Rainfall Values of Stations (mm)**

| Number | Stations | Jan. | Feb. | Mar. | Apr. | May | June | July | Aug. | Sept. | Oct. | Nov. | Dec. |
|---|---|---|---|---|---|---|---|---|---|---|---|---|---|
| | | | | | | | | | | | | **Months** | |
| 1 | Silvan | 104 | 116 | 116 | 84 | 61 | 6 | 0 | 1 | 1 | 19 | 69 | 116 |
| 2 | Siirt | 115 | 107 | 111 | 106 | 66 | 9 | 1 | 0 | 5 | 48 | 86 | 102 |
| 3 | Ahlat | 64 | 67 | 79 | 84 | 62 | 23 | 9 | 4 | 15 | 50 | 68 | 53 |
| 4 | Bitlis | 180 | 156 | 141 | 107 | 65 | 19 | 5 | 3 | 14 | 58 | 97 | 130 |
| 5 | Van | 42 | 35 | 46 | 58 | 41 | 17 | 6 | 3 | 12 | 44 | 49 | 83 |
| 6 | Şırnak | 136 | 121 | 143 | 141 | 64 | 5 | 2 | 1 | 6 | 33 | 83 | 123 |
| 7 | Başkale | 47 | 43 | 69 | 102 | 100 | 42 | 20 | 9 | 16 | 39 | 43 | 38 |
| 8 | Nusaybin | 92 | 62 | 63 | 63 | 40 | 1 | 1 | 0 | 1 | 16 | 43 | 79 |
| 9 | Şemdinli | 106 | 155 | 151 | 195 | 56 | 9 | 3 | 4 | 5 | 77 | 119 | 97 |
| 10 | Hakkari | 102 | 105 | 125 | 146 | 57 | 15 | 3 | 1 | 10 | 26 | 76 | 91 |

In the PWP method percentage weightings are calculated for each of the three adjacent stations constituting subtriangles. PWP method calculations and subareal values are given in Figure 3.33 for January and July.

The results are presented collectively in Table 3.15 for January and July with annual AAR results.

The PWP method yields more reliable results and smaller AAR values depending on the regional variability of the rainfall amounts over the catchment area. For instance, in July, rainfall values have many regional variations. This is due to the semi-arid characteristics of the study area and frequent occurrence of convective rainfalls, which are not expected to cover large areas. It is noted that in July, the Başkale station surroundings appear as the most rainfall-recipient area, which is represented by a subarea of 1,545 km² in the Thiessen method, whereas it is shown as subareas of 3,140 km² and 519 km² in the arithmetic mean and the isohyetal map

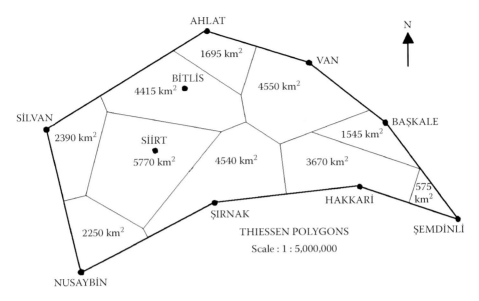

**FIGURE 3.31** Application map of Thiessen method (January and July).

technique, respectively. On the other hand, it is represented by a subarea of 153 km² when the PWP method is considered. Hence, the representation of the most rainfall-intercepting meteorology station with the smallest subarea gives rise to a smaller value of AAR for the catchment area during convective storms.

As the frontal and more homogeneous type of rainfalls are recorded during a specific period, it is expected that the results of the PWP and Thiessen polygon methodologies will approach each other, provided that there are approximate rainfall records at each meteorological station. In practice, whatever the circumstances are, it is preferable to adopt PWP.

### 3.9.6 AUTOMATED AAR CALCULATION

The automation of the triangularization of the AAR calculation technique is very suitable for applications in arid lands, steppe areas, and desert environments. The accuracy of the method is closely related to the validity of the linear rainfall variations. It has been stated by Şen and Eljadid (2000) that the rainfall distribution is rather linear between the meteorology stations considered in the northern part of Libya.

### 3.9.6.1 Data and Study Area

The climatic studies, in general, and rainfall studies in particular, are new products in Libya because there were few meteorological stations before 1969 and even the ones that existed were working in a narrow scope. Due to the huge area (1,755,500 km²) of the country with 90% deserts, it is difficult, if not impossible, to cover all this area by sufficient meteorology stations. Any attempt to study the meteorological variables that exist over Libya is expected to have a significant probability of error.

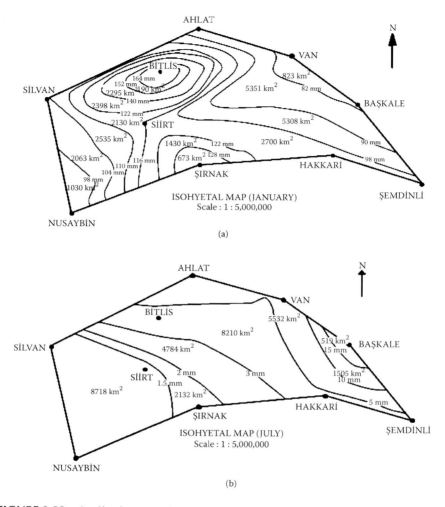

**FIGURE 3.32**  Application map of isohyetal technique, (a) January, (b) July.

On the other hand, the study area contains the Gefara plain in the west corner of the country (11,100 km²) and the Al-Jabal Al-Akhdar mountainous area in the eastern corner along the Mediterranean region (Figure 3.21). These are the most important agricultural locations and the majority of Libya's population live along this coast, which is influenced mainly by the Mediterranean sea climate as mild winters and not excessively hot summers with high humidity and substantial rainfall.

In Libya, as in arid lands, the hydrology of a region depends primarily on its climate, next to its topography and geology. The climate factor which has a great importance in hydrology, is the rainfall and its modes of occurrences (Bruce and Clark, 1966). There is a need for detailed study of this meteorological factor in hydrology studies including the water resources development and management project, agricultural activities, and flood predictions. In these studies the rainfall amounts falling on the specific area are required in volumes.

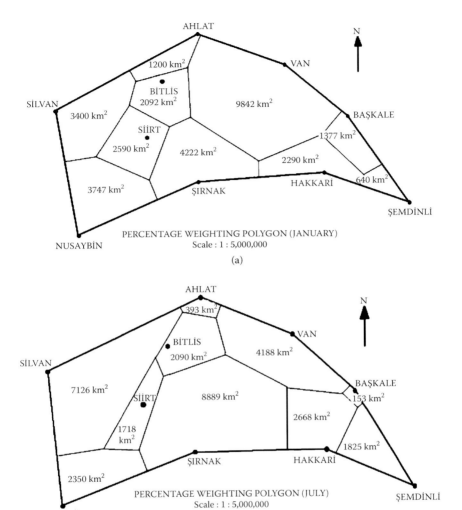

FIGURE 3.33  Application map of PWPs: (a) January, (b) July.

**TABLE 3.15**
**AAR Values in mm for Different Methods**

| Method | January | July | Annual |
|---|---|---|---|
| Arithmetic average | 98.76 | 5.00 | 63.31 |
| Thiessen | 106.30 | 3.94 | 63.57 |
| Isohyetal map | 102.72 | 4.12 | 62.13 |
| Percentage weighting polygon | 89.22 | 2.32 | 54.49 |

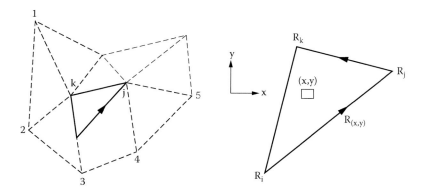

**FIGURE 3.34**   Notations used for a triangular subarea.

### 3.9.6.2   A Rapid Areal Rainfall Calculation

An alternative method for the analysis of rainfall data that is used for computing the volume of rainfall falling on the study area was produced primarily by Akin (1971). The development is analogous to some elementary concepts used in finite element analysis techniques. Consider a region of interest with station locations and rainfall amounts that are plotted on a map, then a series of straight lines are drawn arbitrarily to connect every gauge point with the adjacent gauges. Straight lines are drawn in an counterclockwise direction as shown in Figure 3.34. These straight lines should produce a series of triangles, not necessarily having the same shape.

Each triangle area is known as a subarea, and the corners of the triangles are shown by $(i,j,k)$. Rainfall values at the corners are denoted as $R_i$, $R_j$, $R_k$, and $i, j, k = 1,2, \ldots, n$, where n is the number of subareas. Triangularization of all the stations for the northern part of Libya is shown in Figure 3.35.

On the other hand, Table 3.16 shows subareas generated in this manner for the study area.

It is assumed that the rainfall over the subarea varies linearly between the three-corner gauge points. Thus, at any point $(x,y)$ interior to the $n$-th subarea, the rainfall height $R(x,y)$ is expressed as

$$R(x,y) = \alpha_n + \beta_n x + \gamma_n y \tag{3.40}$$

where $\alpha_n$, $\beta_n$, and $\gamma_n$ are constants related to the gauge measurements at the corners. It is possible to write three simultaneous equations one for each apex (i,j,k) as

$$R(x_i,y_i) = \alpha_n + \beta_n x_i + \gamma_n y_i$$
$$R(x_j,y_j) = \alpha_n + \beta_n x_j + \gamma_n y_j \tag{3.41}$$
$$R(x_k,y_k) = \alpha_n + \beta_n x_k + \gamma_n y_k$$

The solution of constants from these equations in terms of known quantities leads to

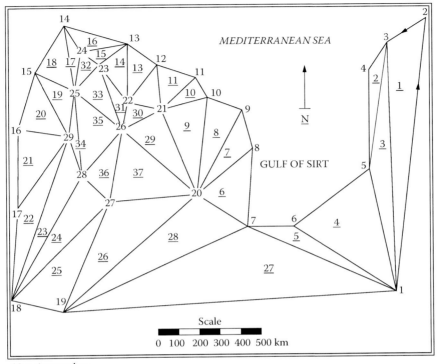

29-point number
37-area number

**FIGURE 3.35** Triangular mesh for northern Libya.

**TABLE 3.16**

**Values of i, j, k for Each Area**

| Subarea Number | i | j | k | Subarea Number | i | j | k | Subarea Number | i | j | k | Subarea Number | i | j | k |
|---|---|---|---|---|---|---|---|---|---|---|---|---|---|---|---|
| 1 | 1 | 2 | 3 | 11 | 11 | 12 | 21 | 21 | 16 | 17 | 29 | 31 | 22 | 23 | 28 |
| 2 | 3 | 4 | 5 | 12 | 12 | 22 | 21 | 22 | 17 | 18 | 29 | 32 | 15 | 25 | 24 |
| 3 | 3 | 5 | 1 | 13 | 12 | 13 | 22 | 23 | 18 | 6 | 29 | 33 | 28 | 23 | 25 |
| 4 | 5 | 6 | 1 | 14 | 13 | 23 | 22 | 24 | 18 | 27 | 26 | 34 | 26 | 25 | 29 |
| 5 | 6 | 7 | 1 | 15 | 13 | 24 | 23 | 25 | 18 | 19 | 27 | 35 | 26 | 28 | 25 |
| 6 | 7 | 8 | 20 | 16 | 13 | 14 | 24 | 26 | 19 | 20 | 27 | 36 | 27 | 28 | 26 |
| 7 | 7 | 8 | 20 | 17 | 14 | 25 | 24 | 27 | 19 | 1 | 7 | 37 | 20 | 28 | 27 |
| 8 | 9 | 10 | 20 | 18 | 14 | 15 | 25 | 28 | 19 | 7 | 20 | | | | |
| 9 | 10 | 21 | 20 | 19 | 15 | 29 | 25 | 29 | 20 | 21 | 25 | | | | |
| 10 | 10 | 11 | 21 | 20 | 15 | 16 | 29 | 30 | 21 | 22 | 28 | | | | |

$$\alpha_n = \frac{a_i R_i + a_j R_j + a_k R_k}{2A_n}$$

$$\beta_n = \frac{b_i R_i + b_j R_j + b_k R_k}{2A_n} \tag{3.42}$$

$$\gamma_n = \frac{c_i R_i + c_j R_j + c_k R_k}{2A_n}$$

where

$$a_i = x_j y_k - x_k y_j$$
$$b_i = y_j - y_k$$
$$c_i = x_k - x_j \tag{3.43}$$

Following a cyclic permutation of $(i,j,k)$ the subarea $A_n$ can be calculated as follows:

$$A_n = \frac{a_i + a_j + a_k}{2} \tag{3.44}$$

The differential volume, $dQ$, of rainfall at any point within the subarea is defined as

$$dQ = R(x,y)dA \tag{3.45}$$

So that the total volume of rainfall associated with the subarea becomes theoretically as

$$Q_n = \iint [\alpha_n + \beta_n x_n + \gamma_n y_n] dx dy \tag{3.46}$$

where the substitution of the above relevant expressions leads after some algebra and to the volume of rainfall for the $n$-th subareas as

$$Q_n = A_n \left[ \frac{\alpha_n + \beta_n x_i + \gamma_n y_i}{3} + \gamma_n \frac{y_i + y_j + y_k}{3} \right] \tag{3.47}$$

In fact, this equation follows directly from Equation 3.41 because $Q_n = \overline{H}_n A_n$ with $\overline{H}_n = (H_i + H_j + H_k)/3$. In the case of $m$ triangular subareas, the total rainfall volume becomes

$$Q = \sum_{i=1}^{n} Q_i \tag{3.48}$$

and the corresponding total area is

$$A = \sum_{i=1}^{n} A_i \tag{3.49}$$

**TABLE 3.17**
**Monthly Average Aerial Rainfall Depths and Volumes**

| Month | Average Depth (mm) | Average Volume ($\times 10^6$ m³) |
|---|---|---|
| January | 28.2 | 97.400 |
| February | 12.7 | 64.889 |
| March | 16.2 | 47.684 |
| April | 4.4 | 16.564 |
| May | 5.5 | 19.980 |
| June | 0.41 | 1.600 |
| July | 0.03 | 0.160 |
| August | 0.17 | 0.800 |
| September | 2.90 | 9.060 |
| October | 7.13 | 23.230 |
| November | 15.7 | 39.600 |
| December | 28.9 | 112.10 |

**TABLE 3.18**
**Seasonal Average Aerial Rainfall Depths and Volumes**

| Season | Average Depth (mm) | Average Volume ($\times 10^6$ m³) |
|---|---|---|
| Winter | 72.800 | 274.389 |
| Spring | 26.100 | 84.228 |
| Summer | 0.613 | 2.560 |
| Autumn | 25.700 | 71.890 |

Finally, the ratio of Equation 3.48 to Equation 3.49 gives the AAR height over *m* subareas as

$$\bar{H} = \frac{Q}{A} \tag{3.50}$$

By means of this procedure, the AAR area and volume are easily calculated if the gauge locations and rainfall amounts are known. Computer software has been developed to calculate the subareas, associated AAR depths, and volumes falling on each subarea, as well as the total area of interest (Eljadid, 1997).

The results are shown for monthly AARs in Table 3.17, whereas the seasonal AAR depths and volumes are presented in Table 3.18.

These values are computed by developed software over the study area as a summation of 37 subareas already shown in Figure 3.34 and Table 3.16.

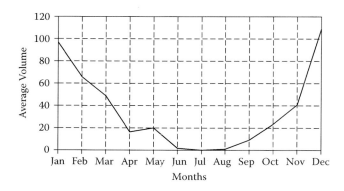

**FIGURE 3.36**   Monthly AAR volume ($\times 10^6\,\mathrm{m}^3$) variation.

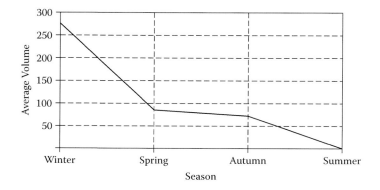

**FIGURE 3.37**   Seasonal AAR volume ($\times 10^6\,\mathrm{m}^3$) variation.

### 3.9.6.3   Volume Calculations

The monthly and seasonal variations with time over the northern part of Libya have been analyzed, and the results are shown in Figures 3.36 and 3.37, respectively. The apparent regional variation in seasonality can be explored further with reference to the short-and long-term records from gauges adjacent to the study area. These figures show that the seasonality is evident for the coastal catchment in December, January, and February. However, December is the predominantly wet month over the study area. On the other hand, it is clear from Figure 3.36 that a basic inverse nonlinear relationship exists between the seasonal AAR volume and the duration, i.e., less rainfall amounts occur more frequently. To be exact, this mode of rainfall distribution is normal according to the spatial variation of rainfall over the northern part of the country (Şen and Eljadid, 1999).

Figure 3.37 shows that the winter season has the dominant wet spell followed by the spring season, then comes the autumn season; however, there is no appreciable rainfall during the summer season.

According to the AAR results, winter is the rainiest season in which the volume of rainfall reaches $275 \times 10^6$ m³. However, rainfall quantities decrease as one moves toward the south, due to the decrease in the Mediterranean climate effect. Winter rainfall is rather little even at its peak; therefore, one cannot depend on it for agricultural and industrial development. The spring rainfall reaches $84 \times 10^6$ m³, which is far less than winter rainfall due to the decrease in number of low pressure systems that are the main cause for rainfall in this season. Spring rainfall does not play a major role in supporting the water requirement for the agricultural and industrial activities in the northern part of Libya. These little quantities make the landscape green, and they can be used as a seasonal source for some agricultural products used in animal husbandry in some years. The autumn season is partially rainy in most of the northern regions. The results show that the northern part of Libya receives rainfall that reaches almost $72 \times 10^6$ m³ during this season. This is due to the rainy air currents which come from the Mediterranean Sea during these periods, and additionally, the sharp decrease in Sahara climate affect most of Libya at this time of the year. The summer is the driest season in the northern part of Libya with $2.56 \times 10^6$ m³ of water, which is a rather significant quantity in this season. The same is due to the Saharan climate, which becomes influential over the country during the summer seasons. This situation does not allow a chance for development of any kind of clouds, which are the major sources of rainfall.

## 3.10   TRIPLE VARIABLE DROUGHT DESCRIPTOR

*Meteorological droughts* are defined whenever rainfall deficit occurs. The *rainfall deficit* may be regarded as the amount of water smaller than a prescribed base (threshold), which in most cases, is considered as 2.5 mm but in arid regions is equal to zero. In humid regions, for hydrological design and planning, droughts of hydrological origin are considered, and because the most significant hydrological variable is the runoff, drought definition is based on the runoff deficit relative to water demand. From an *agricultural* point of view, droughts appear whenever the soil deficit starts, i.e., the plant water demand is greater than the soil moisture availability. A common point to all these different types of drought is a base value for the comparison of either the rainfall, runoff, or soil moisture amounts. Of course, amounts smaller than the adopted base level may occur temporally and spatially, and hence temporal and spatial drought characteristics come into the picture for a drought characteristics analysis. Drought analysis is confined mostly to temporal variations, where time series of concerned variables (rainfall, runoff, or soil deficit) are considered and thereof various drought characteristics such as drought duration, intensity, magnitude, etc., are calculated. Their predictions require different models including statistics, probability theory, and stochastic processes.

The first classical approach involved the evaluation of instantaneously smallest value in an observed hydrologic sequence recorded at a single site (Gumbel, 1958). This method gives information on the instantaneous maximum value of drought magnitude only, without any elaboration on either its duration or areal extent. Yevjevich (1972) presented the first objective definition of hydrologic point drought with applications by Downer et al. (1967), Llamas and Siddiqui (1969), Saldarriaga and Yevjevich (1970), Millan and Yevjevich (1971), Guerrero-Salazar and Yevjevich (1975), and Şen (1976, 1977, 1980b).

In the spatial drought evaluation, drought occurrences at many sites are considered, and with the information and knowledge at individual sites, regional drought characteristics such as drought coverage percentages are evaluated (Tase et al., 1987; Şen, 1980b). Spatial treatment of the droughts is mostly achieved by either probabilistic modeling or spatial analysis such as Kriging, multiple regressions, or trend surface analysis. Although there are also spatio-temporal drought analysis methodologies, they are rather restrictive and have many assumptions in order to simplify the analysis.

A wide variety of drought definitions has been mentioned by Wilhite and Glantz (1985). Perhaps, the main lack in such a common definition arises from the fact that there are several kinds of information needed for drought monitoring. Most of the studies are concentrated on the rainfall shortages as drought implications because rainfall is the most significant input variable for many water-related activities such as water supply, groundwater and reservoir storages, soil moisture, and streamflow. The simplest methodology of temporal drought assessment is the *standardized rainfall index* (SPI), which is used to quantify the rainfall deficit for several time scales, i.e., time averaging periods (McKee et al., 1993). The SPI for a given averaging period of time is physically the difference of rainfall from the mean divided by the standard deviation. Similar to Equation 3.1, the SPI variable $x_i$ is defined for a given series of rainfall $(X_1, X_2, ..., X_n)$ as

$$x_i = \frac{X_i - \overline{X}}{S_x} \tag{3.51}$$

where $\overline{X}$ and $S_x$ are the arithmetic average and the standard deviation, respectively, of the given rainfall series. One of the most significant benefits from this formulation is that the resulting time series include relatively small values that are concentrated around zero with positive and negative deviations. Hence, based on the arithmetic average truncation level, the negative instances within the standardized time series indicate dry spells, and the successive sequence of such dry spells preceded and succeeded by at least one wet spell will indicate the *duration* of the drought, whereas the *magnitude* of the drought is the summation of these negative values. Consequently, the division of magnitude to the drought duration gives the drought *intensity*. SPI does not provide any information on the areal drought extension or objective relation of these drought descriptors to other meteorological or hydrological variables. Therefore, their results are rather restrictive and useful for temporal drought predictions or assessments based on rainfall only.

In practical studies, in order to generalize SPI results, it is necessary that the rainfall series abide with a normal (Gaussian) distribution but, in general, rainfall sequences have non-Gaussian frequency distributions (Sections 3.3 and 3.6). SPI gives information concerning probability of prevailing conditions, percentage of average rainfall and accumulation of rainfall deficits. On the basis of SPI, a drought event is defined for each time scale as a period in which the SPI is continuously negative and the SPI reaches a value of −1.0 or less. A drought begins when the SPI

**TABLE 3.19**
**Drought Categorization by SPI**

| SPI Value Ranges | Drought Categories |
| --- | --- |
| 0 – (0.99) | Mild |
| (−1.00) – (−1.49) | Moderate |
| (−1.50) – (−1.99) | Severe |
| < (−2.00) | Extreme |

first falls below zero and ends when it becomes positive. Droughts are categorized according to Table 3.19.

The SPI has been extensively used recently in the United States for monitoring drought. However, a more intensively used drought index in the United States is due to Palmer (1965), and it is referred to as the Palmer Drought Index (PDI), which is designed to monitor drought related to the soil moisture, especially, for agricultural droughts; it requires temperature data in addition to the rainfall data. Although much of the variations in PDI are derived by rainfall, there are significant roles of temperature and consequent evapotranspiration and humidity in the case of agriculture. However, the application of PDI is difficult in many parts of the world due to unavailability of required data for its application. In a way, PDI combines the effects of rainfall with other variables, and, especially, the temperature.

It is suggested here to relate the rainfall features of drought occurrences to temperature and humidity in a straightforward manner through the triple-variable analysis. The relationship between the SPI and PDI are shown by McKee et al. (1995), and they stated that the correlation reaches its maximum at two locations after 12 months with a magnitude that explains nearly 80% of the variance or which has a correlation coefficient near to 0.9.

Our basic concern is with the rainfall-based droughts, but in addition to rainfall time series, two other closely related records are temperature and humidity. Simultaneous consideration of rainfall, temperature, and humidity presents a mean for drought assessment, provided that their interrelationships are described on a Cartesian coordinate system. The best way of achieving such an interrelationship is through the visualization of rainfall values as mathematically dependent, and temperature and humidity as independent variables. This gives rise to a different drought evaluation technique apart from the temporal and spatial methodologies. It is, in a way, a consideration of rainfall variability related to temperature and humidity variations. There are many benefits from such a triple-variable drought graph, some of which can be given as follows:

1. The variation of rainfall can be depicted according to temperature and relative humidity variations.
2. It is possible to depict the locations of rainfall deficits on the basis of temperature and relative humidity ranges.
3. One can construct arithmetic average and standard deviation variations of rainfall on the basis of temperature and relative humidity.

4. For any given value of relative humidity, it is possible to identify the rainfall variation with temperature.
5. Likewise, rainfall variation with relative humidity can be obtained for any given value of relative humidity.
6. Regions of maximum rainfall can be enclosed, and this gives rise to temperature and relative humidity domain identification for maximum rainfall occurrences, which represent water surpluses.
7. Likewise, minimum rainfall occurrences can be located from triple-variable drought graphs, and correspondingly, temperature and relative humidity ranges can be determined. These occurrences in the triple-variable drought graphs indicate actual rainfall deficits and hence drought concentration locations.
8. It is possible to estimate the rainfall amounts for a given pair of temperature and humidity values. In this manner, one can also estimate drought spells and amounts.

### 3.10.1 APPLICATIONS

The implication of the methodology has been presented for three sites each from different climatology regions of Turkey as shown in Figure 3.38.

Monthly rainfall, temperature and humidity records between 1930 and 2000 are considered in the application. Some features of these sites are indicated in Table 3.20.

Adana lies in the southern region with its typical Mediterranean climatic features. Winters are relatively short with cool periods, but summer seasons have high temperatures and relative humidity. Here, zero SPI line divides the relative humidity–temperature domain in a very stable manner into dry and wet regions (see Figure 3.39). There is a sudden jump in temperature value at almost 50% relative humidity level. High relative humidity values have droughts (SPI < 0) with high temperatures, whereas temperatures drop to almost 5°C with relative humidity ranging from 30 to 50%. This means that at this range of relative humidity, the droughts appear even during very low temperatures.

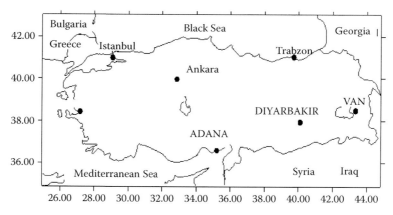

**FIGURE 3.38**  Stations location map.

**TABLE 3.20**
**Different Climatological Sites in Turkey**

| Site Name | Longitude | Latitude | Elevation (m) | Climatic Feature |
|---|---|---|---|---|
| Adana | 3,659 | 3,521 | 27 | Mediterranean |
| Diyarbakir | 3,754 | 4,014 | 677 | Continental |
| Van | 3,827 | 4,319 | 1,661 | Continental |

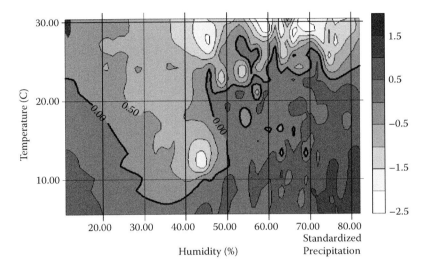

**FIGURE 3.39** Triple-drought signatures for Adana.

The Diyarbakir region lies in the completely semi-arid region of Turkey in the southeastern part with dry and, hence, low relative humidity bands but long periods of sunshine, especially in the summer season. Figure 3.40 is rather similar to the Adana station triple diagram for drought with penetrations of dry SPI values toward the relative humidity range from 45 to 60%. Such a shift is due to semi-maritime effects from the Mediterranean Sea around Diyarbakir. It is possible to say that there is no drought condition below almost 20°C at any relative humidity value except in the aforementioned range.

In the most southeastern part of Turkey, Van station represents a mountainous area and has high elevations with consequently dry air penetrations, short periods of sunshine duration, and snowy periods in winter. The triple diagram for drought identification is shown in Figure 3.41, which has a distinctive pattern compared with each one of the previous stations. First of all, there is on the average no drought effect above almost 55% relative humidity, whatever the temperature value. On the contrary, drought prevalence is observed at a lower relative humidity. Furthermore, especially in this station, the appearances of drought are more due to the relative humidity than the temperature.

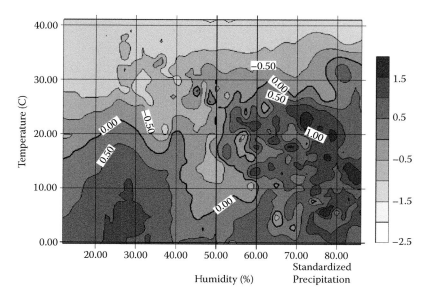

**FIGURE 3.40**    Triple-drought signatures for Diyarbakir.

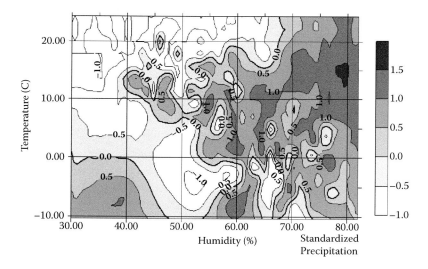

**FIGURE 3.41**    Triple-drought signatures for Van.

## 3.11   JORDAN CLIMATOLOGY

The arid climatic nature, the remarkable length of the dry season, and the insufficient rainfall amounts, in addition to the population growth due to successive immigrations from neighboring countries (mainly Palestine, Lebanon, Syria, Kuwait, and Iraq) has caused an increase in the demand for freshwater, especially in the past three decades, and therefore water scarcity has become a terrible crisis in Jordan

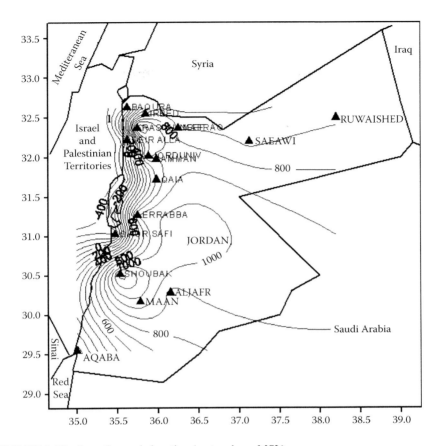

**FIGURE 3.42** Location and elevation (meter above MSL).

(Tarawneh, 2000). The location of stations with topographic contour lines is presented in Figure 3.42.

There are topographic heights less than the Mediterranean Sea level in the western part of Jordan. According to Hershfield (1962), Sharon (1965), and Shehadeh (1976), the frequency distribution of annual rainfall is always close to the normal distribution, even in the driest deserts in the world. In the light of the detailed investigations and illustrations of the rainfall statistical moments in Jordan, the country is divided into three obvious homogeneous rainfall regions (Tarawneh, 2000).

1. The northern region includes the northern heights, western Amman, Irbid, and the extreme northern Jordan Valley (Baqura). The total annual rainfall amount in this region is 400 to 600 mm. The seasonal rainfall distribution is 63, 23, and 14% in winter, spring, and fall, respectively.

2. The second region includes the central part of Jordan (Amman), the southern heights (Shoubak and Rabba), and the northern Jordan Valley (Deir Alla), where the rainfall distributions are as 63, 24, and 12% in winter, spring, and fall, respectively. This region has a total rainfall amount between 250 to 350 mm.

3. The third region consists of the lower locations among and besides the northern and southern heights (Qaia, Wadi Duleil, and Mafraq), the eastern parts (Safawi and Ruwaished), southern and southeastern parts (Ma'an and Al-Jafr) and southern Jordan Valley that extends to Aqaba. The seasonal rainfall distribution is 54, 25, and 21% in winter, spring, and fall seasons, respectively. The total annual rainfall amount in this region varies from 140 to 170 mm in the central west to 70 to 90 mm in the east and 30 to 50 mm in the south.

Either from Mediterranean or polar origin about 28 frontal depressions per year penetrate the eastern Mediterranean region. These depressions center over or near Cyprus, then move toward east-northeast to northeast (11), east (11), southeast (2) and (5) depressions fill up (Meteorological Office, 1962). The majority of these depressions penetrate the area during the winter and spring seasons. The northeastern parts of Jordan have the most exposure to cold and moist air masses, which are usually associated with these systems. Both of the northern and southern heights experience a mean winter relative humidity more than 70% with increases to 76% in Ras Muneef and Queen Alia International Airport (QAIA). The mean winter relative humidity values differ from 61% in the central and southern Jordan valley to 62–66% in the arid lands and to 47%, which is the minimum value that is recorded in Aqaba (Figure 3.43).

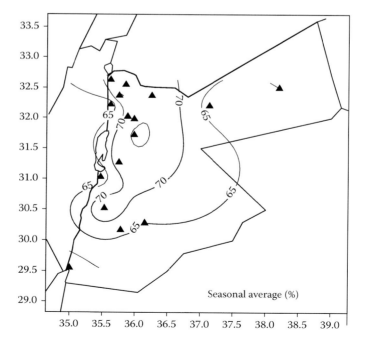

**FIGURE 3.43A**   Seasonal average of winter season relative humidity.

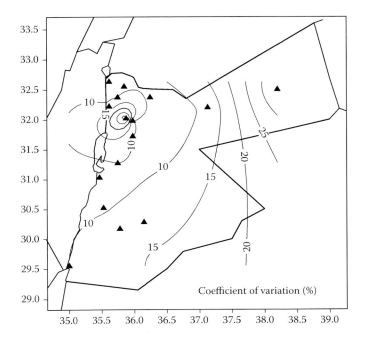

**FIGURE 3.43B**   (Continued). Coefficient of variation of winter season relative humidity.

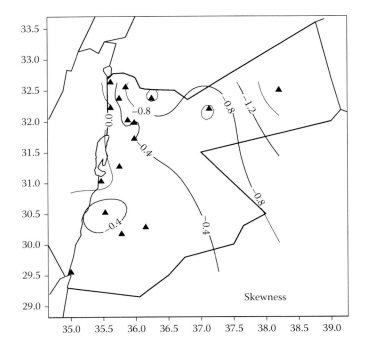

**FIGURE 3.43C**   (Continued). Skewness of winter season relative humidity.

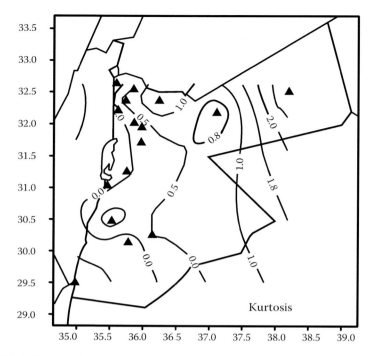

**FIGURE 3.43D**    (Continued). Kurtosis of winter season relative humidity.

Cohen and Stunhill (1996) studied the climate change in the Jordan Valley by analyzing the daily records of maximum and minimum temperatures, rainfall, and global irradiance at three stations located in the northern, central, and southern parts of the Israeli side of the valley. They found no significant trends in annual rainfall. Maximum temperature and diurnal temperature ranges have decreased significantly in the three stations, while trends in minimum temperature were not consistent and not always significant.

The climate change scenarios expect a reduction of rainfall about 20 to 25% in the dry season (April–September) and 10 to 15% in winter time, with temperature decrease about 1.5°C in Jordan during the current half of the century (Ragab and Prudhomme, 2002). In the light of increasing population growth and continuous depletion of the available water resources in a country located at the fringe of the desert, the studies of climatology and climate change are thus important. Vulnerability and adaptation to climate change in Jordan (UNDP, 1999) is one of the comprehensive studies needed by those investigating climate and climate change, environment, water resources, housing and settlement constructions, economy, and agricultural and industrial issues. Abandeh (1999) selected six stations for climate change trend possibility and found that three of them exhibit an increase around 2°C in minimum temperature and a decrease around 1°C in maximum temperature at Amman. In Irbid there is an obvious decrease in maximum temperature and a slight increase in minimum temperature. In Aqaba there is a clear decrease in maximum temperature while no trend has been detected in minimum temperature. There is no noticeable trend in Amman and Irbed rainfall time series but there is a slight decrease in the Aqaba rainfall trend.

## REFERENCES

Abandeh, A., 1999. Climate trends and climate change scenarios, Vulnerability and Adaptation to Climate Change in Jordan. Vol. 1, Project No. JOR/95/G31/IG/99, Al-Shamil Engineering, Amman, Jordan.

Abouammoh, A. M., 1991. The distribution of monthly rainfall intensity at some sites in Saudi Arabia. *Environ. Monitoring Assess.,* Vol. 17, 89–100.

Akin, J. E., 1971. Calculation of areal depth of precipitation. *J. Hydrol.,* Vol. 12, 363–376.

Al-Sefry, S., Şen, Z., Al-Ghamdi, S. A., Al-Ashi, W., and Al-Baradi, W., 2004. Strategic ground water storage of Wadi Fatimah—Makkah region Saudi Arabia. Saudi Geological Survey, Hydrogeology Project Team, Final Report.

Aron, G., Wall, D. J., White, E. L., and Dunn, C. N., 1987. Regional rainfall intensity-duration-frequency curves for Pennsylvania. *Water Resour. Bull.,* 23 (3), 479–485.

Barger, G. L. and Thom, H. C. S., 1949. Evaluation of drought hazard. *Argon. J.,* Vol. 41, 519–527.

Barnes, S. L., 1964. A technique for maximizing details in numerical weather map analysis. *J. App. Meteor.,* Vol. 3, 396–409.

Bayraktar, H., Turalioglu, F. S., and Şen, Z., 2005. The estimation of average areal rainfall by percentage weighting polygon method in Southwestern Anatolia region, Turkey. *Atmos. Res.,* Vol. 73, 149–160.

Bazuhair, S. A., Gohani, A., and Şen, Z., 1997. Determination of monthly wet and dry periods in Saudi Arabia. *Int. J. Climatol.,* Vol. 17, 303–311.

Bell, F. C., 1969. Generalized rainfall-duration-frequency relationships. *J. Hydraul. Div. ASCE,* 95, 311–327.

Benjamin, J. R. and Cornell, C. A., 1960. *Probability Statistics and Decision for Civil Engineers,* McGraw-Hill, New York, p. 684.

Bernard, M. M., 1932. Formulas for rainfall intensities of long durations, *Trans. ASCE,* 96, 592–624.

Berndtsson, R. and Niemczynowicz, J., 1986. Spatial and temporal characteristics of high intensive rainfall in northern Tunisia. *J. Hydrol.,* Vol. 87, No. 3/4, 285–289.

Bruce, J. P. and Clark, R. H. (1966). *Hydrometeorology.* Pergamon Press, New York, 319 pp.

Burlando, P. and Rosso, R. 1996. Scaling and multi-scaling models of depth-duration-frequency curves for storm precipitation. *J Hydrol.,* 187(1–2), 45–64.

Chow, V. T., 1964. The frequency formula for hydrologic frequency analysis. *Trans. AGU* Vol. 32, 231–237.

Cohen, S. and Stunhill, G., 1996. Contemporary climate change in the Jordan Valley. *J. Appl. Meteorol.,* 35, 1051–1058.

Cressman, G. P., 1959. An operational objective analysis system. *Mon. Wea. Rev.,* 87, No. 10, pp. 367–374.

Downer, R., Siddiqui, M. M., and Yevjevich, V., 1967. Applications of Runs to Hydrologic Droughts. Colo. State Univ., Fort Collins, CO, Hydrol. Pap. 23.

Eljadid, A. G., 1997. Hydrometeorological aspects and water resources management in the northern part of Libya. Unpublished Ph.D Thesis, Istanbul Technical University, Meteorology Department, 250 p.

Feller, W., 1967. *An Introduction to Probability Theory and its Application,* Vol. 1, John Wiley & Sons, Inc., New York, p. 509.

Fiddes, D., 1977. Depth-duration-frequency relationships for East Africa rainfall. In: *Proc. Symp. on Floods Hydrol.* (Nairobi, 1975), 57–70. Transport and Road Res. Lab. Supply. Report No. 259.

Fiedler, F.R., 2003. Simple, practical method for determining station weights using Thiessen polygons and isohyetal maps. *J. Hydrol. Eng.* 8 (4), 219–221.

Froehlich, D. C., 1995a. Intermediate-duration-rainfall equations. *J. Hydraulic Eng., ASCE,* 121(10), 751–756.

Froehlich, D. C., 1995b. Long-duration-rainfall intensity equations. *J. Irrigation Drainage Eng.,* 121(3), 248–252.

Froehlich, D. C., 1995c. Short-duration-rainfall intensity equations for drainage design. *J. Irrigation Drainage Eng.,* 121(4), 310–311.

Garcia-Bartual, R. and Schneider, M. 2001. Estimating maximum expected short-duration rainfall intensities from extreme convective storms, *Phys. Chem. Earth (B),* 26(9), 675–681.

Guerrero-Salazar, P. L. A. and Yevjevich, V. (1975) Analysis of Drought Characteristics by the Theory of Runs, Colorado State University, Fort Collins, Hydrology Paper 80.

Gumbel, E. J., 1958. *Statistics of extremes.* Columbia University Press, New York.

Hershfield, D. M., 1962. A note on the variability of annual rainfall. *J. Appl. Meteorol,* Vol. 1, pp. 575–578.

Hevesi, J. A., Istok, J. D., and Flint, A. L., 1992a. Precipitation estimation in mountainous terrain using multivariate geostatistics: Part I. Structural analysis. *J. Appl. Meteorol.* 31, 661–676.

Hevesi, J. A., Istok, J. D., and Flint, A.L., 1992b. Precipitation estimation in mountainous terrain using multivariate geostatistics. Part II : Isohyetal Maps, *J. Appl. Meteorol.,* 31, 661–676.

Keers, J. F. and Wescott, P., 1977. A computer-based model for design rainfall in the United Kingdom. Meteorological Office Scientific Paper, 36, London.

Kedem, B., Chiu, L. S., and Karni, Z., 1990. An analysis of the threshold method for measuring area-average rainfall. *J. Appl. Meteorol.,* 29, 3–20.

Keifer, C. J. and Chu, H. H. 1957. Synthetic storm pattern for drainage design. *J. Hydraul. Div. ASCE,* 83, 1–25.

Koch, G. S. and Link, R. E., 1971. *Statistical Analysis of Geological Data,* Vols. I and II. Dower Publications, New York.

Kohler, M. A., 1949. Double mass curve analysis for testing the consistency of records and for making required adjustments. *Bull. Am. Met. Soc.,* Vol. 30, 188–189.

Koutsoyiannis, D., Kozonis, D., and Manetas, A. 1998. A mathematical framework for studying rainfall intensity-duration-frequency relationships. *J. Hydrol.* 206, 118–135.

Leclerc, G. and Shaake, J. C., 1972. Methodology for assessing potential impact of urban development on urban runoff and the relative efficiency of runoff alternatives: Ralph M. Parsons Laboratory Rept. No. 167, Massachusetts Institute of Technology, Cambridge, MA.

Llamas, J., and Siddiqui, M. M., 1969. Runs of Precipitation Series. Colorado State University, Fort Collins, CO, Hydrology Paper 33.

Maidment, D. R., 1993. *Handbook of Hydrology.* McGraw-Hill, New York.

Mandelbrot, B. B., 1979. *Fractal Geometry of Nature.* Freeman, W. H., San Francisco, 1982.

Martin, R., 1988. A statistical use for characterizing probability distributions with application to rain rate data. *J. Clim. Appl. Met.,* Vol. 28, 354–360.

McKee, T. B., Doesken, N. J., and Kleist, J., 1993. The relationship of drought frequency and duration to time scales. Preprints, *Eighth Conf. on Applied Climatology* (Anaheim, California, USA), 179–184.

McKee, T. B., Doesken, N. J., and Kleist, J., 1995. Drought monitoring with multiple time scales. Preprints, *Ninth Conf. on Applied Climatology* (Dallas, Texas, USA), 233–236.

Meteorological Office, 1962. *Weather in the Mediterranean,* Vol. 1, General Meteorology, 2nd Ed., Her Majesty's Stationary Office. London, U.K., pp. 32–35.

Millian, J. and Yevjevich, V., 1971. Probabilities of Observed Droughts, Hydrology Paper No. 50, Colorado State University, Fort Collins, CO.

Mooley, D. A., 1973. Gamma distribution model for Asia summer monsoon rainfall. *Mon. Weath. Rev.,* Vol. 101, No. 2, 160–176.

Mooley, D. A. and Crutcher, H. L., 1968. An application of Gamma distribution functions to India rainfall. ESSA Tech Report, EDS 5, Silver Springs, MD.

Palmer, W. C., 1965. Meteorol. drought. Research Paper No. 45. U.S. Weather Bureau, NOAA Library and Information Services Division, Washington, D.C., 20852.

Pearson, K., 1957. *Tables of the Incomplete Gamma Function.* Cambridge University Press, Cambridge, MA.

Ragab, R. and Prudhomme, C., 2002. Climate change and water resources management in arid and semi-arid regions: prospective and challenges of the 21st century. *Biosystem Eng.,* 81, 1, 3–34.

Richard, B. D., 1955. *Flood Estimation and Control.* Chapman & Hall Ltd., London, 187.

Saldarriaga, J. and Yevjevich, V., 1970. Application of Run Lengths to Hydrologic Series, Hydrology Paper No. 40, Colorado State University, Fort Collins, CO.

Sasaki, Y., 1960. An objective analysis for determining initial conditions for the primitive equations. Tech. Rep., (Ref. 60-16T), College Station: Texas A&M University.

Sharon, D., 1965. Variability of rainfall in Israel. *Israel Exploration Journal,* Vol. 15, pp. 169–176.

Shehadeh, N., 1976. The variability of rainfall in Jordan. *Dirasat–Humanities,* Vol. III, No. 3, 67–85.

Sherman, C. W., 1932. Frequency and intensity of excessive rainfalls at Boston-Massachusetts. *Trans. ASCE,* 95, 951–960.

Sirdaş, S. and Şen, Z., 2007. Determination of stream-channel infiltration due to flash floods in western Arabian Peninsula. *Hydrol. Eng., ASCE,* in press.

Stevens, M. J. M. and Smulders, P. T., 1979. The estimation of the parameters of Weilbull wind speed distribution for wind energy utilization purposes. *Wind Eng.,* Vol. 3, 132–145.

Summer, G., 1988. *Precipitation Process and Anal.* John Wiley & Sons, New York, 455 pp.

Şen, Z., 1976. Wet and dry period of annual flow series. *J. Hydraul. Eng., ASCE,* 102 (HY 10), 1503–1514.

Şen, Z., 1977. Run-sums of Annual Flow Series, *J. Hydrol.,* Vol. 35, pp. 311–324.

Şen, Z., 1978. Autorun analysis of hydrologic time series. *J. Hydrol.,* Vol. 36, pp. 75–85.

Şen, Z., 1980a. Statistical analysis of hydrologic critical droughts. *J. Hydraul. Eng., ASCE.,* 106 (HY1), 99–115.

Şen, Z., 1980b. Regional drought and flood frequency-analysis-theoretical consideration. *J. Hydrol,* Vol. 46, No.3–4, 265–279.

Şen, Z., 1984. Autorun model for synthetic flow generation. *J. Hydrol.,* 81, 155–170.

Şen, Z., 1997. Objective analysis by cumulative semivariogram technique and its application in Turkey. *J. Appl. Meteorol.* 36 (12): 1712–1724.

Şen, Z., 1998. Average areal precipitation by percentage weighting polygon method. *J. Hydrol. Eng.* 1, 69–72.

Şen, Z. and Eljadid, A. G., 1999. Rainfall distribution functions for Libya and rainfall prediction. *J. Hydrol. Sci.,* Vol. 45, No. 5, 665–680.

Şen, Z. and Eljadid, A. G., 2000. Automated Average Areal Rainfall Calculation in Libya. *Water Resour. Manage.,* Vol. 14, 405–416.

Şen, Z. and Öztopal, A., 2001. Genetic algorithms for the classification and prediction of precipitation occurrence. *Hydrol. Sci. J.,* 46 (2): 255–267.

Tabios III, G. O., and Salas, J. D., 1985. A comparative analysis of techniques for spatial interpolation of precipitation. *Water Resour. Bull.,* 21, 365–380.

Tarawneh, Q., 2000. Water resources environmental and climatic impact assessment with application to Jordan. Unpublished Ph.D. Thesis, Istanbul Technical University, Meteorology Department, 135.

Tase, N., Wilhite, D. A., and Glantz, M. H., 1987. Understanding the drought phenomenon: The role of definitions. Water International, 10, 111–120.

Taylor, C. M. and Lawes, E. F., 1971. Rainfall intensity duration frequency data for stations in East African Meteorological Department, Nairobi Tech. Memo No. 17.

Thiebaux, H. J. and Pedder, M. A., 1987. *Spatial Objective Anal.,* Academic Press, 299 pp.

Thiessen, A. H., 1911. Precipitation averages for large areas. *Mon. Wea. Rev.,* Vol. 39, 1082–1084.

UNDP, 1999. Vulnerability and adaptation to climate change in Jordan. Vol. 1, Project No. JOR/95/G31/IG/99, Al-Shamil Engineering, Amman, Jordan.

Wallis, R. J. and Hosking, J. R. M., 1993. Some statistics useful in regional frequency analysis. *Wat. Resour. Res.,* Vol. 29, No. 2, 271–281.

Walpole, R. E., 1982. *Introduction to Statistics,* Third ed., MacMillan, London.

Walton, W. C., 1969. *Ground Water Resources Evaluation,* McGraw-Hill Book Co., New York.

Wiesner, C. J., 1970. *Hydrometeorology.* Chapman & Hall, London, 232.

Wihl, G. and Nobilis, F. 1975. Die Vertailung und Hohenabhangigkeit von 2.5 und 97.5% der Niderschlagssummen in der Jahrszeiten in Osterreich (The distribution and elevation dependence from 2.5% and 97.5% of seasonal precipitation totals in Austria, in German). *Riv. Ital. Geofis. Sci.* Affini 1, Vol. 1, 134–137.

Wilhite, D. A. and Glantz, M. H., 1985. Understanding of the drought phenomenon: the role and definition. *Water Int.,* 10, 111–120.

WMO (World Meteorological Organization), 1966. Climate Change. Tech. Note No. 79, WMO No. 195, TP 200, I–20. WM, Geneva, Switzerland.

Yevjevich, V., 1967. *An Objective Approach to Definitions and Investigations of Continental Hydrologic Drought,* Hydrology Paper No. 23, Colorado State University, Fort Collins, CO, 43 p.

Yevjevich, V., 1972. Probability and statistics in hydrology. *Water Resour. Publications,* Colorado State University, Fort Collins, CO.

Zahrani, M. I., Saad, G. H., Hawsawi, H. M., Khiyami, H. A., Al-Amawi, F. A., Theban, M., and Şen, Z., 2007. Potential flood hazard in wadi Qanunah, Southwest Saudi Arabia. Saudi Geological Survey, Hydrogeology Project Team, Final Report.

# 4 Runoff and Hydrograph Analysis

## 4.1 GENERAL OVERVIEW

In arid regions people live mostly along the wadi banks and flood plains, depending for their survival on occasional surface water and permanent groundwater resources for domestic and agricultural activities. Infrequent surface water and flood occurrences may cause natural hazards that may threaten property and life. Although these areas are security regions for water supply, they are also dangerous to human life and property.

In any part of the world, the rainfall–runoff relationship plays a fundamental role in many aspects of watershed management such as the determination of the available and sustainable water resources, the design of flood operations and protective measures, and drought management. Although rainfall–runoff modeling is an important aspect in arid and semi-arid areas, there are significant differences from humid climates. The available methods for humid regions are commonly applied almost without distinction to arid regions. A set of different models is available to represent rainfall–runoff relationships, but they have limitations in the hydrologic parameters that are used to describe the rainfall–runoff process in wadi systems (Wheater et al., 1993). Accordingly they must be calibrated and verified based on historical rainfall and flood records if available; otherwise, flood discharge estimations should be achieved through synthetic methods. The factors that affect the rainfall–runoff relationship in a wadi can be divided into two main categories, namely, rainfall and watershed-related quantities.

Field experience indicates that storm rainfall has to exceed a total of about 20 mm before there is any substantial surface runoff. Annual runoff coefficients fluctuate between 0.133 and 0.185, resulting in an overall average of 0.158. Runoff is characterized by spates that last, on average, 12 h from start to finish. These flows are typical torrential catchments having a steep rise and rapid recession. The behavior of flow discharge is parallel to that of the rise–recession pattern. Available data reveal that it takes about 25 min, on average, to reach the target of 100 $m^3$/sec (Nouh, 2006).

The main purpose of this chapter is to review the current methodologies and to propose new alternatives for arid and semi-arid region hydrograph analysis.

## 4.2 HYETOGRAPH–HYDROGRAPH RELATIONSHIP

A *hyetograph* shows the change of rainfall intensity by time, and it has a single peak resulting from a storm rainfall event (Figure 3.9). A *hydrograph* is the graph that shows the change of discharge by time. The former is an input into a drainage basin (wadi), whereas the latter is the output. In any hydrological study the single most significant variable is the discharge at any desired cross-section of a catchment (wadi). For humid regions, there are different analytical, empirical, field, and hybrid techniques for hydrograph identification, which should concentrate on the following natural and rational features. Provided that the following points are valid, these techniques can be used also in arid regions.

1. Availability of surface water in a natural channel with temporal discharge variations depending on meteorological and geological conditions.
2. During nonrainy periods in humid regions, there is a base flow in the channel due to groundwater recharge. Most often such a base flow is missing in arid and semi-arid regions.
3. After the start of the storm rainfall, the discharge in the channel starts to increase.
4. After the storm rainfall cessation, the discharge continues to increase up to a peak discharge then starts to decrease. Hence, after the rainfall stops there is a peak discharge value.
5. After the peak discharge, the recession part of the hydrograph continues for some time and then returns to a natural base flow level, depending on the groundwater recharge from the up-stream parts of the cross-section.

Humid region hydrographs appear both in rainy or nonrainy periods. In arid and semi-arid regions hydrographs are attached with the rainy periods and their after-effects only. In humid regions, one can separate the hydrograph into two parts, namely, *base flow* (nonrainy) and *direct runoff* periods (Figure 4.1a). In arid regions, generally the base flow does not exist and hence the hydrograph is equivalent with the direct flow discharge variations only (Figure 4.1b).

Hydrographs in arid regions are typically characterized by extremely rapid rise of as little as 15 to 30 min. Losses from the flood hydrograph through bed infiltration reduce the flood volume. These *transmission losses* dissipate the flood and obscure the interpretation of observed hydrographs. It is not uncommon that no flood is observed at a gauging station, when further up-stream a flood has been generated and lost to bed infiltration (Şorman and Abdulrazzak, 1993).

On the other hand, the hydrograph is an end result of a hyetograph as the transformation of rainfall into runoff according to watershed features (Figure 4.2).

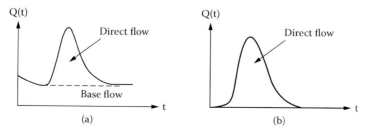

**FIGURE 4.1**   Hydrograph in a (a) humid region, and an (b) arid region.

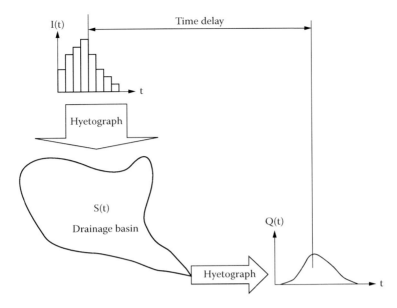

**FIGURE 4.2**  Catchment response to precipitation.

Rainy and nonrainy periods correspond to the rising and recession limbs of the hydrograph, respectively. There is a set of useful information for the planning and design of many water resources that can be derived from a given hydrograph. These are

1. In humid regions, the base flow reflects the contribution of *springs* (ground-water) during nonrainy periods. In general, base flow shows a slight decrease with time in the form of a trend (see Figure 4.3).
2. The area underneath the direct flow hydrograph gives the total volume of surface flow due to a rainfall storm. This volume corresponds to an excess amount of rainfall, which is the remaining rainfall after subtraction of losses including evaporation, infiltration, depression, interception storages, etc.
3. If rainfall continues for long durations, the rising limb of the hydrograph reflects more rainfall volume directly. After the rainfall cessation, this limb continues to rise up to a peak point. The peaks of hyetograph and hydro-graph are not simultaneous, but there is a time delay (lag) in the hydrograph peak with respect to hyetograph peak or centroid (see Figure 4.2).
4. The recession limb reflects the surface sheet flow into the channel after the rainfall cessation (see Chapter 2).

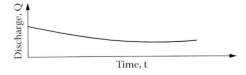

**FIGURE 4.3**  Base flow.

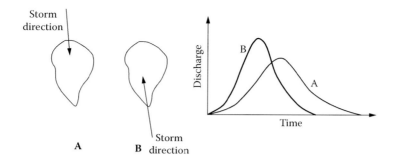

**FIGURE 4.4** Rainfall direction effect on hydrograph.

5. The peak represents the maximum runoff discharge due to hyetograph and reflects the time (*time of concentration*) needed for a raindrop to travel from the farthest point in the up-stream to the outlet.

Although the spatial distribution of rainfall causes variations in hydrograph shape, in practice, the rainfall is considered as uniform. If the center of the storm is close to the basin outlet, then a rapid rise, sharp peak, and rapid recession are observed in the hydrograph. If a larger amount of rainfall occurs in the upper reaches of a basin, then the hydrograph exhibits a lower and delayed peak. The direction of storm movement with respect to orientation of the basin can affect both the magnitude of the peak flow and the duration of surface runoff. Storm direction has an important effect on elongated basins, where up-stream moving storms tend to produce lower peaks with longer durations than storms moving downstream (Figure 4.4).

Thunderstorms produce peak flows on small basins, whereas large cyclonic or frontal-type storms are generally dominant in producing major floods in larger basins. If spatial variability is small, then the variation in the storm intensity over its duration is best presented by a rainfall hyetograph, as commonly applied in design practice. However, the assumption of spatially uniform rainfall is likely to be highly questionable where localized thunderstorm rainfalls predominate.

## 4.3   RATING CURVE

It shows the change of discharge by water level elevation or depth in a given cross-section. It is useful for flood inundation map preparations. In humid regions, rating curves can be obtained by field measurements and subsequent office calculations, whereas in arid regions synthetic and empirical approaches become effective. The following steps are for such a synthetic derivation.

1. A cross-section is selected along the wadi course, especially at one of the stable locations so that the shape of the cross-section does not change with time.
2. The profile of the cross-section is measured through surveying instruments and the geometry of the section is determined.
3. Surface features along the cross-section perimeter are observed and notes are taken visually in the field.

4. The slope, S, of the main channel at the cross-section is measured by considering two points, at 100 m apart toward up-stream and down-stream along the main channel *thalweg* (it is the deepest, continuous line along a watercourse), and their elevations are recorded.

5. After drawing the cross-section with suitable horizontal and vertical scales, the portions of surface features are described as rock, sand, gravel, silty sand, vegetation, etc.

6. For each surface portion a suitable *Manning coefficient*, n, is obtained from available standard tables (Section 4.11).

7. A set of hypothetical depths, D, with reference to the lowest point on the cross-section is chosen, and corresponding hypothetical water areas, A (wet areas), are calculated.

8. For each depth, the *wetted perimeter*, P, is found and then the *hydraulic radius*, R, of the cross-section is calculated as the ratio of the wetted area to the wetted perimeter (Section 2.9).

9. According to the wetted perimeter, the arithmetic average or preferably weighted average of the Manning coefficient is calculated for the cross-section.

10. Manning's formulation is used for calculation of average flow velocity, $\overline{V}$, as in Equation 2.13.

11. The cross-section discharge, Q, is calculated as the multiplication of average velocity by the area as $Q = A\overline{V}$.

12. The plot of discharges versus depths yields to a scatter diagram.

13. A synthetic *rating curve* is a smooth trend through the scatter points as shown in Figure 4.5, which appears in the form of a power function that can be expressed as

$$Q = aD^b \tag{4.1}$$

where *a* and *b* are constants. They can be obtained from the best fit to the scatter of points through a regression approach (Section 4.10.2).

In arid regions, such a synthetic rating curve helps to find discharges corresponding to a set of runoff or flood depth measurements at different times. Hence, the plot of the discharges versus time yields to a compound hydrograph as in Figure 4.6.

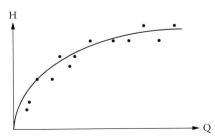

**FIGURE 4.5** Rating curve.

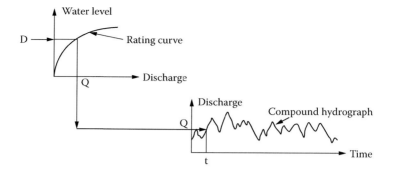

**FIGURE 4.6**   Hydrograph obtained from a rating curve.

## 4.4   CRITICS OF CONVENTIONAL METHODS

The rainfall–runoff relationship plays a key role in any water resources planning, design, operation, and maintenance study. For instance, flood estimations on small drainage basins are required for a number of engineering structures such as *dams*, levees, culverts, and soil conservation purposes. If in a basin designs are of low cost hydraulic structures, then the flood estimation models with large amounts of input data are not warranted. Preferably, parsimonious models are considered with simple basic principles for easy use (Linsley, 1982).

The most frequently used methods in flood estimations on small catchments are the rational method (RM) and the Soil Conservation Service (SCS) method (SCS, 1986). Details of these methods can be obtained readily from the relevant literature (Chow et al., 1988; Linsley, 1982). The main parameter for the RM is the *runoff coefficient*, C, and for the SCS method the *curve number*, CN. These methods are used for design flood discharge estimation provided that the design rainfall information is given.

Pilgrim and Cordery (1993) present the application of the RM as a design procedure. In many applications, C value is considered as constant, but in nature it changes with time and especially in the calculation of the design discharge average recurrence interval, it plays the single most important role. The variation of C with time must be considered in any formulation for finding refined rainfall–runoff conversion mechanism (Kadioğlu and Şen, 2001). Estimation of the C value is difficult and is the major source of uncertainty in many water resources projects. The coefficient must account for all the significant factors affecting the peak flow to average rainfall intensity, not restricted to area and response time. In any water resources design, the Cs are taken from tables based on a set of drainage features (Maidment, 1993). They are chosen in a rather vague manner and largely include personal judgment rather than actual field data. Additionally, various studies show that Cs vary widely from storm to storm, particularly depending on different antecedent wetness and environmental conditions (Hjelmfelt, 1991; Ponce and

Hawkins, 1996; Kadioğlu and Şen, 2001). Generally, the C increases as the average recurrence interval of rainfall increases, thus allowing for nonlinearity in runoff response of the drainage basin. Because considerable judgment and experience are required in selecting satisfactory Cs for a design, there is a need to check values against observed runoff data.

The SCS method is not physically based, and in its formulation there is no time variation. As with C in the RM, CN is also time dependent. Similar problems as in the RM apply for the SCS method. It was found by Wood and Blackburn (1984) that large variations in the CN are required to reproduce observed floods. This implies that actual antecedent moisture condition has a major effect, which is not taken into consideration in the derivation of the method. SCS method popularity is rooted in its convenience, simplicity, authoritative origin, and responsiveness to four readily grasped catchment properties, which are soil type, land use/treatment, surface condition, and antecedent condition (Ponce and Hawkins, 1996).

Williams and Laseur (1976) and Eastgate et al. (1979) have shown that the appropriate values of the C and the CN for use in the estimation of design floods in a given catchment are highly dependent on the manner in which the rainfall duration is determined. The model coefficients are also dependent on the use of soils with higher clay content and farming systems aimed at maximizing soil water storage in those areas for grain crop production. In many catchments, soil surface conditions are not stable over time, management practices may alter, and vegetal cover levels may change over a crop cycle.

In the following sections, few modified methods are proposed for synthetic hydrograph production, which avoid many points of the classical approaches.

## 4.5 INFINITESIMAL RUNOFF COEFFICIENT (IRC)

In nature, the runoff and the rainfall phenomena have spatial and temporal variability. Storm rainfalls of a given average return interval cannot produce the same floods depending on the antecedent conditions of the soil moisture. In general, the saturated catchments yield maximum floods from a storm event than the catchments with low antecedent moisture contents.

Herein, the change of runoff, r, with respect to rainfall, R, during infinitesimally small time duration, dt, is considered as the basis of rainfall–runoff conversion phenomenon. Especially, in arid regions there is a direct relationship between r and R, because most evaporation losses are negligible during the storm rainfall. Hence, it is possible to indicate such a variation mathematically as dr/dR which is referred to as the *infinitesimal runoff coefficient* (IRC). The surface runoff generation from a storm rainfall provides the basic foundations of the rainfall–runoff relationship. Below are the physical implications of the IRC change depending on different hydrological conditions. Runoff is generated by a variety of surface and subsurface flow processes as explained in the following points.

1. It is already stated by Ponce and Hawkins (1996) that saturation overland flow usually occurs during an infrequent storm, or toward the end of a particularly wet season, when the soil is already wet from prior storms. Under a specific set of circumstances, including soil type and texture, the silt entrained by splash erosion may deposit on the surface and create a thin crust that eventually reduces the infiltration rate to a negligible level. Thus, any additional rainfall will be converted to runoff, which is typical in semi-arid and arid regions, where large amounts of surface runoff may take place even through the underlying soil profile, below a relatively thin soil, which remains substantially dry (*Influences,* 1940; Le Bissonnais and Singer, 1993). If, prior to the storm, the catchment is saturated, the whole rainfall will appear in the form of direct runoff with no infiltration. In this case the IRC at any instant will be equal to one, dr/dR = 1.

2. If the catchment has low antecedent moisture levels, then the rainfall will have two components: (a) infiltration, I, and (b) direct runoff, r. On the other hand, the lack of vegetation cover in arid and semi-arid lands removes protection of the soil from raindrop impact, and soil crusting has been shown to lead to a large reduction in infiltration capacity for bare soil conditions (Morin and Benyamini, 1977). In this case, r < R and mathematically, dr/dR < 1.

3. If prior to a storm, the catchment is unsaturated and has very high infiltration rate, then almost all the rainfall will enter the subsurface. This condition implies that the IRC will be very close to zero. Mathematically, as I → ∞, then (dr/dR) → 0.

4. After some time period, an unsaturated catchment attains to completely saturated level, where the IRC becomes equal to 1, i.e., dr/dR = 1. This implies mathematically that as t → ∞, then dr/dR → 1. During this time period the infiltration rate decreases, whereas the runoff rate increases, which means an increase in the IRC.

In general, the IRC variation is confined between 0 and 1 ($0 \leq dr/dR \leq 1$). Such a situation can be expressed by the following mathematical model (Şen, 2007a),

$$\frac{dr}{dR} = 1 - e^{-kR} \qquad (4.2)$$

where *k* is referred to as the *saturation factor,* which represents the catchment saturation condition. Because *r* and *R* are measured in the same unit, Equation 4.2 is dimensionless. In the case of saturated (unsaturated) catchment *k* assumes big (small) values. Although theoretically, $0 < k < \infty$, in physical applications it can assume only finite values. Estimates of *k* should take into account the hydrological soil group, *land use* and treatment classes, and antecedent moisture content.

### 4.5.1 COMPARISON WITH OTHER METHODOLOGIES

Because the model in Equation 4.2 is a way of composing a set of simple hydrological principles, it should be reducible to some of the most commonly used methods.

Herein, its comparison with the RM and SCS methods is presented. Equation 4.2 is valid for small catchment areas where $k$ is assumed to be constant, and its multiplication by drainage area, $A$, leads to the following expression,

$$\frac{dV}{dR} = A(1 - e^{-kR})$$ (4.3)

where $dV = Adr$ is the direct runoff volume increment. Rainfall intensity definition in Equation 3.3 renders Equation 4.3 into the following form:

$$\frac{dV}{dt} = A(1 - e^{-kR})i$$ (4.4)

Because the left-hand side corresponds to surface discharge, $Q$, definition Equation 4.4 can be rewritten as

$$Q = A(1 - e^{-kR})i$$ (4.5)

The *peak discharge*, $Q_{PR}$, from the rainfall occurs after practically a long storm duration mathematically as $R \to \infty$ and hence Equation 4.5 becomes

$$Q_{PR} = Ai$$ (4.6)

Because a certain amount of the rainfall intensity is lost for infiltration the runoff peak discharge, $Q_{Pr}$, is smaller than $Q_{PR}$ and, hence,

$$Q_{Pr} = CAi$$ (4.7)

where $C$ is the runoff coefficient, and this expression is the well-known RM for flood estimation.

On the other hand, the SCS method is based on a simple storm relationship between $R$, actual runoff, $r$, ($r \le R$), *potential maximum retention*, $S$, after runoff begins when $S \ge F$, where $F$ is the *actual retention* after runoff. The initial equation is based on trends observed in data from collected sites; therefore, it is an empirical instead of a physically based equation. Use of a simple water budget principle (NEH, 1999) leads to

$$r = \frac{(R - I_a)^2}{(R - I_a) + S}$$ (4.8)

where $I_a$ is an *initial abstraction* greater than zero, and it can be expressed as a percentage of $S$. In all applications, $I_a = 0.2S$ is adopted, and its substitution into the previous expression gives

$$r = \frac{(R - 0.2S)^2}{R + 0.8S}$$ (4.9)

which is the rainfall–runoff relationship used in the SCS method for estimating direct runoff from storm rainfall. The validity of Equation 4.9 can be checked against the simple hydrological principles that are employed in the derivation of Equation 4.2, and, hence, the following SCS method drawbacks can be identified thus:

1. As a simple hydrology principle, if there is no storm rainfall ($R = 0$), then there is not direct runoff generation ($r = 0$). Substitution of $R = 0$ into Equation 4.9 leads to $r = 0.5S \neq 0$, which implies that without any storm rainfall the direct runoff is equal to the half of the potential maximum initial retention characteristic of the drainage basin.
2. The slope of the rainfall–runoff relationship can be obtained by taking the derivative of both sides in Equation 4.9 with respect to $R$ as

$$\frac{dr}{dR} = \frac{(R-0.2S)(R+1.8S)}{(R+0.8S)^2} \qquad (4.10)$$

which implies that for $S = 0$, $dr/dR = 1$ as in the case of the IRC method. However, when $R = 0$, $dr/dR = -0.5625$, which means that $-0.5625 < dr/dR < 1$. As has been already explained for the methodology developed herein, $0 < dr/dR < 1$:

1. The SCS method neither reveals explicitly the time dependence of the rainfall–runoff relationship nor takes into account explicitly the infiltration characteristics of the drainage basin.
2. The SCS method application needs the CN, ($0 < CN < 100$), which does not have any direct physical implication in hydrological studies. CN is related to $S$ as

$$CN = \frac{1000}{10+S} \qquad (4.11)$$

The selection of CN is a very critical and subjective issue in the application of the SCS method. CN is a transformation of $S$; it is used to make interpolating, averaging, and weighting operations more linear. Depending on the value of $S$ (i.e., CN), SCS method rainfall–runoff, ($R$-$r$), relationships start from different initial values, not from the origin where $r = 0$ for $R = 0$. In order to illustrate this explicitly, it is useful to construct rainfall–runoff graphs by using Equation 4.7 for different CNs as presented in Figure 4.7.

It is obvious from this figure that initial rainfall–runoff values are not physically plausible because they give rise to a negative slope, i.e., to negative IRC values, as explained above. For instance, for small CN numbers, such as $CN = 20$, there is runoff ($r \neq 0$) with no rainfall ($R = 0$). Similar cumulative rainfall–runoff curves can be obtained by integration from $r = 0$ to $r$ simultaneously from $R = 0$ to $R$, which yields

$$r = R - \frac{1}{k}\left(1 - e^{-kR}\right) \qquad (4.12)$$

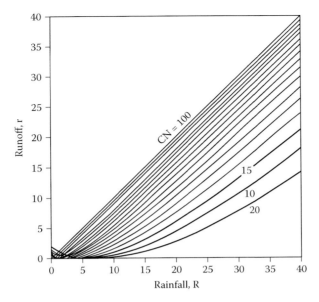

**FIGURE 4.7**   SCS rainfall–runoff relationship.

The graphical representation of this expression is presented in Figure 4.8, which shows a similar rainfall–runoff relationship to Figure 4.2. However, in the IRC method, contrary to Equation 4.10, there are no negative slopes, and all the curves start from the origin as expected, i.e., $r = 0$ for $R = 0$.

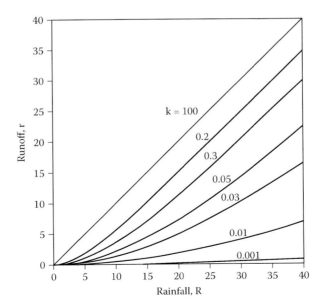

**FIGURE 4.8**   IRC method rainfall–runoff relationship.

Each curve in Figures 4.7 and 4.8 is labeled by different constants as CN and k, respectively. In order to find the correspondence between these constants, it is useful to compare the mathematical expressions of the straight-line portions in both figures. First, the straight-line portion of curves resulting from Equation 4.9 can be evaluated by completing the division operation on the right-hand side which leads to

$$r = R - 1.2S + \frac{S^2}{R + 0.8S} \tag{4.13}$$

For large $R$ values, this expression will have the final straight-line portions of each curve in Figure 4.8 as

$$r = R - 1.2S \tag{4.14}$$

On the other hand, for big $R$ values, Equation 4.12 yields the straight-line portion expression for the curves in Figure 4.8 as

$$r = R - \frac{1}{k} \tag{4.15}$$

The equivalence of Equations 4.14 and 4.15 yields

$$\frac{1}{k} = 1.2S \tag{4.16}$$

Furthermore, elimination of $S$ between Equations 4.11 and 4.16 gives

$$\frac{1}{k} = \frac{1200}{CN} - 12 \tag{4.17}$$

It is possible to deduce logically that in the case of zero potential retention, the CN approaches 100, whereas the value of $k$ increases unboundedly. At the other extreme, theoretically, the upper bound of retention occurs for CN $\rightarrow$ 0 and $k \rightarrow$ 0. Hence, Equation 4.17 is a useful conversion between CN and k. Table 4.1 is prepared from the SCS method according to this principle.

Furthermore, the graphical representation between the CN and corresponding $k$ values is presented in Figure 4.9 on double logarithmic paper.

In order to compare the two methodologies, namely, SCS and IRC, the runoff amounts for the corresponding CN and k values are calculated for two different case studies from Equations 4.11 and 4.17, respectively. In Table 4.2, different rainfall amounts are considered, whereas Table 4.3 presents the results for the same rainfall amounts but different CNs.

The graphical representation of the runoff results from the two methods are given in Figures 4.10 and 4.11 for different and the same rainfall inputs, respectively.

**TABLE 4.1**
**Catchment Descriptions and *k* Values**

| Land Use Description | Cover Description | *k* Values for Hydrologic Soil Group | | | | |
| --- | --- | --- | --- | --- | --- | --- |
| | | A | B | C | D | E |
| Agricultural | Row Crops – Straight Rows + Crop Residue Cover– Good Condition | | 0.148 | 0.25 | 0.38 | 0.472 |
| Commercial | Urban Districts: Commercial and Business | 0.472 | 0.674 | 0.958 | 1.306 | 1.583 |
| Forest | Woods – Good Condition | | 0.0357 | 0.1019 | 0.1944 | 0.279 |
| Grass/Pasture | Pasture, Grassland, or Range – Good Condition | | 0.0533 | 0.1303 | 0.2372 | 0.3333 |
| High-Density Residential | Residential districts by average lot size: 1/8 acre or less | 0.1548 | 0.279 | 0.4722 | 0.75 | 0.9583 |
| Industrial | Urban District: Industrial | 0.2143 | 0.3553 | 0.6111 | 0.8426 | 1.1071 |
| Low-Density Residential | Residential districts by average lot size: 1/2 acre | 0.0278 | 0.0978 | 0.1944 | 0.3333 | 0.4722 |
| Open Spaces | Open Space (lawns, parks, golf courses, cemeteries, etc.). Fair Condition (grass cover 50% to 70%) | | 0.0801 | 0.1855 | 0.3135 | 0.4375 |
| Parking and Paved Spaces | Impervious areas: Paved parking lots, roofs, driveways, etc. (excluding right-of-way) | 8.25 | 4.0833 | 4.0833 | 4.0833 | 4.0833 |
| Residential 1/8 acre | Residential districts by average lot size: 1/8 acre or less | 0.1548 | 0.279 | 0.472 | 0.75 | 0.9583 |
| Residential 1/4 acre | Residential districts by average lot size: 1/4 acre | 0.0511 | 0.1303 | 0.25 | 0.4069 | 0.5577 |
| Residential 1/3 acre | Residential districts by average lot size: 1/3 acre | 0.0357 | 0.1105 | 0.2143 | 0.3553 | 0.5119 |
| Residential 1/2 acre | Residential districts by average lot size: 1/2 acre | 0.0278 | 0.0978 | 0.1944 | 0.3333 | 0.4722 |
| Residential 1 acre | Residential districts by average lot size: 1 acre | 0.0208 | 0.0867 | 0.1771 | 0.3135 | 0.4375 |
| Residential 2 acres | Residential districts by average lot size: 2 acres | 0.0114 | 0.071 | 0.1548 | 0.279 | 0.3796 |
| Water/Wetlands | | | 0 | 0 | 0 | 0 |

It is obvious that both method results are within a less than 10% error limit. This indicates the validity of the IRC method in practical applications. In view of what has been explained previously, it can be readily seen that the IRC method is more viable because it has a physically based derivation.

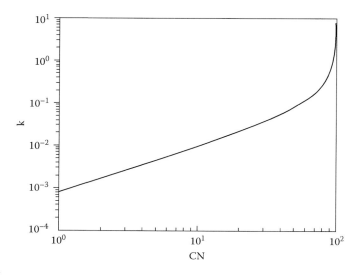

**FIGURE 4.9**  CN and k relationship.

**TABLE 4.2**
**Different Rainfall Inputs**

| Rainfall | CN | Potential Retention | Runoff SCS | Runoff IRS | Error (%) |
|----------|-----|---------------------|------------|------------|-----------|
| 55 | 98 | 4.0833 | 54.756 | 54.755 | 0 |
| 20 | 64 | 0.1480 | 14.540 | 13.590 | 6.5 |
| 45 | 39 | 0.0533 | 30.484 | 27.943 | 8.3 |
| 36 | 12 | 0.0114 | 4.810 | 6.470 | 25.0 |

**TABLE 4.3**
**The Same Rainfall Inputs**

| Rainfall | CN | Potential Retention | Runoff SCS | Runoff IRS | Error (%) |
|----------|-----|---------------------|------------|------------|-----------|
| 36 | 98 | 4.0833 | 35.76 | 35.76 | 0 |
| 36 | 64 | 0.1480 | 30.03 | 29.28 | 2.5 |
| 36 | 39 | 0.0533 | 22.27 | 19.99 | 10.2 |
| 36 | 12 | 0.0114 | 4.81 | 6.47 | 25.0 |

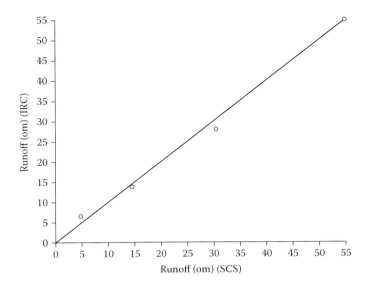

**FIGURE 4.10**   Different rainfall inputs.

Direct implementation of the IRC methodology developed in this section requires records of rainfall and runoff during single or successive storms. Such data is available from NEH (1999), Chapter 16. Both accumulated rainfall and runoff recordings are presented in Table 4.4. The plot of accumulated rainfall and runoff from the second and third columns versus time gives the graphical behavior of these two events as shown in Figure 4.12.

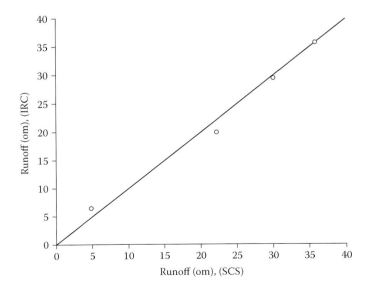

**FIGURE 4.11**   The same rainfall inputs.

**TABLE 4.4**

**Rainfall Tabulated in 0.3-Hour Increments from Plot of Raingage Chart**

| Time (h) (1) | Accumulated Rainfall (in) R (2) | Accumulated Runoff (in) r (3) | Incremental Rainfall (in) dR (4) | Incremental Runoff (in) dr (5) | dr/dR (6) | 1 − dr/dR (7) |
|------|------|------|------|------|------|------|
| 0.0 | 0.00 | 0.00 | 0.00 | 0.00 | 0.00 | 1.00 |
| 0.3 | 0.37 | 0.00 | 0.37 | 0.00 | 0.00 | 1.00 |
| 0.6 | 0.87 | 0.12 | 0.50 | 0.12 | 0.24 | 0.76 |
| 0.9 | 1.40 | 0.39 | 0.53 | 0.27 | 0.51 | 0.49 |
| 1.2 | 1.89 | 0.72 | 0.49 | 0.33 | 0.67 | 0.33 |
| 1.5 | 2.24 | 0.98 | 0.35 | 0.26 | 0.74 | 0.26 |
| 1.8 | 2.48 | 1.16 | 0.24 | 0.18 | 0.75 | 0.25 |
| 2.1 | 2.63 | 1.28 | 0.15 | 0.12 | 0.80 | 0.20 |
| 2.4 | 2.70 | 1.34 | 0.07 | 0.06 | 0.86 | 0.14 |
| 2.7 | 2.70 | 1.34 | 0.00 | 0.00 | 0.00 | 1.00 |
| 3.0 | 2.70 | 1.34 | 0.00 | 0.00 | 0.00 | 1.00 |
| 3.3 | 2.71 | 1.35 | 0.01 | 0.01 | 1.00 | 0.00 |
| 3.6 | 2.77 | 1.40 | 0.07 | 0.05 | 0.71 | 0.29 |
| 3.9 | 2.91 | 1.51 | 0.14 | 0.11 | 0.76 | 0.24 |
| 4.2 | 3.20 | 1.76 | 0.29 | 0.25 | 0.86 | 0.16 |
| 4.5 | 3.62 | 2.12 | 0.42 | 0.36 | 0.86 | 0.16 |
| 4.8 | 4.08 | 2.54 | 0.46 | 0.42 | 0.91 | 0.09 |
| 5.1 | 4.43 | 2.85 | 0.35 | 0.31 | 0.88 | 0.12 |
| 5.4 | 4.70 | 3.09 | 0.27 | 0.24 | 0.89 | 0.11 |
| 5.7 | 4.90 | 3.28 | 0.20 | 0.19 | 0.95 | 0.05 |
| 6.0 | 5.00 | 3.37 | 0.10 | 0.09 | 0.90 | 0.10 |

This figure also shows that there are two storm events and, as expected, the rainfall amounts are always above the runoff records, which show that some portion of rainfall is lost during its transformation into runoff event. The hydrological behavior and response of the watershed is reflected entirely in these records, which are the basic ingredients for *saturation factor*, k, and determination in Equation 4.2. In order to calculate k, Equation 4.2 can be arranged as

$$\log\left(1-\frac{dr}{dR}\right)+kR=0 \tag{4.18}$$

Here, the only unknown is k, and it corresponds to the slope of $\log(1 - dr/dR)$ versus R plot on a semilogarithmic paper. In order to determine k from the given rainfall and runoff records, it is necessary to execute the following steps.

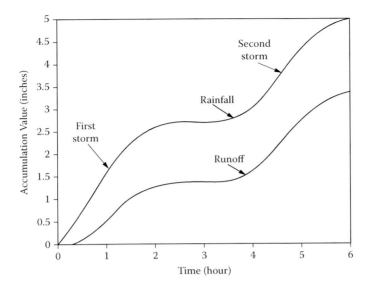

**FIGURE 4.12**  Accumulated rainfall and runoff.

1. Calculate the incremental values of rainfall, $dR$, from the given data, which are presented in column 4 of Table 4.4.
2. Calculate the incremental values of runoff, $dr$, which are given in column 5 of Table 4.4.
3. The division of $dr$ to $dR$ for the same time instances gives values represented in column 6 of the same table.
4. The bracket term in Equation 4.18 is given in the last column of the same table.

Next, it is seen that the second and the last columns in Table 4.4 represent two variables, namely $R$ and $(1 - dr/dR)$. It is obvious from Equation 4.18, mathematically, that there is a reverse linear relationship between these variables on a semilogarithmic paper as in Figure 4.13. One can fit a straight-line through the scatter points, and hence the slope of the fitted straight-line yields the runoff exponent value.

In this figure, 3.7 inches is the accumulated rainfall difference that corresponds to one cycle of logarithmic scale on the vertical axis. The calculations are shown on the same figure and consideration of the two storm events collectively gives $k = 0.62$. The same data has been treated by considering the CN (NEH, 1999), where the CN is adopted as 85. It is interesting to calculate the corresponding saturation factor of this CN from Equation 4.17 as $k = 0.472$, which is not close to the value obtained by the IRC methodology.

There is another way of determining and confirming $k$ by making use of the rainfall–runoff curves as shown in Figure 4.12. For this purpose, rainfall–runoff curves for a set of $k$ values are prepared according to Equation 4.12, and the results are presented in Figure 4.14. In the same figure, the plots of accumulated rainfall and runoff values from Table 4.4 are shown with small circles, which yields $k = 0.6$.

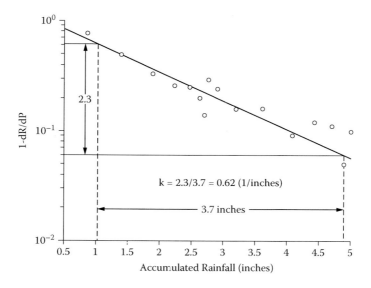

**FIGURE 4.13**   Runoff exponent calculations from two storm events.

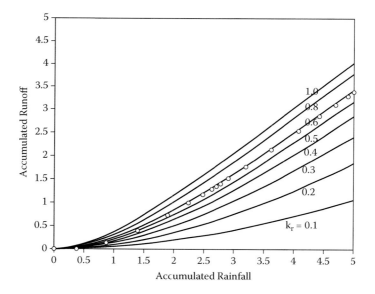

**FIGURE 4.14**   Rainfall–runoff relationship for saturation factor, $k$.

## 4.6   HYDROGRAPH AND UNIT HYDROGRAPH DERIVATION IN ARID REGIONS

In the past two decades, there were several attempts to relate the response of a drainage basin to its geomorphologic or topographic features, using different hypotheses in order to model the damping effect of the drainage network (Rodriguez-Iturbe et al., 1979; Gupta et al., 1980; Karlinger and Troutman, 1985; Mesa and Miffin, 1986; Chutha and Dooge, 1990). Most of these studies are based on classifying the channel network of the drainage basin according to the Strahler (1958) ordering procedure and assuming that the impulse response of each stream area is an exponential distribution function. The assumptions are set forward in order to obtain analytical expressions. Among such assumptions, especially in the model proposed by Rodriguez-Iturbe et al. (1979), the infiltration rate is not taken into consideration initially, but the final hydrograph is then reduced by a factor to compensate for the *infiltration* effects. On the contrary, the following model considers infiltration during the model development because infiltration is one of the most significant hydrologic components especially in arid and semi-arid regions. The resulting impulse response is mathematically equivalent to the response of the conceptual model consisting of parallel and serial linear storage elements (Bras, 1989). It is confirmed through many years of experience that the drainage basin impulse response can be described more adequately by a gamma PDF than a simple exponential PDF (Nash, 1958; Dooge, 1973; Rodriguez-Iturbe et al., 1979; Bras, 1989).

### 4.6.1 TEMPORAL RUNOFF CHANGE WITH RAINFALL

In arid and semi-arid regions, the most important hydrological components during a storm rainfall are the rainfall, runoff, and infiltration rates where evaporation and interception (as there are no significant vegetation covers) are negligible. On the other hand, in natural wadis (catchment areas in arid and semi-arid regions) the surface depression volumes are comparatively very small, and hence they are ignored. Let us concentrate first on the physical implications of the rainfall–runoff ratio change, depending on environmental conditions:

1. If the wadi surface is highly or completely impervious, the whole rainfall rate will be equal to the runoff rate. In this case, all the rainfall appears as the direct runoff with no infiltration. This case also shows that, at any time, instant $dr/dR = 1$. The mathematical implication of this assertion is that the runoff versus rainfall plot appears as a 45° straight-line on the Cartesian coordinate system. Figures 4.7 and 4.8 present 45° lines for CN = 100 and $k = 100$ or very high $k$ values, respectively.

2. In the case of lower or moderate pervious surface and subsurface layers, $dr/dR < 1$, i.e., less than 45° slopes initially. This is already the case for the initial portions of the Soil Conservation Service (SCS; 1971, 1986) rainfall–runoff relationship.

3. For very high infiltration rates, almost all the rainfall will enter the subsurface. This condition implies that the rainfall–runoff relationship will appear very closely to the rainfall axis, i.e., $dr/dR \rightarrow 0$, which implies that the slope is almost equal to zero.

4. Another physical extreme occurs after long time periods because the unsaturated layers of the subsurface becomes completely saturated, which poses a similar condition as in step 1, where asymptotically $dr/dR = 1$ (45° straight lines).

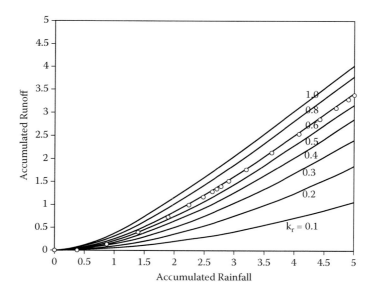

**FIGURE 4.15**  Runoff–rainfall slope variations with time.

5. Joint consideration of all the previous steps rationally leads to theoretical conclusion that $0 \leq dr/dR \leq 1$. A coarse application of these steps is presented by Kadioğlu and Şen (2001) for monthly time durations.

In light of these points, it is possible to end up with the following mathematical model (Şen, 2007b):

$$\frac{dr}{dR} = 1 - e^{-k_s t} \qquad (4.19)$$

where $k_s$ is a parameter representing the soil and subsoil conditions. This is referred to as the *runoff exponent*. The more the imperviousness of the soil, the bigger will be the soil parameter value. In Figure 4.15 various curves resulting from Equation 4.19 are presented with different $k_s$ values.

It is obvious from this figure that, practically, $0 \leq k_s \leq 1$. By definition, Equation 4.19 is valid for unit area. For the whole drainage basin, the area can be considered in the forms of homogeneous subareas, where $k_s$ coefficient is constant. Although in humid regions for large areas, the catchment heterogeneity due to soil characteristics, geology, topographical aspects, vegetation, and land use, etc. is tremendous, this is not the case in the arid zone catchments.

## 4.6.2  INFILTRATION CHANGE WITH TIME

The classic *infiltration rate, f*, expression is given by Horton (1940) as

$$f = f_i + (f_i - f_f)e^{-kt} \qquad (4.20)$$

where $f_i$, and $f_f$, $(f_f < f_i)$ are the initial and final (long-term) infiltration rates, and $k$ is the *infiltration coefficient*, which reflects the vertical water intake capacity of the subsurface. By definition, $f$ is the change of infiltration height by time. Equations 4.19 and 4.20 can be combined by considering the water balance equation simply as

$$R = r + f$$

or, by differentiation,

$$dR = dr + df,$$

the substitution of which into Equation 4.19 leads to

$$\frac{dr}{dr+df}=1-e^{-k_s t}$$

After the necessary arrangements one can obtain,

$$\frac{dr}{df}= e^{k_s t} -1$$

or, finally, in terms of time rates,

$$\frac{dr}{dt}= (e^{k_s t} -1)\frac{dI}{dt}=(e^{k_s t} -1)f' \qquad (4.21)$$

where $f'$ is the *effective infiltration rate* of rain entering subsoil. This is represented by the second term on the right hand-side of Equation 4.20 as

$$f' = (f_i - f_f)e^{-kt} \qquad (4.22)$$

Finally, the substitution of this expression into Equation 4.21 leads to

$$\frac{dr}{dt}=(f_i-f_f)(e^{k_s t} -1)e^{-kt} \qquad (4.23)$$

The graphical representations of this equation for a set of parameters $(k_s, f_f, f_i,$ and $k)$ are shown in Figures 4.16 to 4.18. In each figure, there is a set of curves for different $k$ values. It is obvious that as the $k$ value increases, the runoff rate decreases. Likewise, there is an inverse relationship between the peak ratio value and the parameter sets. On the other hand, the comparison of Figures 4.16 to 4.18 indicates that increase in $k_s$ value causes increase in the peak runoff rates. This is plausible because the increase in $k_s$ is equivalent to the increase in the impervious surface area percentage and hence the increase in the runoff rates. A glance at these curves indicates that

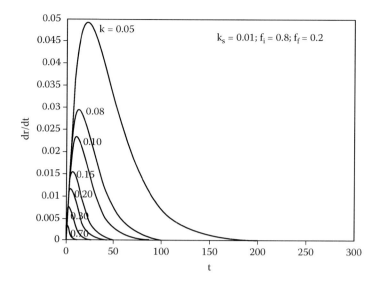

**FIGURE 4.16**   Runoff rate hydrograph.

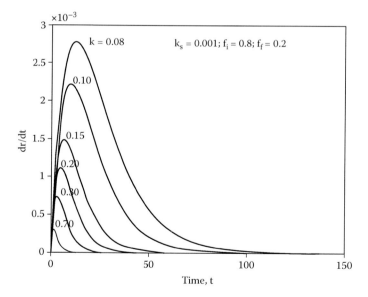

**FIGURE 4.17**   Runoff rate hydrograph.

they have self-similar behaviors (Mandelbrot, 1979). This is tantamount to saying that they all have the same shape at different scales.

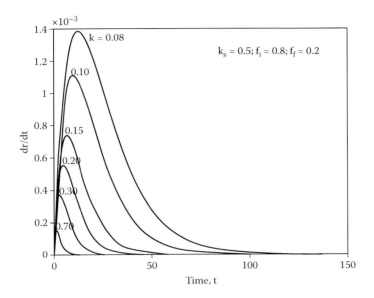

**FIGURE 4.18** Runoff rate hydrograph.

### 4.6.3 HYDROGRAPH FORMULATION

The right-hand side of Equation 4.23 is equal to the *specific discharge* (discharge per area), $q = dQ/A$, which can be obtained as

$$q = \frac{Q}{A} = (f_i - f_f)(e^{k_s t} - 1)e^{-kt} \tag{4.24}$$

This expression constitutes the basis of hydrograph analysis. The hydrological principles in the derivation of Equation 4.24 are not new, but they are combined in an effective manner with logical–rational–physical–mathematical–empirical principles to yield a new hydrograph expression, which is valid especially for arid regions.

The *peak discharge*, $Q_p$, expression can be obtained by setting $dq/dt$ equal to zero which yields after the necessary mathematical operations, first *time-to-peak*, $t_p$ as

$$t_p = \frac{1}{k_s} Ln\left(\frac{k}{k - k_s}\right) \tag{4.25}$$

The substitution of this last expression into Equation 4.24 gives the peak specific discharge expression as

$$q_p = \frac{Q_P}{A} = \frac{k_s (f_i - f_f)}{k - k_s}\left(\frac{k}{k - k_s}\right)^{-k/k_s} \tag{4.26}$$

In this manner, time-to-peak and peak discharge are related to the infiltration ($f_i$, $f_f$ and $k$) and *runoff exponent*, $k_s$ parameters. It is obvious from Equations 4.25 and 4.26 that for physically plausible results, $k$ should be greater than $k_s$ ($k > k_s$).

These expressions replace the empirical equations by Snyder (1938) who has related the time-to-peak and the peak discharge expressions to drainage basin morphologic features such as the drainage area, $A$, longest channel length, $L$, and the centroid distance, $L_c$, between the projection of drainage basin centroid point on the main channel and the outlet (see Chapter 2). With these notations, the classical *Snyder equations* are given as

$$t_p = C_t \, (LL_c)^{0.3} \tag{4.27}$$

and

$$Q_F = C_p \frac{A}{t_p} \tag{4.28}$$

where $C_t$ and $C_p$ are location-dependent parameters which are given as $1.35 < C_t < 1.60$ and $0.23 < C_p < 0.67$ for the Appalachian Mountains in the United States. The major difference between these two sets of equations (Equations 4.25–4.26 and Equations 4.27–4.28) is that former expressions rely on the hydrological aspects rather than morphological parameters as for the Snyder approach.

### 4.6.4  STANDARD DIMENSIONLESS UNIT HYDROGRAPH

Inspection of Figures 4.16 to 4.18 indicates that sets of hydrographs with different parameters have similar shapes. As stated earlier, these hydrographs are *self-similar*, and they can be reduced into a common dimensionless form. Subsequently, each graph's specific discharge ordinates are divided by the maximum specific discharge, $q_P$, and the time values by $t_p$. Whatever the parameter set values, all of the hydrographs collapse on the same standard *dimensionless unit hydrograph* (DUH) whose coordinates are presented in Table 4.5.

The plot of *standard* DUH is shown in Figure 4.19 with the SCS dimensionless UH.

The dimensionless recession limb has greater values at the same dimensionless times than the SCS curve, and hence its recession limb is more durable. Another difference is that the rising limb of the standard DUH has sharper slopes, which indicates that in arid regions flash floods have more potential to occur. Hence, in arid and semi-arid regions, occurrence of flash floods is more possible and more durable than humid regions.

### 4.6.5  COMPARISON WITH OTHER METHODOLOGIES

The SCS method is based on a simple storm relationship between different hydrological variables in Equation 4.9 and similar to Equation 4.11 as

$$S = 25.4 \left( \frac{1,000}{CN} - 10 \right) \tag{4.29}$$

**TABLE 4.5**
**Standard Dimensionless UH Coordinates**

| $t/t_p$ | $q/q_p$ | $t/t_p$ | $q/q_p$ |
|---------|---------|---------|---------|
| 0 | 0 | 3.1 | 0.423 |
| 0.1 | 0.2349 | 3.5 | 0.3269 |
| 0.2 | 0.4273 | 3.9 | 0.2494 |
| 0.3 | 0.5829 | 4 | 0.2327 |
| 0.4 | 0.7068 | 4.5 | 0.1631 |
| 0.6 | 0.8768 | 5 | 0.1129 |
| 0.7 | 0.9303 | 5.5 | 0.0774 |
| 1.1 | 1 | 6 | 0.0526 |
| 1.2 | 0.9921 | 6.5 | 0.0356 |
| 1.5 | 0.9329 | 7 | 0.0239 |
| 1.8 | 0.8423 | 8 | 0.0106 |
| 2.1 | 0.7393 | 9 | 0.0047 |
| 2.4 | 0.6358 | 11 | 0.0008 |
| 2.5 | 0.6023 | 14 | 0 |
| 2.7 | 0.5382 | | |

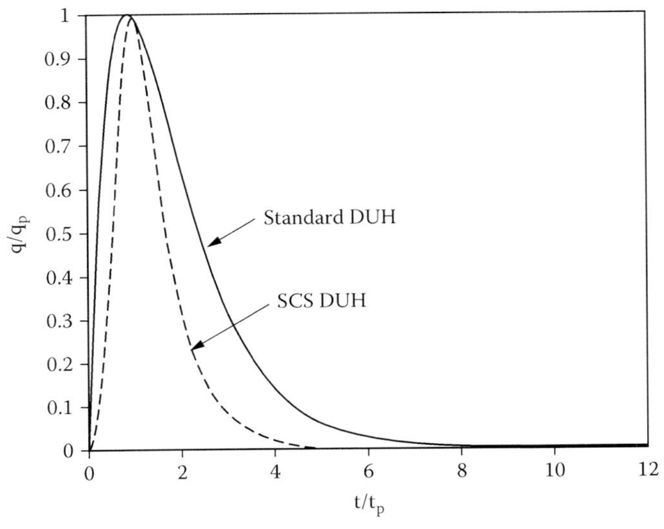

**FIGURE 4.19** Dimensionless unit hydrograph.

where the constant 25.4 is due to SI units. In the SCS method, a triangular UH is used to distribute $r$ runoff depth over time. Peak discharge rates, $q_p$, can be estimated using the following simple equation

$$q_p = \frac{20.8\,Ar}{\left(\dfrac{D}{2} + L_g\right)} \tag{4.30}$$

where $L_g = 0.6t_r$ and $D$ = storm duration in hours equal $2t_r^{0.5}$, in which $t_r$ is the duration of design rainfall and $A$ is the catchment area. In Equation 4.30 there is much interference on an unexplainable basis, such as that the factor 20.8 only applies if the triangular unit hydrograph has a particular geometry, with 37.5% of its volume in the rising limb and a time of recession equal to 1.67 times the time of rise. Unfortunately, in many parts of the world, the SCS method is used but without checking such a condition and, therefore, the results may be questionable. This is due to the fact that the SCS method does not have its unique UH basis, but rather a subjective triangular UH, depending on the choice of CN value. If the UHs are of another shape, a different peak discharge rate would be calculated. On the other hand, as Titmarsh (1989) and Titmarsh et al. (1995) stated, the CN is time dependent, but this point is not taken into consideration in hydrograph determinations.

Because, for infiltration, the soil types are significant, the numerical classification of the $k_s$ parameter will be based on the types of soils. As a preliminary approach the soil classification of SCS into four classes as A, B, C, and D is considered. These classes correspond respectively to "high," "moderate," "slow," and "almost no" infiltration rates. Additionally, each one of these classes is given in terms of outcrop rock types as they appear especially in arid regions (see Table 4.6). Accordingly, the possible average $k_s$ values are presented in the last row. Generally, sand and gravel soils fall within category A. Soils with moderately fine to moderately coarse composition imply B type with moderate runoff and infiltration rates. High-runoff potential soils have type C with slow infiltration rates. Such hydrologic phenomena are expected with moderately fine to very fine texture. Finally, D soil type represents very high (low) runoff (infiltration) rates. It is rational and logical that the smaller $k_s$ values imply big infiltration rates.

The first impression from this table is that in each soil type there are different rock types with different infiltration rates. It is, therefore, recommended that a hydrologist with experience in the field may appreciate more refined values around the global values of $k_s$ as they are presented in this table. This means to say that one should not take the given values in Table 4.6 as crisp $k_s$ but rather use intuition, expertise, and experience in allocating $k_s$ values. For instance, in soil type A, sand and gravel cannot be treated with the same $k_s$ value because gravel will have a comparatively greater infiltration rate than sand and, correspondingly, its $k_s$ value will be

**TABLE 4.6**
**Runoff Exponent Values**

| Soil Type | A | B | C | D |
|---|---|---|---|---|
| Rock outcrops | Gravel, cobbles, boulders, Eolian sand, sand dunes | Quaternary deposits, sand, fracture | Sedimentary rocks, weathered rocks, metamorphic rocks | Clay, silt, intact rock |
| $k_s$ | 0.15 | 0.30 | 0.50 | 0.75 |

relatively greater as an A type of soil. Another point that should be taken into consideration in the use of proposed methodology is the percentage areas of different soil types within the drainage basin. If there are four different subareas—$A_A$, $A_B$, $A_C$, and $A_D$—each with types A, B, C, and D soil, and *runoff exponents* as $k_{sA}$, $k_{sB}$, $k_{sC}$, and $k_{sD}$, respectively, then the average runoff exponent $\bar{k}s$ value should be calculated as

$$\bar{k}_s = \frac{A_A k_{sA} + A_B k_{sB} + A_C k_{sC} + A_D k_{sD}}{A_A + A_B + A_C + A_D} \qquad (4.31)$$

It must be stated at this stage that, in general, the contributing area to the runoff is not the whole catchment area, but only part of it. Many times this is the riparian zone, which is not the case in many arid zone catchments.

### 4.6.6 APPLICATION

The application of the methodology developed herein is presented for annual discharges in Table 4.7, which are taken from the *National Engineering Handbook* (1999). These are the total rainfall and runoff data in inch from a watershed in the United States near Treynor, Iowa.

The final column is calculated as the ratio of runoff to rainfall depth, which is referred to as runoff exponent $k_s$. The relationship between $k_s$ and the CN values calculated in NEH (1999) is given in Figure 4.20.

It is obvious that there is a direct relationship between the CN and $k_s$. The average value of CN is 86, which corresponds to the average $k_s$ as 0.33.

The implementation of the new method is presented for two wadis, namely, Wadi Mathwab and Wadi Dighbij in the Yalamlam basin that lie in the western regions of the Arabian Peninsula along the Red Sea coastal area (see Figure 4.21). Both the regional and local air circulation have a dominant influence on the climate of the region. According to the world climate classification as established by Koppen (Glenn, 1954), the Yalamlam basin can be divided into three main climate types.

1. The hot desert climate that prevails on the Red Sea coast (*Tihamah*).
2. The low latitude semi-arid climate that prevails in the hills.
3. The warm temperate rainy climate with dry winters prevailing in the Scarp Mountains (Taha et al., 1981).

On the Red Sea coast, there are two basic climates, namely, cool to warm and stable air originating from the Mediterranean during the winter period, and warm and moist air due to monsoons coming from the Indian Ocean during the summer.

Generally, rainfall is predominant in the northern mountain areas during winter due to the Mediterranean effect, whereas it is widespread in all regions during spring because of the local diurnal circulation effect. Orographic conditions are clear in winter and spring seasons. In summer, rainfall moves toward the south due to the monsoon flow effect with its southwesterly wind. In fall, the area becomes under the influence of monsoon as well as the local diurnal circulation (Al-Yamani and Şen, 1992).

Extensive field infiltration works in two wadis indicate spatial variability of the measurements within the two catchments with average infiltration coefficients as

**TABLE 4.7**
**Total Rainfall and Runoff Heights**

| Year | Month | Day | Rainfall (in) | Runoff (in) | Peak Discharge (ft³/sec) | CN | $k_s$ |
|------|-------|-----|---------------|-------------|--------------------------|-----|-------|
| 1964 | June  | 22  | 1.18 | 0.58 | 216.8 | 93 | 0.491 |
| 1965 | June  | 29  | 1.30 | 0.64 | 157.0 | 92 | 0.492 |
| 1966 | June  | 26  | 1.04 | 0.40 | 153.0 | 91 | 0.385 |
| 1967 | June  | 20  | 5.71 | 3.76 | 406.0 | 82 | 0.658 |
| 1968 | June  | 13  | 0.97 | 0.28 | 94.0  | 89 | 0.289 |
| 1969 | Aug.  | 20  | 2.23 | 0.17 | 36.9  | 63 | 0.076 |
| 1970 | Aug.  | 2   | 1.92 | 0.70 | 282.4 | 84 | 0.365 |
| 1971 | May   | 18  | 1.10 | 0.73 | 214.0 | 96 | 0.664 |
| 1972 | May   | 5   | 0.62 | 0.29 | 121.0 | 96 | 0.468 |
| 1973 | Sep.  | 26  | 1.25 | 0.28 | 43.7  | 84 | 0.224 |
| 1974 | Aug.  | 17  | 1.12 | 0.10 | 23.5  | 79 | 0.089 |
| 1975 | Aug.  | 29  | 1.66 | 0.30 | 54.2  | 78 | 0.181 |
| 1976 | July  | 17  | 0.56 | 0.02 | 4.2   | 84 | 0.035 |
| 1977 | May   | 8   | 1.06 | 0.43 | 145.4 | 92 | 0.406 |
| 1978 | May   | 19  | 1.12 | 0.20 | 84.1  | 84 | 0.179 |
| 1979 | Mar.  | 18  | 0.93 | 0.54 | 17.2  | 96 | 0.581 |
| 1980 | June  | 15  | 0.83 | 0.34 | 207.0 | 93 | 0.410 |
| 1981 | Aug.  | 1   | 1.63 | 0.33 | 104.0 | 79 | 0.203 |
| 1982 | June  | 14  | 1.35 | 0.50 | 151.0 | 89 | 0.370 |
| 1983 | June  | 13  | 1.78 | 0.41 | 104.0 | 79 | 0.230 |
| 1984 | June  | 12  | 0.76 | 0.45 | 104.0 | 97 | 0.592 |
| 1985 | May   | 14  | 1.26 | 0.22 | 35.6  | 82 | 0.175 |
| 1986 | Apr.  | 27  | 1.94 | 0.75 | 191.0 | 85 | 0.386 |
| 1987 | May   | 26  | 0.86 | 0.38 | 55.0  | 94 | 0.442 |
| 1988 | July  | 15  | 1.96 | 0.03 | 2.8   | 58 | 0.015 |

*Source:* NEH, 1999. *National Engineering Handbook*, Chapter 5, pp. 5–11. With permission.

in columns 4 to 6 of Table 4.8. It is possible to make detailed statistical analysis of these coefficients, but their arithmetic averages are considered herein. The runoff exponent values in column 7 are chosen according to specifications in column 2 by considering the classification guide lines in Table 4.6. The substitution of the infiltration coefficients and the runoff exponent values into Equations 4.25 and 4.26 yields the time-to-peak, $t_p$, and peak discharge, $Q_p$, values as in the columns 8 and 9, respectively, in Table 4.8.

The multiplication of DUH coordinates in Table 4.5 by $t_p$ and $Q_p$ conveniently yields the UHs as in Figures 4.22 and 4.23.

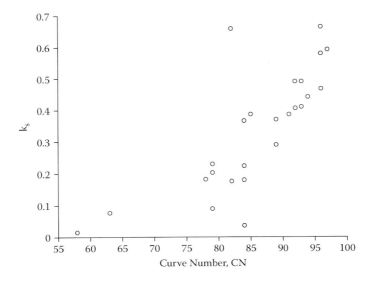

**FIGURE 4.20** Curve number–exponent runoff coefficient scatter diagram.

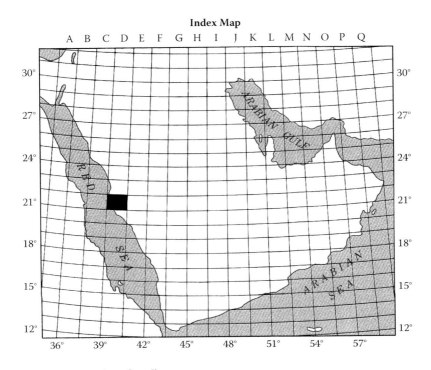

**FIGURE 4.21** Location of wadis.

**TABLE 4.8**

**Infiltration Test Results**

| Wadi Name (1) | Specifications (2) | Area (km²) (3) | $f_i$ (4) | $f_f$ (5) | $k$ (6) | $k_s$ (7) | $t_p$ (8) | $Q_p$ (9) |
|---|---|---|---|---|---|---|---|---|
| Mathwab | Fine-to-medium sand | 15.75 | 1.80 | 1.05 | 0.32 | 0.3 | 9.242 | 216.545 |
| Dighbij | Silt-to-fine sand | 52.51 | 1.00 | 0.54 | 0.23 | 0.2 | 10.18 | 31.913 |

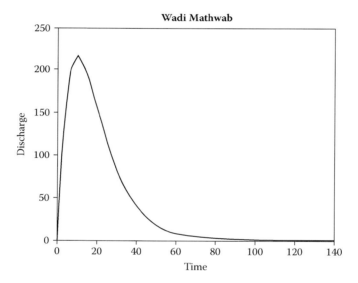

**FIGURE 4.22**    Wadi Mathwab hydrograph.

## 4.7   MODIFIED SNYDER SYNTHETIC HYDROGRAPH

In arid regions, the storm rainfall and resulting runoff measurements are not available and, therefore, the designer is faced with the problem of trying to find the rainfall–runoff relationship by means of artificial, indirect, or synthetic ways. There
are various synthetic unit hydrographs (UHs) derived empirically from the geomorphologic features of the wadi drainage basin configuration on the basis of different
representative parameters that can be obtained from topographic maps. The most
significant geomorphologic parameters that are required in such synthetic methodologies are the drainage basin area, main channel length, slope, the distance of drainage basin centroid to the outlet as in Equations 4.27 and 4.28. However, one should
be careful in the application of these equations because they are developed for humid
regions. Many researchers in arid regions adopt them as if they are applicable in
their study area. This point needs confirmation by some other simple methodologies such as basic rational methods, peak discharge–drainage area relationships, or
by any other available real data. In this section, although Snyder's approach is used

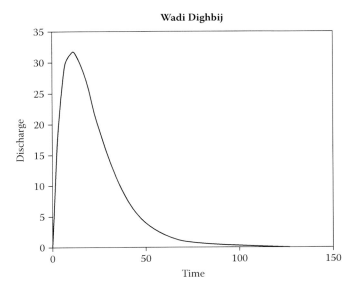

**FIGURE 4.23**  Wadi Dighbij hydrograph.

for confirmation, the direct application of such methodologies for comparison with available measurements, including many drainage basin areas and their relationships to the peak discharge, is found not to be acceptable, prior to some adjustment. So, finally, the Snyder method has been modified according to the regional flood discharge–drainage basin area relationship trend.

Wheater and Brown (1989) presented an analysis of Wadi Ghat, a 597-km$^2$ subcatchment of Wadi Yiba, one of the Saudi Arabian basins in the southwestern part of the Kingdom, where areal rainfall is estimated from five raingages and a classical unit UH analysis is undertaken. A striking illusion of the ambiguity in observed relationships is the relationship between observed rainfall depth and runoff volume, where the greatest runoff volume was apparently generated by the smallest observed rainfall. Goodrich et al. (1997) show that the combined effects of limited storm areal coverage and transmission loss give important differences from humid regions.

It is the purpose of this section to present the necessary modifications to the classical Snyder (1938) model for synthetic hydrograph peak discharge calculations in ungauged watersheds. In general, absence of reliable long-term data and experimental research, and the tendency to employ humid zone experience and modeling, may lead to results that are highly inaccurate.

### 4.7.1  PEAK FLOODS IN ARID REGIONS

There are no *perennial* surface flows in arid regions, and the annual runoff volume is usually concentrated in the form of *flash floods* of short duration but sizable magnitude, which mostly occur during the expected rainy seasons in winter and spring. Due to ground and climate conditions, rainfall is immediately converted to runoff, causing floods or flash floods due to the following:

1. In the upper parts of the wadis, there is almost no soil to trap water. The slopes tend to be steep and the rocks are impervious. Therefore, infiltration losses and retention by filling the depressions are minor.
2. In the foothills of the wadis, the high intensity of the rain seals the surface of bare soil quickly. Consequently, only a shallow depth of soil moisture penetration is achieved before pounding and the onset of surface runoff.

Commonly, flood flows move down the channel network as a *flood wave*, flowing over a bed that is either initially dry or has a small initial flow. Hydrographs are typically characterized by extremely rapid rise times of as little as 15 to 20 min. However, transmission losses from flood hydrograph through bed infiltration are important factors in reducing the flood, and obscure the interpretation of observed hydrographs (Jordan, 1977; Parissopoulos and Weather, 1990; Şorman and Abdulrazzak, 1993; Wheater, 2002). It is not uncommon for a flood to be observed at a gauging station, when further upstream a flood has been generated and lost to bed infiltration.

Several *synthetic hydrograph* derivation methods are available (Snyder, 1938; Clark, 1943; Soil Conservation Service, 1971; 1986), but each has drawbacks in arid region applications because of the implicit assumptions in the derivations and location-specific conditions. Most of the methods in hydrology have two stages: logical and empirical. Any model is a simplified representation of a real hydrological system with a set of logical and empirical equations. The requirement from a model is the reproduction of the catchment scale relationship between storm rainfall and direct runoff.

### 4.7.2 Logical Phase

Logically, there are relationships between the surface flow and catchment characteristics as mentioned in basic textbooks on hydrology (Linsley, 1982; Maidment, 1993). In practical studies, the two most important quantities are time-to-peak discharge and peak discharge, which are related to storm rainfall and drainage basin (wadi) features. A logical phase of hydrograph construction in ungauged drainage basins is already embedded in the classical approach first proposed by Snyder (1938), who suggested three parameters, namely, time-to-peak, $t_P$, (Equation 4.27); peak discharge, $Q_P$, (Equation 4.28); and base time, $t_b$, for the description of a synthetic hydrograph such as

$$t_b = 3 + 3\left(\frac{t_P}{24}\right) \qquad (4.32)$$

It is stated that because the base length of the hydrograph is always greater than three days (Raudkici, 1979), Snyder's method is applicable to fairly large catchments only (Langbein et al., 1947; Taylor and Schwarz, 1952; Gray, 1961). In arid regions, the geomorphologic wadi characteristics are the only available quantitative data that can be used to estimate the peak discharge or derivation of the flood hydrograph.

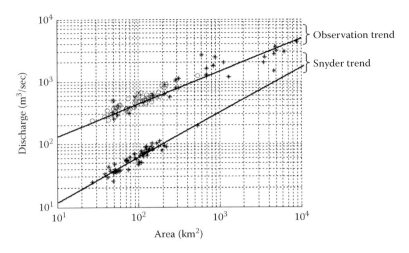

**FIGURE 4.24**  Peak discharge–catchment area relationship.

### 4.7.3 EMPIRICAL PHASE

Many researchers all over the world use the Snyder synthetic hydrograph methodology without checking its validity for arid regions. It is necessary to check calculated peak discharges with the actual measurements from the field. For this purpose, the peak discharge–catchment area relationship on a double logarithmic paper given by Al-Suba'i (1991) is used here. He collected all the available surface runoff peak discharge observations from a set of wadis in the southeastern Arabian Peninsula and plotted them on a double logarithmic paper as the peak discharge ($Q_p$) versus basin area (A) shown in Figure 4.24.

The observation trend appears on the double logarithmic paper as a straight line, which implies mathematically that it is a power function and after the application of classical regression technique average peak discharge, $\overline{Q}_p$, is related to A as

$$\overline{Q}_p = 43A^{0.522} \qquad (4.33)$$

This observed trend is shown in Figure 4.24 as the observation trend. However, observed $Q_p$'s have random deviations from this trend line. On the other hand, the classical Snyder method is used for the Snyder peak discharge, $Q_S$, calculations from Wadi Baysh in the southwestern part of the Arabian Peninsula. This wadi has 54 sub-wadis and the necessary geomorphologic quantities are given in the first five columns of Table 4.9, where the first column includes the sub-basin number. Climate and hydrologic conditions in the study area have been explained by many researchers and some detailed regional studies on the subject include the works by Italconsult (1973), Al-Qurashi (1981), Al-Jerash (1983, 1985, 1988), Şen (1983), Nouh (1987, 1988a,b), Al-Suba'i (1991) and Şen and Al-Suba'i (2002). Further publications on arid and semi-arid region conditions are presented in two reports by FAO (1981) and UNESCO (2002).

## TABLE 4.9
## Wadi Baysh Sub-Basin Features

| Sub-Basin | Area (km²) | Lc (km) | L (km) | So | $Q_s$ (m³/sec) | Sub-Basin | Area (km²) | Lc (km) | L (km) | So | $Q_s$ (m³/sec) |
|---|---|---|---|---|---|---|---|---|---|---|---|
| 1 | 146.50 | 6.00 | 17.86 | 0.048 | 104.01 | 28 | 153.00 | 10.80 | 23.60 | 0.058 | 83.78 |
| 2 | 134.00 | 8.00 | 19.73 | 0.058 | 84.72 | 29 | 146.80 | 10.49 | 24.44 | 0.045 | 80.24 |
| 3 | 100.90 | 11.30 | 28.57 | 0.064 | 51.47 | 30 | 51.21 | 4.70 | 8.98 | 0.068 | 48.10 |
| 4 | 51.61 | 6.29 | 13.52 | 0.098 | 39.28 | 31 | 112.50 | 6.81 | 21.98 | 0.051 | 72.28 |
| 5 | 176.30 | 11.01 | 27.68 | 0.052 | 91.50 | 32 | 168.90 | 8.30 | 21.87 | 0.079 | 102.38 |
| 6 | 27.06 | 5.00 | 9.17 | 0.079 | 24.79 | 33 | 44.77 | 4.20 | 14.32 | 0.039 | 37.81 |
| 7 | 135.30 | 10.49 | 17.94 | 0.059 | 81.14 | 34 | 49.21 | 7.81 | 15.03 | 0.031 | 34.01 |
| 8 | 181.40 | 9.99 | 17.60 | 0.067 | 111.02 | 35 | 52.45 | 7.39 | 14.18 | 0.074 | 37.50 |
| 9 | 78.22 | 5.70 | 13.45 | 0.115 | 61.42 | 36 | 90.70 | 8.30 | 16.86 | 0.085 | 59.44 |
| 10 | 69.96 | 5.00 | 13.90 | 0.047 | 56.57 | 37 | 108.30 | 14.19 | 30.70 | 0.051 | 50.49 |
| 11 | 118.70 | 8.00 | 21.96 | 0.041 | 72.67 | 38 | 74.42 | 11.07 | 25.68 | 0.032 | 39.44 |
| 12 | 58.15 | 7.50 | 20.97 | 0.044 | 36.80 | 39 | 93.51 | 10.61 | 16.89 | 0.007 | 56.92 |
| 13 | 113.80 | 9.00 | 18.70 | 0.079 | 70.58 | 40 | 53.70 | 8.30 | 16.17 | 0.041 | 35.64 |
| 14 | 126.30 | 12.49 | 23.38 | 0.071 | 66.39 | 41 | 41.48 | 6.89 | 15.60 | 0.018 | 29.43 |
| 15 | 104.90 | 8.40 | 18.54 | 0.074 | 66.59 | 42 | 535.90 | 19.01 | 49.50 | 0.002 | 198.32 |
| 16 | 146.10 | 11.01 | 22.06 | 0.065 | 81.17 | 43 | 49.01 | 11.30 | 27.06 | 0.021 | 25.41 |
| 17 | 123.20 | 6.20 | 17.71 | 0.046 | 86.86 | 44 | 112.40 | 9.00 | 26.19 | 0.023 | 63.01 |
| 18 | 57.41 | 8.00 | 16.31 | 0.039 | 38.43 | 45 | 60.27 | 9.99 | 15.27 | 0.016 | 38.50 |
| 19 | 144.50 | 7.90 | 19.84 | 0.033 | 91.54 | 46 | 98.06 | 9.65 | 16.20 | 0.022 | 38.21 |
| 20 | 128.20 | 8.80 | 23.38 | 0.042 | 74.85 | 47 | 53.73 | 6.70 | 17.64 | 0.029 | 37.06 |
| 21 | 99.57 | 14.31 | 29.46 | 0.026 | 46.88 | 48 | 43.15 | 4.01 | 8.29 | 0.095 | 43.54 |
| 22 | 132.10 | 7.90 | 22.16 | 0.044 | 80.96 | 49 | 89.15 | 6.00 | 13.81 | 0.076 | 68.37 |
| 23 | 71.45 | 4.01 | 20.31 | 0.040 | 55.10 | 50 | 107.70 | 6.00 | 18.29 | 0.055 | 75.92 |
| 24 | 125.30 | 8.30 | 18.04 | 0.055 | 80.47 | 51 | 38.30 | 5.60 | 9.48 | 0.080 | 33.58 |
| 25 | 219.60 | 16.70 | 41.53 | 0.037 | 89.05 | 52 | 71.45 | 6.29 | 14.63 | 0.056 | 53.10 |
| 26 | 194.70 | 16.71 | 42.15 | 0.026 | 78.60 | 53 | 124.00 | 7.81 | 12.07 | 0.087 | 91.52 |
| 27 | 207.60 | 13.70 | 35.25 | 0.025 | 93.85 | 54 | 73.32 | 7.00 | 13.33 | 0.078 | 54.27 |

The intermediate calculations for $Q_s$ values are not presented here, but for Equations 4.27 and 4.28 the coefficients are taken as $C_t = 2$ and $C_p = 0.62$, which are the arithmetic average values as given by Snyder. The plot of the area versus Snyder discharge from Table 4.9 is also shown in Figure 4.24. It is obvious that the Snyder method underestimates all the discharges, but there is a general trend also in this case and it is shown as the Snyder trend in the same figure. This trend line provides average Snyder discharges, and it is also in the form of a power function, and with the data at hand, its general expression becomes

$$\overline{Q}_s = 2.19A^{0.73} \tag{4.34}$$

## 4.7.4  DISCHARGE ADJUSTMENT

Hence, the classical Snyder approach does not yield proper peak discharge values for Wadi Baysh. It is necessary to modify it in such a way that the discharge predictions fall around the observation trend. For the modification, the first step is to calculate the discharge for any desired sub-basin by using Equation 4.33. In this manner, the discharges are obtained on the observation trend only. It is necessary to add to this observation trend value the errors (positive or negative) so that the scatter of predicted peak discharges can also be obtained. For this purpose, Snyder discharge errors, $e_s$, from the Snyder trend are calculated as the difference between the Snyder peak discharge and the Snyder trend value as

$$e_s = Q_s - \overline{Q}_s = Q_s - 2.19A^{0.73}$$

On the right-hand side all the values are known from Table 4.9. Hence, there will be positive and negative Snyder errors. Each such error will have its corresponding observation error, $e_o$, from the observation trend. It is assumed that the ratio between $\overline{Q}_p$ and $\overline{Q}_s$ is equal to the error ratios as,

$$e_O = \frac{\overline{Q}_P}{\overline{Q}_S} e_s$$

Hence, the final peak discharge values, $Q_p$, can be calculated as the summation of $\overline{Q}_p$ and the corresponding error from Equation 4.36 which yields

$$Q_P = \overline{Q}_P \pm \frac{\overline{Q}_P}{\overline{Q}_S} e_S = \overline{Q}_P \left( 1 \pm \frac{1}{\overline{Q}_S} e_S \right) \tag{4.35}$$

The application of this equation to each wadi led to final peak discharge values as shown around the observation trend with small circles in Figure 4.24. Hence, the peak discharge (flood) estimations accord with the observed peak discharges from the previous studies as presented in this figure.

Logically, there should be a direct relationship between $L$ and $L_c$, which has not been taken into consideration by Snyder or any other researcher. In practical studies, the determination of $L_c$ is time consuming. Since the values of $L$ and $L_c$ are available in Table 4.9 their plots appear as a straight line on a double logarithmic paper as in Figure 4.25.

The mathematical form of this relationship is again in a power form as

$$L_c = 0.2682L^{1.198} \tag{4.36}$$

The substitution of this expression into Equation 4.27 leads, after simple algebraic calculations, to

$$t_p = C_{st}L^{2/3} \tag{4.37}$$

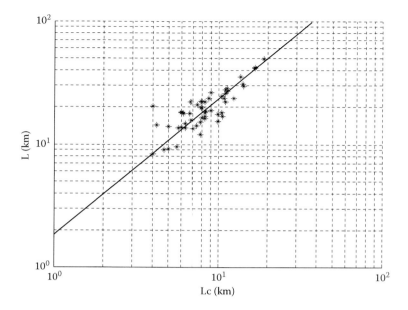

**FIGURE 4.25**   Channel lengths relationship.

where $C_{st}$ is new coefficient. This is a simpler expression than the Snyder formulation, and it needs the length of the main channel only.

## 4.8   DIMENSIONLESS UNIT HYDROGRAPH (DUH)

Synthetic and dimensionless unit hydrographs (DUHs) are derived from watershed characteristics rather than measured rainfall and runoff data, which are not frequently available in arid and semi-arid zones. For instance, Snyder (1938) and Espey and Winslow (1974) methods utilize empirical equations for estimating the salient points of the hydrograph. On the other hand, the two-parameter gamma PDF is most commonly used in various forms depending on the smooth and single peak valued hydrographs. For instance, using the concept of *instantaneous* UH, Nash (1958, 1959) and Dooge (1959) derived the Gamma PDF-like hydrographs from the cascade of linear reservoirs. Depending on data availability, Croley (1980) and Aron and White (1982) derived the two parameters of the gamma PDF, which are useful in deriving a DUH. For ungauged catchments such as the wadis in the arid regions, the two-parameter gamma PDF provides the complete shape of the UH. Singh (2000) and Bhunya et al. (2003) stated that it is possible to obtain the parameter estimations from geomorphologic characteristics of the catchment.

The basic assumption is that the whole hydrograph ordinates are dependent on the sub-basin area; therefore, the shape of the hydrograph is expected to be similar to the relative frequency distribution of the catchment areas. Figure 4.26 shows the relative frequency distribution of wadi sub-catchments. The frequency distribution is skewed to the right, which fits a two-parameter gamma PDF, with shape and scale parameters as 3.314 and 3.436, respectively.

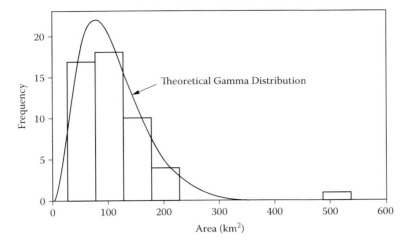

**FIGURE 4.26**   The two-parameter gamma PDF.

After the gamma PDF assumption, the DUH coordinates are presented in Table 4.10. In order to find the DUH, the frequency and area are divided by the mode and its corresponding area, respectively. The graphical representations of this approach, SGS, and SCS DUHs are given in Figure 4.27. The work presented in this section has been prepared as a technical report (Khiyami, et al., 2005) for the Saudi Geological Survey (SGS); accordingly, the newly derived DUH is referred to as SGS UH (Şen, 2007b).

**TABLE 4.10**
**Dimensionless SGS UH Coordinates**

| $t/t_b$ | $q/q_b$ | $t/t_b$ | $q/q_b$ | $t/t_b$ | $q/q_b$ | $t/t_b$ | $q/q_b$ |
|---------|---------|---------|---------|---------|---------|---------|---------|
| 0 | 0 | 3.4974 | 0.0558 | 1.8135 | 0.6019 | 5.3109 | 0.0022 |
| 0.1295 | 0.0662 | 3.6269 | 0.0449 | 1.9430 | 0.5232 | 5.4404 | 0.0017 |
| 0.2591 | 0.2439 | 3.7565 | 0.0361 | 2.0725 | 0.4500 | 5.5699 | 0.0013 |
| 0.3886 | 0.4617 | 3.8860 | 0.0289 | 2.2021 | 0.3836 | 5.6995 | 0.0011 |
| 0.5181 | 0.6655 | 4.0155 | 0.0231 | 2.3316 | 0.3244 | 5.8290 | 0.0008 |
| 0.6477 | 0.8263 | 4.1451 | 0.0184 | 2.4611 | 0.2724 | 5.9585 | 0.0006 |
| 0.7772 | 0.9335 | 4.2746 | 0.0147 | 2.5907 | 0.2272 | 6.0881 | 0.0005 |
| 0.9067 | 0.9880 | 4.4041 | 0.0117 | 2.7202 | 0.1885 | 6.2176 | 0.0004 |
| 1.0000 | 1.0000 | 4.5337 | 0.0092 | 2.8497 | 0.1555 | 6.3472 | 0.0003 |
| 1.1658 | 0.9701 | 4.6632 | 0.0073 | 2.9793 | 0.1277 | 6.4767 | 0.0002 |
| 1.2953 | 0.9171 | 4.7927 | 0.0058 | 3.1088 | 0.1044 | 6.6062 | 0.0002 |
| 1.4249 | 0.8471 | 4.9223 | 0.0045 | 3.2383 | 0.0850 | 6.7358 | 0.0001 |
| 1.5544 | 0.7676 | 5.0518 | 0.0036 | 3.3679 | 0.0690 | | |
| 1.6839 | 0.6844 | 5.1813 | 0.0028 | | | | |

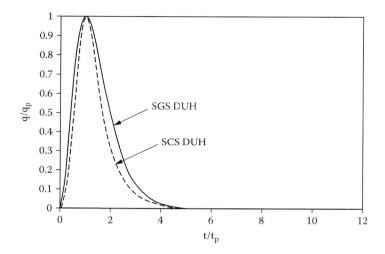

**FIGURE 4.27**   SCS and SGS DUHs.

It is obvious that the SGS DUH follows closely the SCS DUH, but SGS has a longer high-discharge flow tail than SCS. Conversely, SGS graph has more discharge values along the recession limb. These are the features that make the use of the SCS approach slightly biased for arid regions. Besides, it has similar features with the standard DUH in Figure 4.19.

### 4.8.1   Applications

The SGS methodology is applied to Wadi Baysh sub-basin No. 42, which has the maximum area. The application is presented in two steps. The first step concerns time-to-peak and discharge coefficients as they appear in Equations 4.27 and 4.28, respectively. It is assumed that the peak in the runoff hydrograph is taken as equivalent to the storm rainfall average duration. There were 111 individual storm rainfall events during the 1980–1991 period over the Wadi Baysh area with the rainfall duration frequency distribution as in Figure 4.28.

It is obvious that this frequency distribution has an exponential shape similar to explanations in Chapter 3. The average storm rainfall duration is 1.36 h with a standard deviation of 1.58 h. After all the aforementioned adjustments, it is possible to redefine the new dimensionless peak discharge coefficient, $C_{sp}$, and time coefficient, $C_{st}$. The time coefficient can be calculated from Equation 4.35 as

$$C_{st} = \frac{t_p}{L^{2/3}} \tag{4.38}$$

The discharge coefficient can be obtained by the elimination of $Q_P$ between Equations 4.28 and 4.35, which lead to

$$C_{sp} = \frac{\overline{Q}_P t_p}{A} \left( 1 \pm \frac{1}{\overline{Q}_s} e_s \right) \tag{4.39}$$

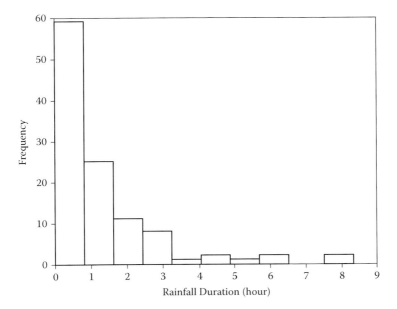

**FIGURE 4.28** Storm rainfall duration frequency distribution.

Substitution of the relevant quantities into Equations 4.38 and 4.39 yield the SGS model parameters as in Table 4.11.

Although the possibility of a meaningful relationship between these coefficients and various geomorphologic features of sub-basins was sought, in reality they are independent from each other. Therefore, it is decided to look at the PDFs of these parameters and the results appeared as gamma PDFs with parameters given in Figures 4.29 and 4.30, respectively.

The sub-basin No. 42 has representative values as collectively presented in Table 4.12.

Application of the present methodology includes the following steps.

1. Determine the geomorphologic features of the drainage basin, namely the catchment area, $A$, and main channel length, $L$.
2. Calculate the coefficients $C_{st}$ and $C_{sp}$.
3. Calculate time-to-peak, $t_p$, from Equation 4.35 and peak discharge, $Q_P$, from Equation 4.28.
4. Multiply the SGS DUH coordinates from Table 4.10 by $t_p$, and $Q_P$ to obtain the hydrograph for the concerned wadi. Here, it is necessary to multiply $t/t_p$ and $q/q_p$ columns by 11.72 h and 1239.82 m³/h. The resultant hydrograph is shown in Figure 4.31.
5. Check whether the area under this hydrograph is equal to the drainage area, $A$, for the UH.

The area under the hydrograph is found as 24773.93 m³ = 0.0248 × 10⁶m³, which is the amount of direct runoff water volume over the whole sub-basin. According

**TABLE 4.11**
**Time-to-Peak and Discharge Coefficients**

| Basin No. | $C_{sp}$ | $C_{st}$ | Basin No. | $C_{sp}$ | $C_{st}$ |
|---|---|---|---|---|---|
| 1 | 1.23 | 5.40 | 28 | 1.32 | 8.79 |
| 2 | 1.29 | 7.28 | 29 | 1.29 | 9.10 |
| 3 | 1.25 | 13.83 | 30 | 1.47 | 6.77 |
| 4 | 1.38 | 10.26 | 31 | 1.18 | 8.04 |
| 5 | 1.25 | 8.80 | 32 | 1.26 | 6.66 |
| 6 | 1.48 | 11.48 | 33 | 1.20 | 9.27 |
| 7 | 1.45 | 8.03 | 34 | 1.41 | 12.89 |
| 8 | 1.44 | 6.19 | 35 | 1.42 | 11.48 |
| 9 | 1.34 | 7.06 | 36 | 1.38 | 9.07 |
| 10 | 1.27 | 7.24 | 37 | 1.30 | 15.70 |
| 11 | 1.24 | 8.50 | 38 | 1.29 | 16.09 |
| 12 | 1.24 | 13.56 | 39 | 1.49 | 10.28 |
| 13 | 1.36 | 8.55 | 40 | 1.40 | 13.09 |
| 14 | 1.39 | 11.01 | 41 | 1.34 | 13.90 |
| 15 | 1.34 | 8.67 | 42 | 1.19 | 7.53 |
| 16 | 1.36 | 8.84 | 43 | 1.27 | 22.97 |
| 17 | 1.24 | 6.23 | 44 | 1.20 | 10.56 |
| 18 | 1.38 | 12.24 | 45 | 1.51 | 12.97 |
| 19 | 1.28 | 6.85 | 46 | 1.36 | 11.56 |
| 20 | 1.25 | 8.83 | 47 | 1.27 | 12.12 |
| 21 | 1.33 | 16.39 | 48 | 1.44 | 6.67 |
| 22 | 1.23 | 7.83 | 49 | 1.35 | 6.71 |
| 23 | 1.04 | 7.83 | 50 | 1.22 | 6.90 |
| 24 | 1.35 | 7.42 | 51 | 1.52 | 9.66 |
| 25 | 1.22 | 12.23 | 52 | 1.34 | 8.43 |
| 26 | 1.22 | 13.50 | 53 | 1.53 | 5.66 |
| 27 | 1.22 | 10.26 | 54 | 1.43 | 8.34 |

to UH definition, the basin area should be covered by 1 cm = 0.01 m = $10^{-2}$ m of uniform water over the whole area, ($539.9 \times 10^6 m^2$), which gives the expected runoff water over the drainage basin as $539.9 \times 10^6 \times 10^{-2} = 5.399 \times 10^6 m^3$, and it is not equal to $0.0248 \times 10^6 m^3$. The division is equal to almost 217, which indicates that the hydrograph ordinates in Figure 4.31 must be multiplied by this number in order to obtain the valid UH. After the necessary operations, the UH for sub-basin No. 42 results as in Figure 4.32.

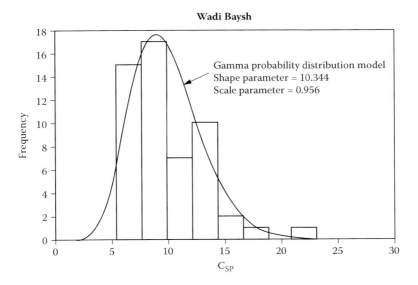

**FIGURE 4.29**  Frequency distribution models for peak discharge coefficients.

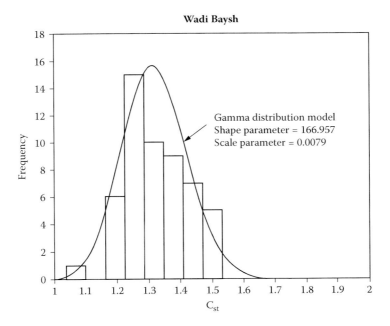

**FIGURE 4.30**  Frequency distribution models for time-to-peak coefficients.

**TABLE 4.12**

**Representative UH derivation**

| Basin No. | Area (km²) | $C_p$ | $t_p$ (h) | $C_{st}$ | Q (m³/h) | Rainfall (×10⁻⁶ m) |
|-----------|-----------|-------|-----------|----------|----------|-----------------------|
| 42        | 535.9     | 1.19  | 11.7      | 7.53     | 1239.82  | 2.31                  |

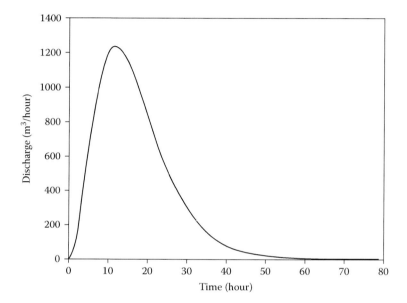

**FIGURE 4.31**    Wadi Baysh sub-basin No. 42 hydrograph.

## 4.9    FLOOD ESTIMATION IN UNGAUGED WADIS

It is economically impractical and physically difficult to gauge all streams in a country. Consequently, many formulas for estimating flood magnitude and total runoff have been published to date. Some of these are based on theoretical considerations, whereas others are purely empirical. However, most of these formulas vary to the extent they provide for major factors causing floods. In general, a basin discharge is a function of climatic and watershed characteristics. The main climatic variable is defined by the rainfall amount, its intensity, distribution, and duration, whereas the watershed features include the drainage area, shape, slope, etc., which affect surface runoff. The latter factors represent the system structure and act as an operator to convert a time sequence of naturally occurring precipitation into a time sequence of runoff (Seyhan, 1976). This interaction is not exclusive between watershed and climatic factors alone, but also takes place among watershed factors themselves. It becomes even more complicated as the distribution of the vegetation, geological formations, soil condition, and spatial and temporal variation of climatic factors are considered.

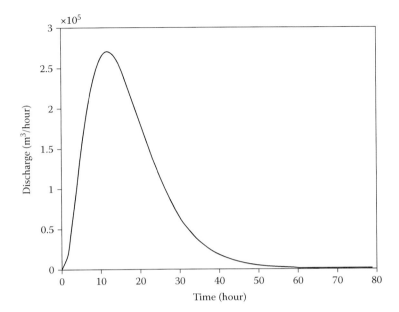

**FIGURE 4.32**  Wadi Baysh sub-basin No. 42 UH.

Eventually, a combination of these processes will determine the percentage of rainfall that ends up as runoff after several passages of storages and transfers.

Problems in the analysis of flood records from arid areas include those of gauging and measurements, low or zero annual maxima in a number of years, and the suitability or otherwise of the mean annual flood as the appropriate scaling factor for the dimensionless curves (Farquharson et al., 1992). In addition, annual *flood magnitude* and *frequency* data are needed for dam siting as well as for dimensioning of the hydraulic structures for flood control and irrigation purposes. In the following, these quantities will be evaluated for the wadis on which the dams are proposed (Şen and Al-Suba'i, 2002).

### 4.9.1  Area-Flood Discharge Method

Some methods have been followed to estimate the *runoff coefficient, C,* and volume for the catchments of ungauged streams. The longest wadi of the western Arabian Peninsula, Wadi Baysh, is intersected by several other westward flowing steep and narrow wadis originating in the Asir mountains and discharging their waters into the Red Sea (see Figure 4.33).

Calculation of the mean values of $C$ for the catchments of the four gauged streams in and around the study area is shown in Table 4.13.

The $C$ value is found to vary from 0.048 to 0.078. Figure 4.34 shows a plot of log C against log A (catchment area). This suggests a straight-line relationship of the form:

$$C = A^{-0.359} \qquad (4.40)$$

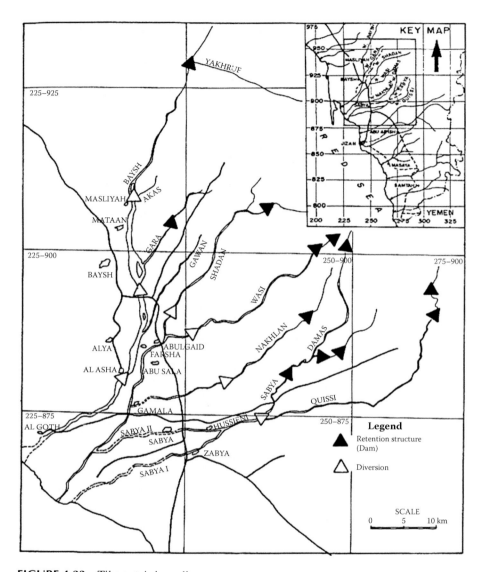

**FIGURE 4.33** Tihamat Asir wadis.

and this is assumed to be valid for the range of catchment areas considered in the region.

Hence, Equation 4.40 has been used to estimate C for the catchments of the ungauged wadis, as shown in Table 4.14. On the other hand, it is possible, for given catchment areas, to estimate annual precipitation amounts from the depth–area lines in Figure 4.35, which are shown in the third column of Table 4.14.

Additionally, *annual rainfall volumes*, $V_A$, for the entire catchment of each wadi are calculated by the multiplication of the second and third columns, and the results

**TABLE 4.13**

**Annual Surface Runoff and Catchment Runoff Coefficient of Gauged Wadis in Tihamat Asir Region**

| Year | Wadi Baysh | Wadi Damas | Wadi Jizan | Wadi Khulab |
|------|-----------|-----------|-----------|------------|
| 1970 | 22.15 | 34.85 | 39.93 | — |
| 1971 | 85.50 | 22.35 | 18.70 | — |
| 1972 | 134.09 | 24.30 | 41.46 | 55.67 |
| 1973 | 45.83 | 11.03 | — | 17.91 |
| 1974 | 87.18 | 32.80 | 83.92 | 32.90 |
| 1975 | 76.89 | 82.06 | 63.92 | 38.33 |
| 1976 | 45.38 | 59.06 | 57.60 | 35.14 |
| 1977 | 36.64 | 52.37 | 67.64 | 43.90 |
| 1978 | 35.09 | 55.96 | 98.44 | 46.66 |
| 1979 | 40.09 | 24.11 | 58.38 | 28.07 |
| 1980 | 37.99 | 17.95 | 43.26 | 13.16 |
| 1981 | 95.76 | 51.60 | 94.69 | — |
| 1982 | 121.16 | 66.85 | 78.90 | — |
| 1983 | 150.91 | 42.53 | 38.39 | 23.58 |
| 1984 | 66.08 | 7.90 | 26.50 | 12.42 |
| 1985 | 110.14 | 62.64 | 60.11 | — |
| 1986 | 102.24 | 38.42 | 40.03 | — |
| Total ($10^6$ m²) | 1293.00 | 687.00 | 911.00 | 347.00 |
| Average ($10^6$ m³) | 76.00 | 40.00 | 57.00 | 32.00 |
| Catchment area (km²) | 4652.00 | 1108.00 | 1430.00 | 900.00 |
| Annual volume of precip. ($10^6$ m³) | 1586.00 | 570.00 | 755.00 | 405.00 |
| Runoff coeff. | 0.048 | 0.071 | 0.076 | 0.078 |

are entered in the fourth column of the same table. The final *annual runoff volume*, $V_R$, calculations are given in the last column as the multiplication of annual rainfall volume by the corresponding calculated mean runoff coefficients. In a similar manner, by using only the catchment upstream areas of the proposed dam sites, mean annual runoff volume at dam sites can be estimated.

The annual volume of the wadi flows in the study region varies widely. Figure 4.36 shows the calculated and measured instantaneous peak discharge values available for wadis in the southwestern Arabian Peninsula including Yemen.

The mean annual flood volume, $V_T$, and discharge, $Q_T$, with *recurrence intervals* of 50, 100, and 200 years are considered significant for *sediment yield* evaluations, reservoir capacity determinations, and *flood control* and irrigation purposes.

In order to determine the values of $V_T$ and $Q_T$ for the wadis under consideration, the *region curve* is presented by Nouh (1988b) and reproduced in Figure 4.37 for

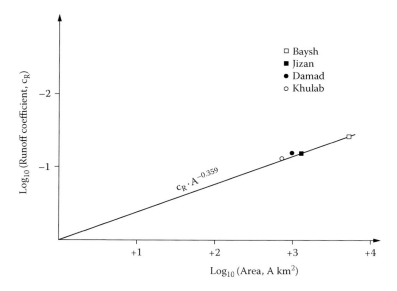

**FIGURE 4.34**   Runoff coefficients as a function of area.

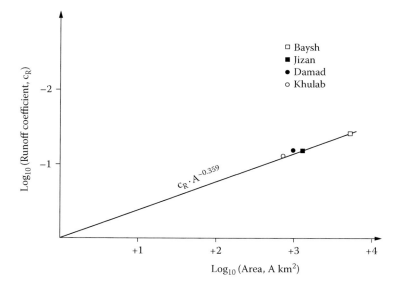

**FIGURE 4.35**   Depth–area curves for mean annual rainfall.

prediction of flood frequency in the southwest Arabian Peninsula, together with the mean annual discharge $Q$ calculated from the values of $V_A$ in Table 4.14. The results are displayed in Table 4.15.

**TABLE 4.14**
**Runoff Coefficients and Mean Annual Runoff Volumes**

| Wadi Name | Catchment Area (A, km²) | Mean Annual Precipitation (P, mm) | Precipitation Volume (V_A, 10³ m³) | Mean Runoff Coefficient, (C) | | Runoff Volume (V_R, 10⁶ m³) | |
|---|---|---|---|---|---|---|---|
| | | | | Measured | Calculated | Measured | Calculated |
| Baysh | 4,652 | 342 | 1,591 | 0.05 | 0.05 | 76 | 76 |
| Ikas | 68 | 370 | 25 | 0.22 | — | — | 6 |
| Qura | 309 | 423 | 131 | — | 0.13 | — | 17 |
| Shahdan | 213 | 510 | 109 | — | 0.15 | — | 16 |
| Was'a | 324 | 446 | 145 | — | 0.13 | — | 18 |
| Sabya | 675 | 453 | 306 | — | 0.10 | 40 | 29 |
| Damad | 1,108 | 515 | 571 | 0.07 | 0.08 | 57 | 46 |
| Jizan | 1,430 | 528 | 755 | 0.07 | 0.07 | — | 58 |

*Note:* Calculations are done according to the following equations. The volume results are taken as round million cubic meters.

$V_A = AP$
$V_R = V_A C$

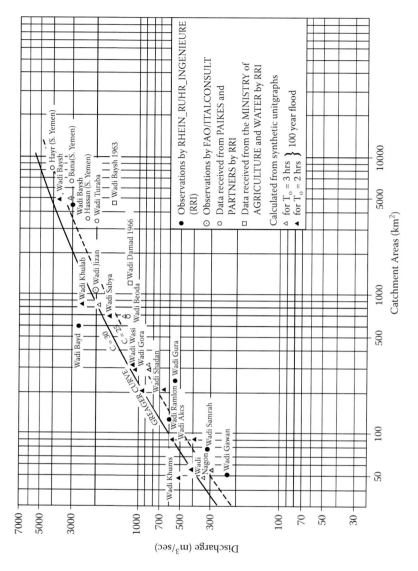

**FIGURE 4.36**  Calculated and measured peak discharges of wadis.

## 4.9.2  KINEMATICS WAVE METHOD OF FLOOD HYDROGRAPH

Richard (1955) developed an empirical method, based on rational formula, compromising between extreme difficulty and undue simplicity. This approach enables the construction of a complete storm hydrograph that requires the collection of available rainfall records pertaining to individual storms, as well as the variables defining the watershed characteristics.

In almost all the empirical flood prediction studies, the following assumptions are listed as simplifying factors in formulations:

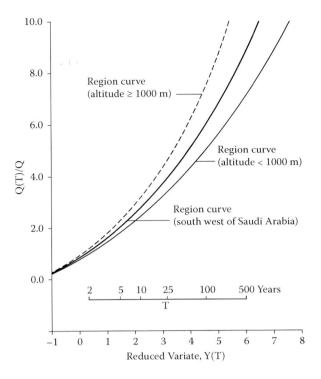

**FIGURE 4.37**   Regional curves for rainfall of flood frequency.

**TABLE 4.15**
**Expected Mean Annual Flood Discharge, $Q_T$, and Mean Annual Flood Volume, $V_T$ ($10^6$ m³) for Different Return Periods**

| Wadi | Mean Annual Flood Discharge Q | 25 Years $Q_T$ | $V_T$ | 50 Years $Q_T$ | $V_T$ | 100 Years $Q_T$ | $V_T$ | 200 Years $Q_T$ | $V_T$ |
|---|---|---|---|---|---|---|---|---|---|
| Baysh | 880.00 | 3,187 | 275 | 3,724 | 322 | 4,948 | 428 | 5,414 | 468 |
| Ikas | 0.64 | 232 | 20 | 271 | 23 | 360 | 31 | 394 | 34 |
| Qura | 192.00 | 69 | 60 | 81 | 70 | 1,080 | 93 | 1,182 | 102 |
| Shahdan | 182.00 | 660 | 57 | 771 | 67 | 1,024 | 89 | 1,121 | 97 |
| Was'a | 209.00 | 756 | 65 | 884 | 76 | 1,157 | 102 | 1,286 | 111 |
| Sabya | 340.00 | 1,230 | 106 | 1,437 | 124 | 1,909 | 165 | 2,089 | 181 |
| Damad | 520.00 | 1,910 | 165 | 2,232 | 193 | 1,966 | 256 | 3,247 | 280 |
| Jizan | 672.00 | 2,434 | 210 | 2,844 | 246 | 3,779 | 327 | 4,136 | 357 |

1. The storm producing the flood is stationary and covers the whole catchment.
2. Its duration is equal to the time of concentration, which is an idealized concept and is defined as the time taken for a drop of water falling on the most remote point of a drainage basin to reach an outlet; remoteness relates to time of travel rather than distance (Maidment, 1993).
3. The rainfall is uniformly distributed over the catchment and throughout the storm period.
4. The storm runoff coefficient is uniform over the catchment and during the storm period.
5. The slope is uniform over the catchment.

Possible departures from these assumptions are also provided in the formulation, which can be applied successfully for storms with durations less than the time of concentration, and even by taking the average runoff coefficients and the catchment features. Furthermore, it can be applied to small as well as large catchments for the determination of maximum intensity, time of concentration, and volume of flood (Sirdaş and Şen, 2007).

Assume that a catchment of length $L$ and unit width $W = 1$ is exposed to a uniform and continuous rainfall intensity, $\bar{i}$. Let the slope of the catchment be $S$. With the continuation of the storm, rainfall will commence to fall over the whole catchment. At the beginning, the contribution to the outlet will be only from the rain falling in its immediate vicinity. The water from the remote parts of the watershed arrives at the outlet during the time of concentration, $t_c$ (Chapter 2). However, at the upstream points of the watershed, the water flows away during this time. Meanwhile, at a point further down the catchment, the water passing off will be replaced from upstream. At the outlet the depth, $D$, of water at time, $t$, will be as

$$D = \bar{C}\bar{i}t \qquad (4.41)$$

where $C$ denotes the runoff coefficient. This result is correct, based on the analysis of flow using the kinematics wave approximation, at least for times up to $t_c$, beyond which the depth stays constant as long as rain continues at its previous steady and uniform rate over the catchment. Now the velocity of flow in a channel can be expressed by the classical Chezy formula as follows (Chow, 1959):

$$V = K\sqrt{RS} \qquad (4.42)$$

where $R$ denotes the hydraulic mean radius, and $K$ is a coefficient. In Section 3.6 of Chapter 3 it is shown that for almost all the cross-sections in arid regions, $W \gg D$ and as a result $R \approx D$. Taking this point into consideration, it is possible to rewrite Equation 4.42 as

$$V = K\sqrt{DS} \qquad (4.43)$$

The depth will increase gradually and, in time, $t_c$, it will cause the maximum velocity, $V_{max}$, which can be found by substituting Equation 4.41 into Equation 4.43 as

$$V_{max} = K\sqrt{\bar{C}\bar{i}t_c S} \qquad (4.44)$$

However, the average velocity, $\bar{V}$, is given mathematically as

$$\bar{V} = K\frac{1}{t_c}\int_0^{t_c}\sqrt{\bar{C}\bar{i}t_c S}\,dt$$

By taking integration, it becomes

$$\bar{V} = K\frac{2}{3}\sqrt{\bar{C}\bar{i}t_c S} \qquad (4.45)$$

Comparison of this expression with Equation 4.44 yields

$$\bar{V} = \frac{2}{3}V_{max} \qquad (4.46)$$

By considering Equation 2.11 from Chapter 2, one can write that

$$t_c = \frac{L}{\bar{V}}$$

or by considering Equation 4.45,

$$t_c = \frac{3L}{2K\sqrt{\bar{C}\bar{i}t_c S}} \qquad (4.47)$$

The solution of this expression for $t_c$ leads to

$$t_c = \left[\frac{L^2}{K^2\bar{C}\bar{i}S}\right]^{\frac{1}{3}} \qquad (4.48)$$

which results when the whole catchment contributes to the flow. Furthermore, a similar expression to Equation 4.48 is given by Morel-Seytoux (1988) and here

$$C = \frac{2}{3}K$$

Substitution of $\bar{i}$ from Equation 3.6 into Equation 4.48 gives

$$\frac{t_c^3}{t_c+1} = \frac{9}{4K^2}\cdot\frac{L^2}{CSRf(a)} \qquad (4.49)$$

Time of concentration can be determined from this expression, provided that other terms are obtained by appropriate office and field techniques. Equation 4.49 shows that $t_c$ is a function of rainfall characteristics in terms of $R$, $C$, $K$, $a$, $L$, and $S$.

### 4.9.2.1  Flood Hydrograph

According to the aforementioned formulations the peak discharge, $Q_P$, will be reached when the whole catchment area, $A$, contributes, and in this case, simply, the rational formulation gives

$$Q_p = C\bar{i}A \qquad (4.50)$$

If the interest is in the flood volume, $F_V$, then it can be obtained from the basic discharge definition (volume per time) as,

$$F_v = 3600C\bar{i}At_c \qquad (4.51)$$

where $F_V$ is in m$^3$, $\bar{i}$ is in mm/hr, $A$ is in km$^2$, and finally $t_c$ is in hr.

If only a part, say, $A_r$, of the catchment is contributing, then the discharge, $Q_r$, is proportional to this partial area. As shown in Figure 4.38, $A_r$ is that part, which is intercepted by an arc with a center point at outlet, $O$, and radius $r_1$, so that the time $t_1$ required by water to flow this distance is given by

$$t_1 = \frac{r_1}{V_1}$$

Here, $t_1$ is similar to Equation 4.48 and can be written as

$$t_1 = \left[ \frac{r_1^2}{K^2 C\bar{i}S} \right]^{\frac{1}{3}}$$

Hence, its ratio to Equation 4.48 gives

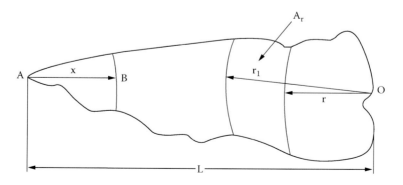

**FIGURE 4.38**  Partial area of the basin.

$$t_1 = t_c \sqrt[3]{\left(\frac{r_1}{L}\right)^2}$$ (4.52)

Likewise, the ratio of the discharge $Q_r$ to Equation 4.50 leads to

$$Q_{r1} = Q_p \frac{A_r}{A}$$ (4.53)

Hence, it is possible to consider successive radii $r_1$, $r_2$, ..., $L$ and calculate corresponding discharges $Q_1$, $Q_2$, ..., $Q_p$ in order to define the rising limb of the hydrograph. At the moment $Q_p$ is reached by assuming that the rainfall duration is equal to the time of the concentration, the rainfall ceases and, in the meantime, the flood commences to decline.

The curve of falling limb is considered as the reverse of the curve of the rising limb. When the rainfall ceases, the water commences to recede from the upper point, $A$, of the catchment (see Figure 4.38). Assume that at time $t_1$ measured from the cessation of the rainfall, the water has receded a distance $x$ from the upstream to some point $B$, as shown in the same figure. Hence, the area that still contributes is equal to $A-A_x$, where $A_x$ is the area intercepted by an arc of radius $x$ and center at $A$ (see Figure 4.38). So, the flood discharge can be calculated as

$$Q_x = Q_p \left(1 - \frac{A_x}{A}\right)$$ (4.54)

and the time for falling limb of hydrograph can be found similarly to Equation 4.52 as

$$t_x = t_c \sqrt[3]{\left(\frac{x}{L}\right)^2}$$ (4.55)

By taking a number of $x$ values from zero to $L$, and finding the corresponding values of $A_x$, the flood discharge for falling limb can be calculated from Equation 4.54 and the corresponding time from Equation 4.55, thus enabling the curve of the falling limb to be constructed.

## 4.10 CROSS-SECTION FLOOD DISCHARGE

In humid regions with records of necessary hydrologic cycle component measurements, it is possible to write down the water balance equation and continue calculations without data availability. As this is not the case in arid regions, a digression must be considered in the calculations, which are mostly based on empirical relations. Especially, in flood assessments in a cross-section for the purpose of flood *inundation map* preparation, the following simple system components are important.

1. The runoff discharge, $Q$, and preferably peak discharge, $Q_p$, that will result from the storm rainfall event over the wadi area.

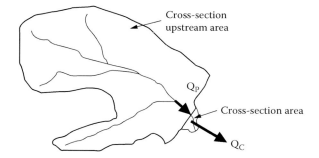

**FIGURE 4.39**   Catchment flood system.

2. The cross-sectional discharge, $Q_C$, is the amount of water that is allowed to pass through the section.
3. If $Q_P > Q_C$ then there is a possibility of *flood hazard*, otherwise the cross-sectional area will assimilate the surface flow peak discharge.
4. Another trend of study includes the preparation of flood inundation map including different risk levels.

It is possible to visualize the catchment flood system as in Figure 4.39, which is based on two geometrical areas, namely the upstream catchment area from the cross-section and the cross-sectional area.

The conceptual model for the cross-sectional flood system can be viewed in two different ways, either as the *overall flood hazard* (Figure 4.40a) or *gradational flood hazard* (Figure 4.40b), corresponding to different risk levels. Overall flood assessment requires peak discharge calculation only once, and it is regarded as the most hazardous situation at and around the cross-sections. However, in the case of gradational flood assessment a series of subsequent flood discharge calculations, each attached with a certain risk level, is calculated and then its flow area coverage is determined within the given cross-sectional area by considering the cross-sectional geometry. It is therefore necessary to have the profile geometrical shape of some control cross-sections along the wadi channel.

In the former case there is no risk attachment and the interest lies in whether $Q_P > Q_C$ only. This is referred to as the conventional flood prediction and assessment, which has different techniques as have been explained in many textbooks

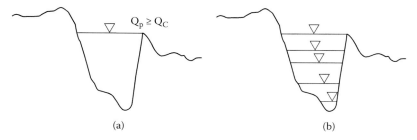

**FIGURE 4.40**   Flood interests: (a) overall flood, (b) gradational flood.

(Maidment, 1993). Most of these techniques are based on a long sequence of runoff measurements, which can then be treated by different PDFs such as Gumbel, Pearson, and Log-Pearson II, etc., but they have not been considered in this book because series of runoff measurements are not available in arid regions.

### 4.10.1 CONTROL CROSS-SECTIONS

Because the flood *hazard potential* is more common in the downstream part of a catchment, most of the cross-sections are taken at downstream locations. In the selection of any cross-section the following guidelines must be considered.

1. It is preferable to select three best representative cross-sections within each wadi sub-basin. If this is not possible due to time and budget restrictions, then most often two cross-sections are selected, one in the upstream and the other in the downstream portion of the sub-basin.
2. The selection of cross-sections should be more intensified at the downstream wadi channels. Also a preliminary reconnaissance inspection should be made on a quick field trip, if possible. This will help to identify the most flood-inundation subjected areas by expert viewing, with information support from the local settlers.
3. The cross-sections must be selected on rather stable portions of the wadi course and, if possible, at more or less straight-line locations. This is a requirement for the simplification of calculations.

The cross sections are also selected at convenient places such as at possible dam sites, downstream of confluence points, significant section areal change locations, bridges, etc.

In order to determine the cross-section, geometry field surveying applications must be carried out across each section perpendicular to the wadi longitudinal direction. For this purpose, the surveying instrument is located over a point almost in the middle of the section and preferably at the lowest elevation within the wadi course, then a sequence of right and left elevation readings are taken at a set of points with appropriate distances from the instrument location. The measurement points are selected during the field observations as points of surface texture changes. For instance, along the cross-section periphery, if there is a change from fine sand to gravel or to vegetation or to rock, then a measurement point is taken. Finally, the plot of distances versus elevations leads to the profile of the cross-section as schematically shown in Figure 4.41.

In this figure, L and R show left- and right-hand sides, respectively, as we look downstream in the field. The material variations in the cross-section periphery are important for the friction losses between the water and natural wadi bed. It is not only the total length of the wetter perimeter that is needed in the calculations, but also the nature of the material at different sectors along the perimeter. The rougher the material is, the more will be the friction and hydraulic losses and, accordingly, reductions in the flow velocity (Section 4.11). In Figure 4.42 the wetted perimeter of a representative cross-section is exposed with different material compositions. It should be noted that the total length of the wetted perimeter changes, depending on the water level in the cross-section.

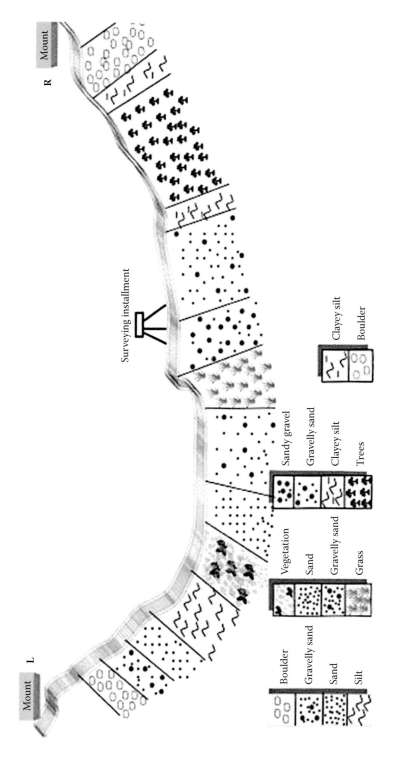

**FIGURE 4.41** Representative cross-section profile (Courtesy of the Saudi Geological Survey).

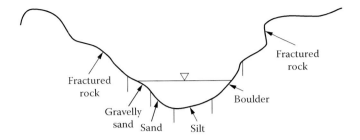

**FIGURE 4.42**    Wetted perimeter in CS-2.

Manning's equation (Equation 2.10) is used for the average velocity calculation for a set of depths, D, and the discharge, Q, calculation results according to steps in Section 4.3 are given for four cross-sections, CS-1, CS-2, CS-3, and CS-4, in one of the wadis (Table 4.16).

Although it is possible to choose very frequent depths, in practice several depths are sufficient for the general description of the cross-section rating curve.

### TABLE 4.16
### Basic Factors and Discharge Calculations

| Depth, D (m) | Discharge, Q (m³/sec) |
|---|---|
| Cross-Section 1 (CS-1) | |
| 1 | 113.74 |
| 4 | 406.16 |
| 11 | 859.09 |
| 17 | 1,037.05 |
| Cross-Section 2 (CS-2) | |
| 1 | 214.19 |
| 3 | 361.23 |
| 6 | 427.05 |
| 14 | 718.57 |
| Cross-Section 3 (CS-3) | |
| 1 | 76.97 |
| 3 | 175.98 |
| 9 | 396.66 |
| 13 | 541.02 |
| Cross-Section 4 (CS-4) | |
| 1 | 75.47 |
| 3 | 177.39 |
| 6 | 244.08 |
| 13 | 468.72 |

**TABLE 4.17**
**Synthetic Rating Curve Parameters**

| Cross-Section Number | Rating Curve Parameters | |
|---|---|---|
| | a | b |
| CS-1 | 113.83 | 0.843 |
| CS-2 | 214.19 | 0.385 |
| CS-3 | 76.97 | 0.746 |
| CS-4 | 75.47 | 0.655 |

### 4.10.2  DISCHARGE–STAGE RELATIONSHIP

For flood inundation mapping, it is necessary to know the discharge change with depth in each cross-section. The relationship between the Q and D in a given cross-section is referred to as the *discharge–stage relationship* or *rating curve* (see Section 4.3). Its general expression is given in Equation 4.1 as a power law. It is possible to find the parameters a and b from the given data plots on double logarithmic paper in Figure 4.43. Invariably, the scatter of points on the double logarithmic paper appears as straight lines.

The coefficients of the cross-section discharge–depth power law equation are presented in Table 4.17.

## 4.11  CURVE NUMBER AND MANNING'S COEFFICIENT IN ARID REGIONS

Flood peak discharge estimation in any hydrological study and its application in arid and semi-arid regions require basic data, which is missing due to various difficulties including remote rural and undeveloped areas, expertise deficit, necessary economical and instrument investments, and many other local hindrances. It is, therefore, necessary to indulge toward indirect and almost empirical formulations to serve hydrological and engineering purposes. In such approaches, besides the rainfall regime over the catchment (wadi, in arid regions) and morphological features, two other major factors are the *land use* and the *soil type* groups. Although, there are different tables for humid regions for the classification of these two factors, unfortunately, they are either complete or missing for arid regions. The classification should be also based mainly on geological features, because in many parts of arid regions there is neither vegetation cover nor large-scale urban developments, but extensive geologic rock outcrops only.

In arid regions, most often than soil, rock types and outcrops play a more significant role in runoff calculation assessments. The application of surface runoff prediction is very sensitive to CN, and especially the fixation of the initial abstraction ratio as 0.2 preempts a regionalization based on geological and climatic setting (Chen, 1982). The CN method is basically an infiltration loss model, although it may also account for interception and surface storage losses through its initial abstraction feature. Long-term losses such as evaporation and evapotranspiration are not

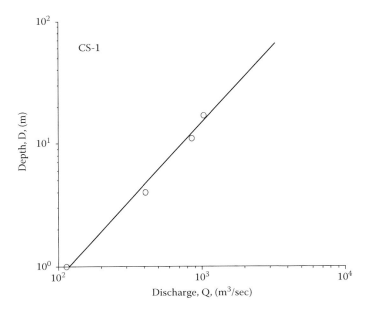

**FIGURE 4.43A**   Representative rating curves.

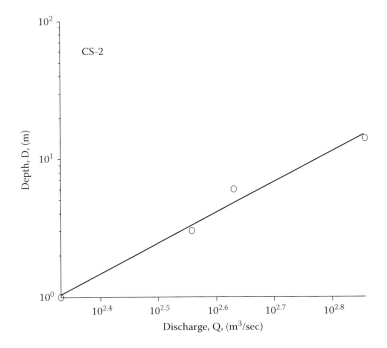

**FIGURE 4.43B**   (Continued).

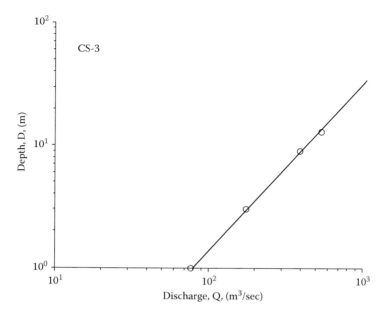

**FIGURE 4.43C**   (Continued).

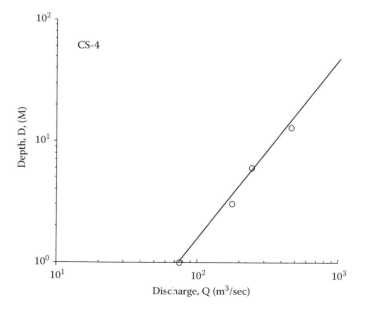

**FIGURE 4.43D**   (Continued).

accounted for by this model. This method converts storm rainfall depth into direct runoff volume (Hawkins, 1993). It is basically a temporally lumped model, where surface phenomenon has importance. Surface phenomena include processes such as crust development and hydrophobic soil layer, which render the soil surface impermeable, promoting surface runoff. The application of the CN to arid region basins must be done with a great deal of care, considerably more than now used by many engineering organizations. The SCS method is a widely used method where data for site-specific analysis CNs are not available.

On the other hand, stream channel discharge calculations require a convenient roughness (Manning) coefficient depending on the land use in the area. Although much research has been done on roughness coefficient in humid areas by considering land use and soil classification, again there is a need concerning the roughness values for arid region flood plains, which include mostly rock outcrops. This coefficient reflects the resistance of material (soil, rock, vegetation, etc.) against the water flow in channels. Hence, it is affected by land use activities such as the pattern of the surface natural or man-made features. In the sparsely vegetated flood plains of arid regions, the major roughness is caused by geological surface texture of the channel and flood plain.

Suggested values for the Manning coefficient, n, tabulated according to factors that affect roughness, can be found in Chow (1959) and Streeter (1971). Roughness characteristics of natural channels are given by Barnes (1967). Although much research has been done to determine roughness coefficients for open-channel flow (Carter, et al., 1963), less has been done for arid region flood plain roughness coefficients, which are typically very different from those for humid area approaches.

### 4.11.1  ROUGHNESS COEFFICIENT

In arid and semi-arid regions partial flood and almost all sheet flows take place over rock outcrops. There is a tendency to regard the selection of roughness coefficients either arbitrarily or intuitively. However, specific procedures could be used to determine the values for roughness coefficients in wadi channels and flood plains. The roughness values for channels are determined by reevaluating the effects of certain roughness factors in the channels (Barnes, 1967). The other method involves the evaluation of the geological texture of the flood plain to determine the Manning roughness coefficient value. This latter method is particularly suited to handle roughness for geologically controlled flood plains. Here, in addition to what is available in the literature, the expert views are also incorporated for a better presentation.

In general, the cross-sections have irregular geometric shapes typical of many natural channels, or they can have a regular shape (triangular, trapezoidal, or semicircular) given by hydraulic structure designers. The flow may be confined to one channel, and especially during floods, it may occur in the compound channels consisting of channel and *flood plain* sections. Irregular cross-section may be divided typically into subsections, where major roughness or geometric changes occur, such as at the junctions of course gravel, sand, and fractured rock as shown in Figure 4.42.

Subsections should reflect representative conditions in the reach rather than only at the cross-section. Roughness coefficients are determined separately for each subsection. The most important factor that affects the selection of channel roughness value in arid regions are the geology, type. and size of the materials that compose the bed and banks of the channel in addition to residential, agricultural, forest, industrial, etc. areas, as in humid regions. A sand channel is defined as a channel in which the bed has an unlimited supply of sand, which ranges in grain size from 0.062 to 2 mm. Resistance to flow varies greatly in sand channels because the bed material moves easily and takes on different configurations or bed forms, which are functions of flow velocity, grain size, bed shear, and temperature.

Resistance to flood flows in channels and flood plains is represented by a roughness coefficient, which naturally reflects the reaction of the earth surface in touch with overflow. The bigger this coefficient is, the smaller the surface water velocity, and hence proportionally smaller is the discharge through the cross-section. The velocity depends on both the geometric features of the cross-section area as well as the material, which comes into contact with water along the wetted perimeter. In many parts of arid regions, the *land use* is not effective artificially as land surface is modified by human, except in small scales, but most frequently natural land surfaces with geology start to play the dominant role, especially in rural and undeveloped areas. If there are human-altered land use places such as residential, agricultural, and industrial areas, then the universal Manning roughness coefficient can be employed from relevant tables. What is not easily available is the Manning roughness coefficient for geological surfaces depending in general on rock types such as sedimentary, volcanic, and metamorphic and, in particular, in the subclassification of each rock from a hydrological point of view. The roughness coefficients are presented in Table 4.18 for common use in arid regions depending on the rock outcrop types.

### 4.11.2   Soil and Rock Classifications

From a hydrological point of view, soils and rocks can be classified according to their infiltration capabilities. Although this will depend on the grain size, fracture intensity, and aperture in many cases, it is better to group the soils and rocks linguistically from two points of view, namely, infiltration, f, and surface direct runoff (discharge, Q). Soils and geology of the wadi primarily influence the groundwater component and the *transmission losses* (Chapter 5). Hydrologic soil groups are classified based on the soil drainage potential. High infiltration rates reduce the surface runoff; high permeability, combined with high transmissivities, substantially enhances the base flow component. According to SCS, soils are grouped into four hydrologic soil groups based on their water drainage potential as $A_S$, $B_S$, $C_S$, and $D_S$. Definitions of the classes for soils are as follows (SCS, 1986).

$A_S$: Soils with low runoff potential. Soils having high infiltration rates even when thoroughly wetted and consisting chiefly of deep well-drained to excessively well-drained sands or gravels.

$B_S$: Soils having moderate infiltration rates even when thoroughly wetted and consisting chiefly of moderately deep to deep, moderately well-drained to well-drained soils with moderately fine to moderately coarse textures.

**TABLE 4.18**
**Curve Number (each value under the soil types indicates the curve number)**

| Geological Description | | Manning Roughness Coefficient | Hydrological Rock Groups | | | |
|---|---|---|---|---|---|---|
| | | | $A_F$ | $B_F$ | $C_F$ | $D_F$ |
| | | Geological Description and Land Use | | | | |
| Sedimentary | Coarse gravel | 0.028 | 75 | 85 | 90 | 95 |
| | Medium gravel | 0.026 | 65 | 75 | 80 | 90 |
| | Fine gravel | 0.024 | 50 | 65 | 75 | 85 |
| | Sand | 0.022 | 40 | 60 | 70 | 80 |
| | Silt | 0.020 | 10 | 35 | 70 | 80 |
| | Sandstone | 0.050 | 45 | 60 | 70 | 80 |
| | Limestone | 0.040 | 60 | 65 | 75 | 90 |
| Volcanic | Granite | 0.025 | 30 | 40 | 60 | 95 |
| | Diorite | 0.020 | 40 | 45 | 65 | 90 |
| | Gabbro | 0.018 | 50 | 55 | 75 | 85 |
| | Basalt | 0.017 | 60 | 65 | 70 | 80 |
| Metamorphic | Slate | 0.015 | 55 | 65 | 70 | 80 |
| | Schist | 0.020 | 60 | 65 | 80 | 90 |
| | Gneiss | 0.023 | 65 | 70 | 85 | 90 |
| | Marble | 0.030 | 65 | 75 | 85 | 90 |

$C_S$: Soils having slow infiltration rates even when thoroughly wetted and consisting chiefly of soils with a layer that impedes downward movement of water, or soils with moderately fine to fine textures.

$D_S$: Soils with high runoff potential. Soils having very slow infiltration rates even when thoroughly wetted and consisting chiefly of clay soils with a high swelling potential, soils with a permanent high water table, soils with a clay pan or clay layer at or near the surface, and shallow soils over nearly impervious material.

Figure 4.44 indicates these four classification regions, under three linguistic groups of infiltration, I, and flood discharge, Q, as low, moderate and high.

It is interesting that in the classical grouping of soils there are missing groups such as high infiltration with moderate discharge, etc.

Similar classification of rock fracturing will be taken into consideration in the determination of CN by considering rock outcrops, which constitute the surface features in arid and semi-arid regions, where there are no forests and significant vegetation covers. The rock classification for the CN determination should be based on the following descriptions of the rocks. In such a classification, the knowledge presented by Şen (1996) can be taken into consideration. Similar to soil classification, the rock classification can be described as follows:

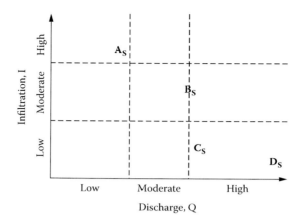

**FIGURE 4.44**   Hydrologic soil groups.

$A_F$: Rocks with high fracture patterns and significant apertures lead to low runoff potential. These rocks have high infiltration rates with extensive fractures vertically, and hence such regions are excessively well-drained.

$B_F$: Rocks having moderate fracture degrees with moderate infiltration rates even when thoroughly wetted and consisting chiefly of moderately deep to deep, moderately well-drained to well-drained rocks with moderately fine to moderately coarse apertures.

$C_F$: Rocks having weak fracture degree, and hence slow infiltration rates, even when thoroughly wetted and consisting chiefly of rocks with moderately fine to fine fracture texture and apertures.

$D_F$: Rocks with very low, in fact, negligible fracture degree, and hence high runoff potential. These rocks have very slow infiltration rates even when thoroughly wetted. Rocks with a permanent high water table may also behave in this manner.

According to this classification, the CNs are presented in Table 4.18 together with the Manning roughness coefficients.

## 4.13   FLOOD HAZARDS

Among all environmental hazards, flood is the most common to many societies all over the world. The main reasons are the widespread geographical distribution of river valleys in humid regions or wadi courses in arid and semi-arid regions and low lying coasts together with their long-standing attractions for human settlement, the surface and groundwater resources availabilities. Although in many cases the threat is limited to comparatively well-defined floodplain and low-lying areas such as estuaries, no country is immune from flood hazards (Smith, 1992).

On the other hand, there are benefits in nonflood-prone areas as they are necessary parts of environmental and wadi catchment ecosystems; they help to maintain a wide range of wetland habitats; they maintain fertile soil by silt deposition and flush-

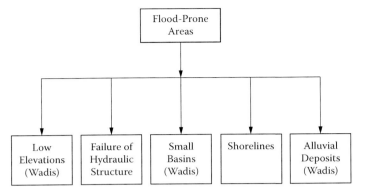

**FIGURE 4.45**   Flood-prone environments.

ing salts from the surface layers; and they provide water for natural irrigation and for fisheries as protein source. It is possible to state that in normal years with balanced hydrological conditions floods bring benefit to a society rather than destruction. In arid region countries, sudden floods in the form of flash occurrences are the most dangerous hazards in the middlestream portion of the wadi system with *inundation risks* in the downstream. The most flood-prone environments are presented in Figure 4.45.

Low-lying areas suffer the most from the floodings and *inundation hazards*. Thousands of people live in these low lying areas due to groundwater availability and transportation facilities. The significance of small basins is due to the occurrence of *flash floods* because during an intensive storm rainfall the basin receives more than it could transfer as surface water in a short period of time. Flash floods occur more in arid and semi-arid regions where there are favorable conditions of steep topography, weak vegetation, and short duration but high-intensity rainfall. Especially, narrow wadis and settlement centers generate rapid runoff due to surface water speed increase and reduction in the surface layer permeability as a result of urbanization. *Alluvium fans* support urban development due to their groundwater potentiality but, at the same time and especially in arid regions, they create a special type of flash flood potential. Alluvial fans are risk-prone environments because the drainage channels can meander unpredictably across the relatively steep slopes, bringing high velocity flows, which are charged intensively with sediment. Their speeds may reach 5 to 10 m/sec.

Although floods vary from year to year, there has been increasing concern that the number and the magnitude of floods have been increasing in the past few years. This is a more valid statement for arid and semi-arid regions because the recent climate change effects gave rise to extensive rainfalls in some parts of the world, leading to unprecedented flooding. Despite growing investment in alleviation schemes, flood losses continue to increase, which could be attributed to a variety of factors. These factors can be gathered into three broad categories:

1. Physical causes related mostly to atmospheric events following the increase in the frequency and magnitude of rainfall and consequent flood events.

2. Wadi causes are related to the surface features of the drainage basin in terms of geomorphology, geology, vegetation, etc. The necessary ingredients are the catchment area, slope, drainage density, main channel length, time of concentration, etc.

3. Human causes due to increase in the vulnerability by greater floodplain occupancy. In spite of the fact that floodplains are one of the most topographically obvious regions of all hazard-prone environments, widespread invasion has occurred as a result of countless individual decisions rooted in the belief that the location benefits outweigh the risk.

Appreciation of these factors is as important as flood hydrology in understanding flood hazards. In arid regions, most flood hazards occur as a result of the second point. Flood protection studies and mitigations started in the beginning of the 20th century with three distinctive evolution stages (Smith, 1992):

1. Hydraulic structural stage (1930s–1960s): Engineering structures are built, such as flood protection dams, levees, successive dikes, diversion canals, etc. During these studies the basic flood generation reasons are identified, and the peak flood discharge calculation procedures, algorithms and formulations are established.

2. Floodplain management stage (1960s–1980s): During these decades most emphasis is given to the combination of mitigation measures such as early flood warning, land use planning, etc.

3. Flood mitigation stage (1980s onward): Measures, especially land use control, were taken by considering different flood scenarios.

Sudden changes in the common behavior of hydrological events such as flooding or flash flooding with progressive and long-lasting results may make many human activities dangerous. These affect both the way a disaster is identified and managed. In order to reduce such dangers, the following immediate or long-term crisis points can be considered for solutions.

1. Short-term effects, for example drowning or injuries during flooding.

2. Medium-term effects, such as progressive food shortages or epidemics following a flooding.

3. Long-term problems, such as epidemics and severe lack of food and drinkable water.

If there are no external assistances or precautions, then hydrological events may turn into an emergency case as a disaster. One working definition of a disaster is that it causes at least 10 deaths or results in an appeal for outside assistance. Climate change appears to be responsible for at least some of this increase, and although global warming has been acknowledged, the term is generally misleading because it does not convey key elements involving water.

On the other hand, wadi channels in arid regions are part of the hydrologic cycle, with surface water transport after the evaporation and infiltration losses from the upstream to downstream areas. The convergence of different wadi branches downstream with their cumulative amounts of surface water causes surface water volume increase downstream, which may consequently lead to *flood inundation* of lower areas. If the region is drained by a single wadi or wadi system, it is called a *drainage basin* or *watershed*. The surface area of the drainage basin collects the meteorologically inborn rainfall water and leads it to the low-lying points within the drainage basin, referred to as stream or river in humid regions but as wadi in arid regions.

*Flash floods* are events that occur in many parts of the world including arid regions, and they may cause potential hazards to human life and property. These floods may rise rapidly due to impervious hard rock catchments and move along the sand- and gravel-filled wadis, which are normally very dry. The flood speeds are usually faster than a person can escape from the rough channels. Flash floods normally reach the sea or are lost in the inland deserts. However, they also help to fill the wadi alluviums that later provide *groundwater recharge* for local agricultural lands or partially for the nearby cities.

Although a number of water balance studies have been conducted for a variety of watersheds throughout the world (e.g., Flerchinger and Cooley, 2000; Scanlan, 1994; Yin and Brooks, 1992; Kattelmann and Elder, 1997; Mather, 1979) the rainfall–runoff, in addition to the water balance of the semi-arid range land, present some interesting challenges. These watersheds, which are dominated by precipitation and evaporation, exhibit a high degree of variability in vegetation communities on scales much smaller than addressed by most hydrologic modeling. Thus, arid region catchments pose a unique set of problems for hydrologists modeling high evaporation rates and studying spatially varying plant communities associated with changes in soil and effective precipitation, and intermittent areal flow which lags only a few months in the early spring.

Most of the alluvium aquifers in arid regions occur in the wadis (dry valleys), which provide depressions for deposition and occasional surface runoff occurrences. The groundwater reservoirs in these wadis are directly related to flash floods. Groundwater resources in arid region aquifers are depleted through pumping or by natural subsurface flow into the sea. However, it is replenished during floods following adequate rainfalls. These replenishments of groundwater depend on the local climatological and geomorphologic conditions, in addition to the geological composition of the area.

Flash floods are not uncommon in arid regions and present a potential hazard to life, personal property, and structures such as small dams, bridges, culverts, wells, and dykes along the wadi courses. After a short period of intensive rainfall, flash floods are formed rapidly, and they flow down over extremely dry or nearly dry water courses at speeds more than 1. 5 m/sec faster than a person can escape from the rough and sandy wadi channels (Dein, 1985). Although flash floods are among the most catastrophic phenomena, the volume of the infiltration from floods is a major source of groundwater replenishment to aquifers hydraulically connected with water courses on the surface. Moreover, this volume of water could be increased significantly by impounding the floods with surface dams or successive dykes. Importance of flood studies, other than dealing with surface and subsurface water interactions, includes focus on flood influences on engineering structures, such as dams, bridges, culverts and spillways.

The absence of detailed records on major floods is noticeable in most wadis. In general, comparatively sufficiently recorded data exist on normal rainfall. The set of available rainfall data together with the drainage basin characteristics facilitate the use of empirical equations to estimate relevant flood discharges. As explained by Parks and Sutcliffe (1987) the problem of flood measurement is more acute in arid areas than elsewhere. Floods are in general flashy and hence, the problem of defining the peak level accurately by water level record or maximum level is aggravated by siltation of inlet pipes (Farquharson et al., 1992).

Countries can reduce the effects of floods and droughts in two main ways:

1. Prepare for a disaster such as with the flood inundation map preparation guide for potential wadis.
2. Mitigate disaster by different approaches (outside the scope of this report).

## REFERENCES

Al-Jerash, M., 1983. Models of estimating average annual rainfall over the west of Saudi Arabia. *J. Fac. Arts and Humanities*, King Abdulaziz University, Jeddah, Saudi Arabia 3, 107–152.

Al-Jerash, M., 1985. Climatic subdivisions in Saudi Arabia: an application of principal component analysis. *J. Climate* 5, 307–323

Al-Jerash, M., 1988. Climatic water balance in Saudi Arabia: an application of Thornthwaite-Mather model. *J. Fac. Arts and Humanities*, King Abdulaziz University, Jeddah, Saudi Arabia 5, 1–62.

Al-Qurashi, M., 1981. Synoptic climatology of the rainfall in the south-western region of Saudi Arabia. Unpublished M.Sc. Thesis, Western Michigan University, Michigan, U.S.

Al-Suba'i, K., 1991. Erosion-sedimentation and seismic considerations for dam siting in the central Tihamat Asir region. Unpublished Ph.D. Thesis. King Abdulaziz University, Faculty of Earth Sciences, Jeddah, Kingdom of Saudi Arabia.

Al-Yamani, M. S. and Şen, Z., 1992. Regional variation of monthly rainfall amounts in the Kingdom of Saudi Arabia. *J. King Abdulaziz University: Faculty of Earth Sciences* 6: 113–133.

Aron, G. and White, E. L., 1982. Fitting a Gamma distribution over a synthetic unit hydrograph. *Water Resour. Bull.*, Vol. 18, No. 1, 95–98.

Barnes, H. H., Jr., 1967. *Roughness Characteristics of Natural Channels,* U.S. Geological Survey Water-Supply Paper 1849, 213 p.

Bhunya, P. K., Mishra, S. K., and Berndtsson, R., 2003. Simplified two-parameter Gamma distribution for derivation of synthetic unit hydrograph. *J. Hydrologic. Eng.*, ASCE, Vol. 8, No. 4, 226–230.

Bras, R. L., 1989. *Hydrology: An Introduction to Hydrologic Sciences.* Addison-Wesley: Reading, MA.

Carter, R. W., Einstein, H. A., Hinds, J., Powell, R. W., and Silberman, E., 1963. Friction factors in open channels, progress report of the task force on friction factors in open channels of the Committee on Hydro-mechanics of the Hydraulics Division: Proceedings, American Society of Civil Engineers, *Journal of the Hydraulics Division*, 89, No. HY2, pt. 1, pp. 97–143.

Chen, C-L., 1982. An evaluation of the mathematics and physical significance of the soil conservation service curve number procedure for estimating runoff volume. In *Rainfall–Runoff Relationship*, Singh, V.P. (Ed.). Water Resources Publications: Littleton, CO, pp. 80–161.

Chow, V. T., 1959. *Open-Channel Hydraulics:* New York, McGraw-Hill Book Co., 680 p.

Chow, V., Maidment, D., and Mays, L., 1988. *Applied Hydrology:* McGraw-Hill, New York, 572 p.

Chutha, P. and Dooge, J. C. I., 1990. The shape parameter of the geomorphic unit hydrograph. *J. Hydrol.* 117(1–4): 81–97.

Clark, C. O., 1943. Storage and unit hydrograph. *Proc. Am. Soc. Civ. Eng.,* 9, pp. 1333–1360.

Croley, II, T. E., 1980. Gamma synthetic hydrographs. *J. Hydrol.,* Vol. 47, 41–52.

Dein, M. A., 1985. Estimation of floods and recharge volumes in wadis Fatimah, Naaman and Turabah. Unpublished M.Sc. Thesis, Faculty of Earth Sciences, King Abdulaziz University, Jeddah, Saudi Arabia, 127 p.

Dooge, J. C. I., 1959. A general theory of the unit hydrograph. *J. Geophys. Res.* Vol. 64, No. 2, 241–256.

Dooge, J. C. I., 1973. *Linear Theory of Hydrologic Systems. ARS Technical Bulletin* No. 1468. U.S. Department of Agriculture, Washington, DC.

Eastgate, W. I., Swartz, G. L., and Briggs, H. J., 1979. Estimation of runoff rates for small Queensland catchments. *Tech. Bull.* No. 15, Div. Land Utilization, Queensland Dept. of Primary Industries, Brisbane, Australia.

Espey, W. H., Jr., and Winslow, D. E., 1974. Urban flood frequency characteristics. *J. Hydraul. Div.,* ASCE, Vol. 100, No. 2, 279–293.

FAO, 1981. Arid Zone Hydrology for Agricultural Development. Rome, Irrigation and Drainage Paper, 37, 271 p.

Farquharson, F. A. K., Meigh, J. R., and Sutcliffe, J. V., 1992. Regional flood frequency analysis in arid and semi-arid areas. *J. Hydrol.* 138, 487–501.

Flerchinger, G. N. and Cooley, K. R., 2000. A ten-year water balance of a mountainous semi-arid watershed, *J. Hydrol.,* 237, No. 1–2, 86–99.

Glenn, T., 1954. *An Introduction to Climate,* 3rd ed. McGraw-Hill: New York.

Goodrich, D. C., Lane, L. J., Shillito, R. M., Miller, S. N., Syed, K. H., and Woolhiser, D. A., 1997. Linearity of basin response as a function of scale in a semi-arid watershed. *Water Resour. Res.,* Vol. 33, No. 12, pp. 2951–2965.

Gray, D. M., 1961. Interrelationship of watershed characteristics. *J. Geophys. Res.,* 66, 1215–1223.

Gupta, V. K., Waymire, E., and Wang, E. C. T., 1980. A representation of an instantaneous unit hydrograph from geomorphology. *Water Resour. Res.,* 16(5): 855–862.

Hawkins, R. H., 1993. Asymptotic determination of runoff curve numbers from data. *J. Irrigation and Drainage Eng.,* ASCE 119(2): 334–345.

Hjelmfelt, A. T., Jr., 1991. Investigation of curve number procedure. *J. Hydrol. Eng., ASCE* 117(6): 725–737.

Horton, R. E., 1940. An approach toward a physical interpretation of infiltration capacity. *Soil Science Society of Am. J.,* 5: 399–417.

"Influences" of vegetation and watershed treatment on runoff, silting, and streamflow. (1940). Misc. Pub. No. 397. USDA, Washington, D.C.

Italconsult, 1973. Climate in area VI south, Hydrological special paper No. 2, Saudi Arabian Ministry of Agriculture and Water.

Jordan, P. R., 1977. Streamflow transmission losses in Western Kansas. *J. Hydraul. Div., ASCE,* 108, HY8, 905–919.

Kadioğlu, M. and Şen, Z., 2001. Monthly precipitation–runoff polygons and mean runoff coefficients. *Hydrol. Sci. J.* 46(1): 3–11.

Karlinger, M. R. and Troutman, B. M., 1985. Assessment of the instantaneous unit hydrograph derived from the theory of topologically random network. *Water Resour. Res.* 21(11): 1693–1702.

Kattelmann, R. and Elder, K., 1997. Hydrological characteristics and water balance of an alpine basin in the Sierra Nevada. *Water Resor. Res.*, 27, 1553–1562.

Khiyami, H. A., Şen, Z., Al-Harthy, S. C., Al-Ammawi, F. A., Al-Balkhi, A. B., Al-Zahrani, M. I., and Al-Hawsawy, H. M., 2005. *Flood hazard evaluation in wadi Hali and wadi Yibah. Technical Report.* Saudi Geological Survey, 138 p.

Langbein, W. B. et al., 1947. Topographic characteristics of drainage basins. Water supply paper 968-C, U.S. Geological Survey, 125–155.

Le Bissonnais, Y. and Singer, M. J., 1993. Seal formation, runoff, and interrill erosion from seven California soils. *Soil Sci. Soc. of Am. J.*, 57, 224–229.

Linsley, R. K., 1982. Rainfall–runoff models—an overview. In *Proc. Int. Symposium on Rainfall–Runoff Relationship*, Singh, V. P. (Ed.). Water Resources Publications: Littleton, CO.

Maidment, D. R., 1993. *Handbook of Hydrology*, McGraw-Hill, Inc., New York.

Mandelbrot, B. B., 1979. *Fractal Geometry of Nature.* W.H. Freeman: San Francisco.

Mather, J. R., 1979. Use of the climatic water budget to estimate streamflow. Water Resources Center. University of Delaware, Technical Research Report, July, Dept. Geography Newark, DE, 528.

Mesa, O. J. and Miffin, E. R., 1986. On the relative role of hill slope and network geometry in hydrological response. In *Scale Problem in Hydrology*, Gupta, V. K., Reidel, D. and Hingham, M. A. (Eds.), 1–17.

Morel-Seytoux, H. J., 1988. Introduction to overland flow theory, *Proc. 8th Annual Hydrology Days Publications*, 57 Seiby Lane, Atherton, CA 94027, 260–293.

Morin, J. and Benyamini, Y., 1977. Rainfall infiltration into bare soil. *Water Resour. Res.*, Vol. 13, No. 5, pp. 813–817.

Nash, J. E., 1957. Determining runoff from rainfall. *Proc. Inst. Civ. Eng.*, Dublin, Ireland.

Nash, J. E., 1958. The form of instantaneous unit hydrograph. In *General Assembly of Toronto, 3–14 September 1957, Volume III, Surface Water, Prevision, Evaporation. IAHS Publication* No. 45. IAHS Press: Wallingford, 114–121.

Nash, J. E., 1959. Synthetic determination of unit hydrograph parameters. *J. Geophys. Res.*, Vol. 64, No. 1, 111–115.

NEH, 1999. *National Engineering Handbook*, Chapter 5, pp 5–11.

Nouh, M., 1987. Analysis of rainfall in southwest region of Saudi Arabia. *Proc. Instn. Civil Eng.*, 83(2), 339–349.

Nouh, M., 1988a. Estimate of floods in Saudi Arabia derived from regional equations. *J. Eng. Sci.*, King Saud University, 14 (1), pp. 1–26.

Nouh, M., 1988b. On the prediction of flood frequency in Saudi Arabia. *Proc. Instn. Civil Eng.*, 85(2), 121–144.

Nouh, M., 2006. Wadi flow in the Arabian Gulf states. *Hydrol. Process.*, Vol. 20, 2393–2413.

Parks, Y. P. and Sutcliffe, J. V., 1987. The development of hydrological yield assessment in NE Botswana, *Brit. Hydrol. Soc. Symp.*, 14–16 September, Hull, Brit. Hydrol. Soc. I.C.E., London, 12.1–12.11.

Parrisopoulos, G. A., and Wheater, H. S., 1990. Numerical study of the effects of layers on unsaturated-saturated two-dimensional flow. *Water Resour. Manage.*, 4, 97–22.

Pilgrim, D. H. and Cordery, I., 1993. *Flood Runoff. Handbook of Hydrology*, Maidment, D. R. (Ed.). McGraw-Hill: New York, 9.1–9.42.

Ponce, V. M. and Hawkins, R. H., 1996. Runoff curve number: has it reached maturity? *J. Hydrol. Eng.*, ASCE 1(1): 11–19.

Raudkici, A. I., 1979. *Hydrology: An Advanced Introduction to Hydrological Processes and Modeling,* Pergamon, New York.

Richard, B. D., 1955. *Flood Estimation and Control.* Chapman & Hall Ltd., London, 187.

Rodriguez-Iturbe, I., Devoto, G., and Valdes, J. B., 1979. Discharge response analysis and hydrologic similarity: the interrelation between the geomorphologic IUH and the storm characteristics. *Water Resour. Res.,* 15(6): 1435–1444.

Scanlan, B. R., 1994. Water and heat flux in desert soils: 1, Field studies. *Water Resour. Res.,* 30, 709–719.

Seyhan, E., 1976. Regression of morphometrical variables with Synthetic Hydrograph parameters. Series B, No. 65, 18 p. Geographic Institute, Utrecht, The Netherlands.

Singh, S. K., 2000. Transmuting synthetic hydrographs into Gamma distribution. *J. Hydrol. Eng., ASCE,* Vol. 5, No. 4, 380–385.

Sirdaş, S. and Şen, Z., 2007. Determination of stream-channel infiltration due to flash floods in western Arabian Peninsula. *Hydrol. Eng., ASCE,* (in press).

Smith, K., 1992. *Environmental Hazards Assessing Risk and Reducing Disaster.* Routledge, London, 324 p.

Snyder, F. F., 1938. Synthetic unit hydrographs. *Transactions American Geophysical Union* 19: 447–454.

Strahler, A. N., 1958. Dimensional analysis applied to fluvially eroded landforms. *Bull. Geol. Soc. Am.,* 69, 279–300.

Soil Conservation Service (SCS), 1971. *National Engineering Handbook, Section 4: Hydrology.* USDA: Springfield, VA.

Soil Conservation Service (SCS), 1986. *Urban Hydrology for Small Watersheds,* Technical Report 55. USDA: Springfield, VA.

Streeter, V.L., 1971. *Fluid Mechanics:* McGraw-Hill Book Co., New York, 5th ed. 705 p.

Şen, Z., 1983. Hydrology of Saudi Arabia. In *Symposium on Water Resources in Saudi Arabia,* Riyadh, A68–A94.

Şen, Z., 1996. Theoretical RQD-porosity-conductivity-aperture charts. *International J. Rock Mechanics and Mining Sciences & Geomechanics Abstracts,* 33 (2): 173–177.

Şen, Z., 2007a. Modified hydrograph method for arid regions. *Hydrol. Process.* 21, (in print).

Şen, Z., 2007b. Hydrograph and unit hydrograph derivation in arid regions. *Hydrol. Process.,* 21, 1006–1014.

Şen, Z., 2007c. Instantaneous runoff coefficient variation and flood estimation model. *ASCE, Hydrol. Eng.,* (in print).

Şen, Z. and Al-Suba'i, K., 2002. Hydrologic considerations for dam siting in arid regions: a Saudi Arabian study, *Hydrol. Sci. J.,* 47 (1), 1–19.

Şorman, A. U. and Abdulrazzak, M. J., 1993. Infiltration-recharge through wadi beds in arid regions, *Hydrol. Sci. J.,* 38, 3, 173–186.

Taha, M., Harb S., Nagib, M., and Tantawy, A., 1981. The climate of the Near East. In *World Survey of Climatology, vol. 9, Climates of Southern and Western Asia,* Takahashi, K. and Arakawa, H. (Eds). Elsevier: New York, 183–255.

Taylor, A. B. and Schwarz, H. E., 1952. Unit hydrograph lag and peak flow related to basin characteristics, *Trans. Am. Geophys. Union,* 33, 235–346.

Titmarsh, G. W., 1989. *Flood estimation for small agricultural catchments.* Darling Downs Region, Queensland, Eng. thesis, University of New South Wales at Sydney, Australia.

Titmarsh, G. W., Cordery, I., and Pilgrim, D. H., 1995. Calibration procedures for rational and USSCS design flood methods. *J. Hydraulic Eng., ASCE,* 121(1): 61–70.

UNESCO, 2002. Hydrology of Wadi Systems. IHP Regional Network on Wadi Hydrology in the Arab Region. IHP-V, Technical Documents in Hydrology, No. 55, Paris, 39 p.

Wheater, H. S., 2002. Hydrological processes in arid and semi-arid areas. *Int. Hydrol. Programme*, Hydrology of Wadi Systems. IHP Regional Network on Wadi Hydrology in the Arab Region, Technical Documents in Hydrology, No. 55, UNESCO, Paris, pp. 5–22.

Wheater, H. S. and Brown, R. P. C., 1989. Limitations of design hydrographs in arid areas – an illustration from southwest Saudi Arabia. *Proc. 2nd Natl. BHS Symp.*, 3.49–3.56.

Wheater, H. S., Jakeman, A. J., and Beven, K. J., 1993. Progress and directions in rainfall-runoff modeling. In: *Modeling Change in Environmental Systems*, Wiley, Chichester.

Williams, J. R. and Laseur, V., 1976. Water yield model using SCS curve numbers. *J. Hydrol. Eng., ASCE*, 102(9): 1241–1253.

Wood, M. K. and Blackburn, W. H., 1984. An evaluation of the hydrologic soil groups as used in the SCS runoff method on rangelands. *Water Resour. Bull.*, Vol. 20, No. 3, 379–389.

Yin, Z. Y. and Brooks, G. A., 1992. Evaporation in the Oketenokee Swamp watershed: a comparison of temperature and water balance methods. *J. Hydrol.*, 131, 293–312.

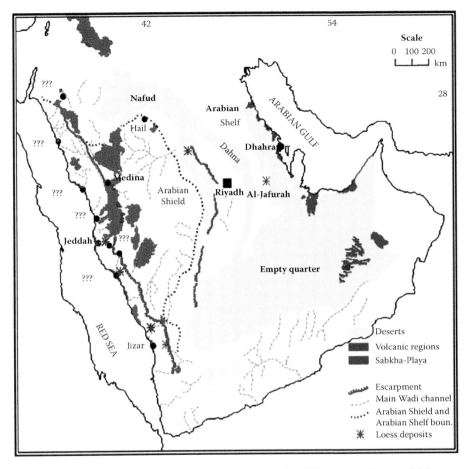

**FIGURE 1.3**  Arabian Peninsula features (courtesy of Saudi Geological Survey, SGS).

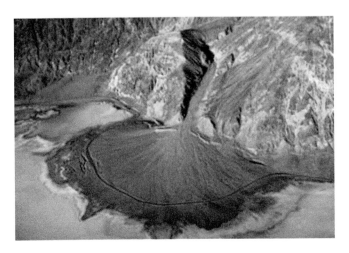

**FIGURE 2.6**  Alluvial fans.

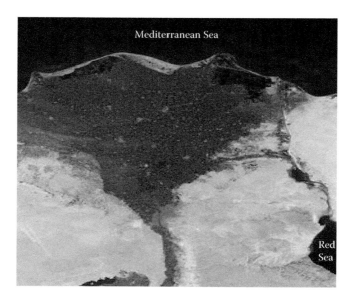

**FIGURE 2.9**    Nile River delta formation.

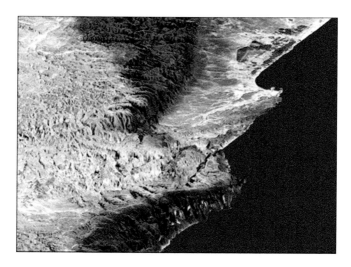

**FIGURE 2.10**    Coastal plain deposits.

# 5 Groundwater Recharge and Evaluation

## 5.1 GENERAL OVERVIEW

Water at varying depths below the earth surface in saturated zones of the geological layers is called a *groundwater reservoir*. Groundwater is the most important hydrologic cycle component in arid regions. If there is no impermeable layer above, then the upper level of the saturated zone is the *water table*, above which the unsaturated (aeration) zone exists, including soil moisture and vadose water.

Surface water structures are not preferred in arid regions due to excessive evaporation losses, except for short-duration impoundment. The ratio of extensive surface area to the volume of small reservoirs leads to high evaporation loss. Micro-storage facilities lose, on average, 50% of their impoundment to evaporation in arid and semi-arid areas (Gleick, 1993; Sakthivadivel et al., 1997). Groundwater storage in *subsurface dams* or *aquifers* provides water supply resources for many months and years with little or no evaporation loss. It also has the advantage that storage can be near or directly under the point of use and immediately available through wells and pumps on demand. The development of tube well technology offered extra facilities for groundwater abstraction in addition to conventional large diameter hand-dug wells, especially in arid regions.

This chapter will emphasize groundwater recharge, aquifer parameter determination from field work, and tests, in addition to chemical constituents in the light of geological, hydrological, and practical field discussions for arid and semi-arid regions.

## 5.2 GEOLOGY AND GROUNDWATER

Structures playing different roles in groundwater quantity and quality variations include the following: groundwater reservoirs occurring in igneous, sedimentary, and metamorphic rock; voids between minerals and grains; and joints, fractures, and faults. The distribution and composition of rocks affect the availability and chemical constituents of groundwater. It is, therefore, inevitable that any groundwater study should include in its early stages a detailed geological prospecting of the area for a more accurate appreciation of the geometrical extension of the aquifer. In general, a geological study should include a lithological phase covering mineral composition, grain size, sorting and packing; a stratigraphic phase describing the age, unconformities, and geometrical relationships between different lithologies; and a study of structural features such as fissures, joints, fractures, folds, and faults. Collection of this information gives a rather clear picture of the subsurface geology, leading to a better understanding of various water bearing formation distributions.

In nature, as rainwater reaches the ground surface, it is almost homogeneous but its penetration into the rocks gives a heterogeneous appearance due to the haphazard distribution of minerals in the geological formations. The infiltrating water from the wadi seeks its way with difficulty toward groundwater storage and then moves rather horizontally toward outlet points such as wells or springs at small velocities. During such a journey, it gets in touch with different types of minerals and rocks that give the groundwater its characteristic quality. In general, the zones close to the outcrops have comparatively good quality water, which becomes rather brackish and saline as it gets closer to the abstraction zone. The quantity and especially the continuity of groundwater are more dependent on hydrological principles than geology.

The geological setup and various features in a region are vital steps in obtaining a better understanding of groundwater occurrence, movement, recharge, and discharge areas. In general, these studies can be initiated by rather inexpensive field reconnaissance trips on surface geology inspections that are succeeded by subsurface investigations of various degrees of detail, depending on the significance of the study, financial resources availability, and stipulated time. In order to have a clear idea of the subsurface geological setting and its relation to groundwater, either cross-sections in some preferred directions or three-dimensional pictures of the subsurface geology (fence diagrams) are necessary, and these assist in making meaningful correlative interpretations (Fetter, 1994).

Any groundwater study with incomplete geological information leads to erroneous conclusions. It is, therefore, advised strongly that hydrological studies be preceded by geological studies so as to achieve a good understanding of groundwater occurrence, distribution, availability, and abstraction.

## 5.3   GROUNDWATER AND RECHARGE

Groundwater recharge depends on several factors such as infiltration capacity, stochastic characteristics of precipitation, and climate factors. The spatial and temporal distribution of the rainfall mainly controls the natural *groundwater recharge*. In arid regions, recharge occurs through the *ephemeral* streams that flow through the wadi course, but most of the water is absorbed in the unsaturated zone before reaching the aquifer. In semi-arid regions, the recharge is irregular and occurs only in periods of heavy rainfall. In humid regions, recharge is mainly in the winter period.

Groundwater recharge has a random behavior depending on the sporadic, irregular, chaotic, and complex features of storm rainfall occurrences, geological composition, and geomorphologic features. In areas of *perennial* surface water flow there are continuous infiltration and consequent groundwater recharge along the main flow channel in all seasons, especially in summer. In arid regions, the recharge possibilities are available in an intermittent manner due to occasional storm rainfalls and their ephemeral stream flow for some seasons. Recharge is irregular and occurs during periods of intense rainfall. Otherwise, rainfall only wets the top soil of the unsaturated zone without any water reaching groundwater storage.

Groundwater recharge can be defined simply as the downward flow of water reaching the groundwater table via the soil and unsaturated zone. Simmers et al. (1997) stated that a clear distinction should be made between the *potential* recharge and the *actual recharge*. In potential recharge, water is available for recharge from the soil zone but it may migrate to another destination due to the dynamics of the unsaturated zone. For example, in areas of high water table, when potential recharge (excess of precipitation over evapotranspiration) cannot enter the groundwater, it becomes runoff. In actual recharge, the water reaches the water table and adds to the groundwater reservoir.

Similar to the need for defining the potential and actual evaporation, it is necessary to make distinctions between the potential and actual recharge rates. Potential recharge is the type where, if there is a continuous surface water supply, then the recharge takes place through the infiltration and following percolation processes until the whole unsaturated layer is saturated. In such a process there is a continuous groundwater *hydraulic head* increase. However, the *actual* recharge occurs only after sufficient storm rainfall occurrences, sufficient in the sense that after the soil and unsaturated zone wetting there should be an actual contribution to groundwater storage, which must cause groundwater hydraulic head rise to a certain extent. Hence, actual recharge is always smaller than the potential alternative.

### 5.3.1 Transmission Losses

The relationship between wadi flow *transmission losses* and groundwater recharge will depend on the underlying geology. Once infiltration has taken place, the alluvium underlying the wadi bed is effective in minimizing evaporation loss through capillary rise (the coarse structure of alluvial deposits minimizes capillary effects). Hellwig (1973) found that dropping the water table below 60 cm in sand with a mean diameter of 0.53 mm effectively prevented evaporation losses, and Sorey and Matlock (1969) reported that measured evaporation rates from stream bed sand are lower than those reported for irrigated soils.

Parrisopoulos and Wheater (1991) combined 2-D simulation of unsaturated wadi-bed response with the Deardorff (1977) empirical model of bare soil evaporation to show that evaporation losses are not in general significant for the water balance or water table response in short-term simulation (for periods up to 10 d). However, the influence of vapor diffusion is not explicitly represented, and long-term losses are not well understood. Andersen et al. (1998) showed that losses are high when the alluvial aquifer is fully saturated, but are small once the water table drops below the surface.

Şorman and Abdulrazzak (1993) provided an analysis of groundwater rise due to transmission loss for an experimental reach in Wadi Tabalah, southwest Saudi Arabia and estimated that on the average 75% of bed infiltration reaches the water table. There is, in general, little information available to relate flood transmission loss to groundwater recharge. The differences between the two are expected to be small, but will depend on residual moisture stored in the unsaturated zone and its subsequent drying characteristics. If the water table approaches the surface, relatively large evaporation losses may occur. In the study of the sand rivers of Botswana, it was expected that recharge of the alluvial river beds would involve complex unsaturated zone response. Observations showed that the first flood of the wet season was sufficient to fully recharge the alluvial river bed aquifer. This storage is topped in subsequent floods, and depleted by evaporation when the water table is near-surface, but in many sections sufficient water remained throughout the dry season to provide adequate sustainable water supplies for rural villages.

## 5.3.2 TYPES OF RECHARGE

Major sources of groundwater recharge in arid regions include rainfall infiltration or direct recharge, natural or induced infiltration from surface water, infiltration of flood water through wadi beds, river recharge, inter-aquifer flows, and irrigation and urbanization recharge. *Direct recharge* is the entrance of rainwater at the place of surface without transformation to the surface runoff. Such areas are often in the upstream parts of wadis, where the rainfall occurs most frequently. However, in the case of local convective rainfall events direct recharge may also occur in any part of the wadi. *Indirect recharge* is due to runoff water, which occurs mostly outside the rainfall influence area. For instance, the Euphrates and Tigris rivers frequently have rainfall events in Turkey, but as a consequence of runoff they support groundwater recharge in Syria and Iraq, which are away from the rainfall influence areas.

Among the spatial recharge types are *sheet flow*, depressions, joints, pondings, river and wadi channels, artificial spreading areas, etc. The recharge types in arid regions together with the mechanism are presented by Lloyd (1986) as in Figure 5.1.

Allison (1988) and Foster (1988) used the terminology of local (or diffuse) and localized recharge for direct and indirect recharge. Direct recharge can be defined as water added to the aquifer through the unsaturated zone by direct percolation of rainfall at the spot where it falls. Indirect recharge occurs where water fulfills the soil moisture deficits and evapotranspiration process before reaching the groundwater reservoir.

Indirect recharge occurs from percolation to the aquifer following surface water runoff (surface water category) and localization (localized category) in joints, pondings, and lakes or through the wadi beds. Indirect recharge produced as a result of infiltration during flood pulses is considered as the most important contribution to the groundwater table in wadi channels. Small local floods merely compensate for soil moisture deficits and evapotranspiration, particularly during the dry season and, therefore, the amount of water that goes to the water table will not be a significant contribution.

The most extensive, rich, and significant recharge in arid regions is indirect recharge, which spreads floodwaters over thousands of square kilometers on both sides along the main wadi channel. At times, in many places both direct and indirect recharges occur simultaneously. The calculation of direct recharge is comparatively easier due to the following reasons.

1. The earth surface area giving rise to recharge is smaller than indirect recharge and, hence, estimations are more reliable.
2. Due to smaller influence areas, the geometrical composition of direct recharge areas are more homogeneous.
3. As rainwater directly reaches groundwater storage in the simplest and shortest way possible, its quality is more similar to rainwater composition than to groundwater composition.

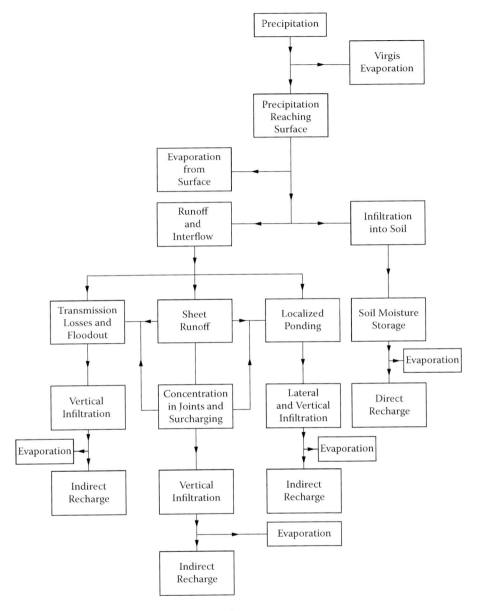

**FIGURE 5.1**   Recharge mechanisms in arid areas.

4. The water movement is almost vertical, so there is less probability that it
   will come into contact with lateral geological variations.

Because the indirect recharge area lies along the wadi channel and extends from
upstream to the outlet point downstream, its calculation is more difficult; additional

care and preservations are required for accurate results in making estimations. The locations where indirect recharge occurs are as follows:

1. Through distinct fractures, fissures, and *solution cavities* in *hard rocks* (igneous and metamorphic) and limestone (*soft rock*) during sheet overflow.
2. Depressions over the drainage basin area that first appear as small lakes with surface water storage. The stored water recharges the aquifer partially; the remainder is lost as evaporation.
3. Along the main wadi channel on both sides, depending on the inundation area, wide at some places and very narrow at others. The wider the spread area, the more is the infiltration and recharge to groundwater storage.
4. In general, *alluvial fans, piedmonts, depressions,* and *subsidence* near the foothills and contact zones are major indirect recharge locations. In these areas, indirect recharge takes place through the fractures and fissures within the mountains that lead groundwater into the low-lying alluvium-fill wadis and to the down-slope water over the mountain cliffs.

In addition to the above-mentioned recharge types, there are also different causes and mechanisms that play roles in the recharge process. Among such dominant mechanical factors are surface features (topography, geomorphology, runoff, depression dimensions, vegetation), atmospheric processes (temperature, evaporation, humidity, solar irradiation, wind speed), land use (agriculture, transportation roads, urban areas), drainage pattern (streams, creeks, rivers, sub-basins), geology (soil type, rock type, fracturing), and unsaturated zone (granular composition, thickness, effective porosity). The role of these factors change from humid to arid regions, and in arid regions the basic mechanisms are indirect recharge in wadi channels, depressions, limestones and *sabkhahs*, volcanic rocks, *sand dunes* and contact lines between different lithologies. Basmaci and Hussain (1988) pointed out that groundwater recharge is a complex function of meteorological, geomorphologic, and hydrologic characteristic in a given region. However, they said that this function differs significantly from humid to arid climates.

According to Scanlon et al. (2002), the definitions of direct and indirect recharge are oversimplified as the *subsurface recharge* is not explicitly considered, whereas the so-called preferred pathways are a common phenomenon with even direct recharge. Apparently, in many locations a combination of both direct and indirect recharge occurs. Comparison of direct and indirect recharge leads to the following points:

1. Estimates of direct recharge can be more reliable than indirect recharge.
2. With increasing *aridity*, direct recharge becomes less significant, whereas indirect recharge happens more in terms of *total recharge* to an aquifer.
3. Recharge occurs to some extent even in the most arid regions, although increasing aridity decreases the net downward flux with greater time variability.
4. Successful groundwater recharge estimation depends on, first, identifying the probable flow mechanism and important features influencing recharge for a given locality.

Lerner (1990) defined the category of localized recharge as intermediate between direct (at the same place where precipitation falls) and indirect (along mainstream channels). In this category, water takes some horizontal movement before reaching the aquifer. He pointed out that this type of recharge is the most dominant in arid and semi-arid areas and, unfortunately, the most difficult for estimation. This type of recharge occurs as follows:

1. *Hard rock* terrain: Recharge is through distinct fissures in weathered, bare hard rock or limestone terrain.
2. *Topographic depressions*: The main concentration of recharge was discovered in such depressions even with highly permeable material (e.g., Freeze and Banner, 1970; Harhash, 1980; Rehm et al., 1982).
3. *Minor wadis*: This type of recharge is a characteristic feature of arid and semi-arid regions where recharge occurs due to ephemeral flows generated after the storms.
4. Mountain front recharge: Many alluvial aquifers including alluvial fans, *piedmont plains,* and *subsidence* basins are categorized in this type of recharge. The recharge system mainly works along the mountain boundary and consists of two main constituents: (a) subsurface influx from the mountain mass to the basin fill, and (b) runoff infiltration in defined and undefined channels and from mountain slopes.

Rushton (1988) listed a wide range of factors that control the actual groundwater recharge mechanism as follows:

1. Surface processes: Topography, precipitation, surface runoff, ponding of water, cropping pattern, actual evapotranspiration.
2. Irrigation: Nature of irrigation scheduling, losses from canals, irrigation fields, and watercourses.
3. Rivers: Rivers flowing into and leaving the study area and gaining from or losing water to the aquifers.
4. Soil zone: Nature of soil, depth, hydraulic properties, variability of the soil spatially and with depth, cracking and swelling.
5. Unsaturated zone: Flow mechanism through unsaturated zone; zones with various hydraulic conductivities.
6. Aquifer: Water acceptance capacity of the aquifer, variation of aquifer condition with time.

Dincer (1980) identified the following recharge conditions, which are most likely to occur in arid zones:

1. Runoff recharge: This is the most common type of recharge in arid zones. He subdivided the runoff recharge into three groups.
    a.  Recharge to the wadi alluvium.
    b.  Recharge to the karst limestone.
    c.  Infiltration through *sabkhahs.*

2. Groundwater recharge in volcanic rocks: Recharge occurs by the infiltration of rainfall in young basaltic rocks or areas covered by pumice.
3. Groundwater recharge in sand dunes: Recharge occurs through the infiltration of the rainfall. These areas are considered as the most favorable areas for the groundwater recharge.

### 5.3.3 Recharge Estimation Methods

Groundwater recharge estimation is one of the most significant hydrological component calculations, especially in arid region countries where the spatial and temporal rainfall variations are sporadic and irregular. Even in developing and developed countries, it has priority in water resources evaluations. Groundwater abstraction in excess of recharge gives ways for groundwater level falls in many parts of the world. This has led to excessive costs in water pumping in addition to reduced rate of abstraction. If proper *groundwater management* is not planned for the future by taking into consideration the recharge possibilities, then groundwater storage may be depleted in a short time. It is, therefore, necessary to have reliable groundwater recharge rate estimations in order to delineate the recharge and well field areas, to control regional groundwater movement, and to keep a balance between water supply and recharge (see Chapter 6).

Although theoretically there are physical, chemical, and hydrological methods for recharge estimation, in practical studies depending on the scale of the work, there arise various difficulties such as those due to temporal and spatial variation of rainfall, heterogeneous geological lithology and stratigraphy, and the effects of measuring instrumentation on the surface flow and, consequently, to infiltration rates. In any recharge estimation method, the important factor is the recharge rate, which represents the volume of water from a unit area during a unit time. In this definition, the unit area and time interval imports difficulties due to topographical, geological, meteorological, hydrological, vegetation pattern, and spatial and temporal variations. There are a number of factors as given below, which make the recharge estimation a difficult one.

1. Variation in the temporal and spatial distribution of rainfall.
2. Redistribution of infiltration and percolation flux by lateral flow near instrumented site, thereby providing inaccurate results from observation wells.
3. Flow away from the regional water table may be poorly defined.
4. Storm water runoff may be an important factor in the redistribution of rainfall from the region of higher elevation and more humid conditions. Isotopically similar water may have reached sampling wells by quite different routes.

Natural groundwater recharge flux determination in arid and semi-arid regions is the most difficult to quantify because of large spatial and temporal variation in rainfall, climate, soils, soil moisture, topography, geologic setting, vegetation patterns, etc. Recharge occurs intermittently in arid regions, usually during periods of storms followed by long periods with little or no rainfall and recharge. There are two main methods for determining recharge flux, namely, direct method and indirect method.

### 5.3.3.1   Direct Method

Although there are lysimeter instruments for direct recharge measurements, their construction and management are difficult to control, and they are costly. Hence, in practical studies their use is almost impossible, but they are used in some parts of the world just for research purposes. It is therefore necessary to use indirect methods for such estimations. Recharge rates are difficult to measure directly because the measurement must be at depth, and it requires construction of lysimeters, which is expensive. The direct method is not suitable for arid regions because of low average rainfall.

### 5.3.3.2   Indirect Method

These may not be as reliable as direct methods but due to their simplicity, economy, and quick result-finding, they are preferable in practical studies with reliable estimations. Among such techniques are various balance approaches such as surface water balance, measurement of saturation and water potential within the vadose zone, groundwater budget analysis, water table response after rainfall, groundwater discharge within well-defined groundwater basin, numerical groundwater flow model, measurement of environmental isotopes, and use of chemical mass balance methods such as with chloride.

## 5.4   TECHNIQUES FOR GROUNDWATER RECHARGE ESTIMATION

The groundwater recharge calculations have two major components: the *infiltration rate* and the *infiltration area*. Additionally, in any groundwater recharge problem, especially in arid regions, the following series of questions are important.

1. What is the capacity of groundwater reservoir to accept recharge? The more saturated the reservoir, the less capacity for future recharge; the extra water must be conserved in some other way.
2. How much water can accommodate the unsaturated zone? The permeability and porosity coefficients of the medium are important, and low permeability does not allow high recharge rates.
3. Are there other potential recharge possibilities nearby or at some distances, so that the extra water can be saved for future use?
4. What are the dimensions of potential recharge within the wadi area and preferably in the wadi channel?
5. Should actual and potential recharge rates be separated from each other?

In arid and semi-arid regions, the groundwater recharge calculations are mostly achieved by isotopic water composition identification (Athavale and Rangarajan, 1990; Rushton, 1990; Zimmerman et al., 1967; Munich, 1968a,b) or a *chloride mass balance* (CMB) approach (Edmunds et al., 1990; Houston, 1990).

The first groundwater study in the Arabian Peninsula was done by Dincer et al. (1974) who concluded that for recharge to occur in the sand dunes, the rainfall amount should be over 50 mm. This conclusion was obtained after the stable isotopes study on groundwater samples collected from the area. Caro and Eagleson (1981) applied a water balance approach for groundwater recharge estimation in Minjur and Wasia aquifers in the central Arabian Peninsula with values ranging from 10 to 24 mm and from 1 to 6 mm. respectively. Ghurm and Basmaci (1983) stated that the groundwater recharge in the upstream of wadis may reach up to 41%. Al-Kabir (1985) noted that deuterium excess of groundwater samples collected from wadis showed recharge from high intensity summer rainfalls. Abdulrazzak et al. (1989) used water balance approach to Wadi Talabah in the southwestern region of the Arabian Peninsula, and they estimated that only 3% of the rainfall occurs as runoff out of which 75% contributes to groundwater. Finally, Bazuhair and Wood (1996) used CBM for groundwater recharge estimations in the wadi aquifers within the western region of the Arabian Peninsula. There are several techniques that have been applied for different hydrogeological provinces by many workers for estimating natural recharge in the arid and semi-arid regions.

Athavale and Rangarajan (1990) used the tritium injection method to estimate natural recharge in the *hard rock* (granite, basalt, and indurated Pre-Cambrian sediments) regions of southern India. This artificial tritium tracer technique is based on the assumption of the piston flow model (Zimmerman et al., 1967; Munich, 1968a,b) for water movement in the unsaturated zone. In this technique, tritiated water is injected at a certain depth in the unsaturated zone, and the vertical movement this tracer undergoes is investigated during the next hydrogeological cycle. After a certain time interval, the moisture content and tracer concentration are measured from various depth intervals. The moisture content of the soil column between injection and displacement depths is the measure of recharge to groundwater.

Edmunds et al. (1990) applied the solute (chloride) profile techniques for groundwater recharge estimation in the semi-arid and arid regions of Cyprus and Central Sudan. The unsaturated zone chloride profiles results of recharge from Cyprus (a mean annual recharge of 50 mm/year with mean annual rainfall equal to 420 mm) and Central Sudan (recharge is 1 mm/year as a result of mean annual rainfall equal to 180 mm) were in good agreement with the tritium profile results obtained from Cyprus and the unsaturated zones chloride profile results from adjacent areas of the Central Sudan. It is concluded that the chloride profiles are a cheap and effective tool for estimating direct recharge in porous lithologies of semi-arid regions and also for investigating recharge history.

Houston (1990) estimated the groundwater recharge on the basement rocks of Zimbabwe by using the ratio of chloride in the rainfall to that in groundwater with the following assumptions.

1. The principal source of chloride in groundwater is from the atmosphere.
2. No contribution comes from evaporate sources or any host rocks (granite or gneiss).

He also estimated a groundwater recharge rate of 2 to 3% of rainfall in the granite aquifer and 0.5 to 1% of rainfall in the gneissic aquifer. He attributed the low recharge rates in the gneissic aquifer to addition of chloride from another source. A plot of chloride (mg/l) in groundwater versus mean annual rainfall shows that the groundwater from gneissic areas tends to have higher chloride content compared to granite. Thus, for a particular rock type area a significant relationship exists between rainfall and chloride contents.

Rushton (1990) used various tracer methods to estimate the recharge flux of the Mehsana alluvial aquifer In India. He used two alternative methods, namely, peak and total tritium, for estimating the recharge flux. From both of these methods similar results are obtained for two areas. The recharge values vary between 3 to 4% of the rainfall. His methods are based on the assumption that the tritium and water both move at the same rate and a piston flow mechanism, in which the younger layer of moisture replaces the older one, pushes down an equal amount of water immediately below. This process continues until the last such layer is added to the water table.

Lloyd (1986) discussed various aspects of the precipitation or flood flow recharge in arid regions. He summarized that the recharge from rainfall is rare, and runoff is the main source of recharge. He also concluded that there is no single solution to overcome this problem. Foster (1988) suggested that the selection of a given method should be based on the hydrological environment and vegetation patterns, and type of recharge, that is, diffused or localized. Gee and Hillel (1988) reviewed various recharge estimation methods used in arid regions. They discussed various flaws of the water balance and soil moisture models. They recommended direct measurement of recharge, that is, construction of lysimeters and the tracer techniques. Allison (1988) made a reasonably comprehensive discussion of the various physical, chemical, and isotopic techniques for the estimation of groundwater recharge. He summarized that the chemical and isotopic methods of recharge estimation can be used more successfully than the physical-based method in arid and semi-arid regions.

Based on lithologic and morphologic criteria Torrent and Sauveplan (1977) mentioned that in the eastern parts of the Arabian Peninsula along the Red Sea coast, there are two types of potential aquifers, namely, generalized and discontinuous. The first one is characterized by a certain continuity of characteristics in space, although this can obviously vary with the heterogeneity of the aquifer material and thickness changes of the formations such as wadi alluvium, surface and coastal deposits, and *basalt flows*. The second type such as granular crystalline rocks (e.g., granitoids) and nongranular volcanic or metamorphic rocks are characterized by little or no original permeability but with secondary permeability acquired either through physico-chemical alteration or through fissuring and fracturing. The factors that govern the productivity are as follows:

1. For generalized aquifers:
   a. Permeability of terrain, which can vary with space
   b. Thickness of aquifer
2. For discontinuous aquifers:
   a. Intensity of fracturing
   b. Depth of weathering

Resources from the above aquifers, derived from recharge or renewable resources, consist of that part of the rainfall that infiltrates downward to supply the aquifers. This infiltration can occur directly, immediately following a period of rainfall or at a later period from surface waters concentrated by runoff. These resources are dependent primarily on rainfall and secondarily on series of factors such a lithology, morphology of the drainage basin, fracturing or weathering, depth of the water table, and vegetation.

For wadi alluvium the morphology of the drainage basin is the most important secondary factor because it determines the flow characteristics of the runoff, its ability to gather, and consequently its capacity for infiltration. Rainfall, runoff, evapotranspiration, and infiltration are the essential parameters of the hydrogeological regime of a drainage basin. With a prior knowledge of these, it is possible to calculate theoretically, the effective infiltration likely to reach the aquifers or the renewable parameters of a given area. Unfortunately, evapotranspiration is a difficult parameter to measure because it results from the sum of two phenomena:

1. Evaporation, either directly from the ground or deferred until collection after runoff
2. Consumption and transpiration by vegetation

## 5.5   CHLORIDE MASS BALANCE (CMB) METHODS

This method was used first by Eriksson and Khunaksem (1969). Later Allison and Hughes (1983) recognized the significance of this method and modified it to use in the unsaturated zone. Since then, others such as Stone (1984) and Wood and Sanford (1995) have used the CBM for groundwater recharge estimation in arid and semi-arid zones. The method assumes that in groundwater the principal chloride source is from atmosphere, and there is no chloride contribution from host rocks (evaporates, granite and gneiss),

$$\text{Recharge}\,(\text{mm}) = \frac{\text{Rainfall}\,(\text{mm}) \times \text{Rainfall}\,\text{Cl}\,(\text{mg/l})}{\text{Groundwater}\,\text{Cl}\,(\text{mg/l})}$$

In many arid and semi-arid regions, there are plans for strategic exploitation and use of groundwater resources. This makes the recharge estimation more important for such arid and semi-arid regions of the country. Consideration of these difficulties, especially during the past decade, led many researchers to use the simple method of CMB. The chloride ion is used in chemical recharge studies because of its conservative nature; the ion neither leaches from nor is absorbed by sediments particles. This is based on the assumption that the chloride concentrations in the rainfall and the recharge areas are in steady-state balance, i.e., input is equal to output without chloride storage change during a specific time period, often one month or year. The CMB method yields underestimated recharge rates in nonirrigated areas when compared with other methods, whereas it gives similar results for irrigated areas (Grismer et al., 2000; Flint et al., 2002; Russo et al., 2003).

It has been shown by Wood and Sanford (1995) that the CMB method can yield regional rates of recharge under certain conditions and assumptions. Bazuhair and Wood (1996) stated that the CMB method yields groundwater recharge rates that have been integrated spatially over the watershed for tens of thousands of years. Various applications of the CMB method have been presented for different parts of the world (Eriksson and Khunakasem, 1969; Eriksson, 1976; Allison and Hughes, 1983; Grismer et al., 2000; Edmunds et al., 1990; Harrington et al., 2002; Scanlon et al., 2002). However, it is not clear how temporal and spatial variations are integrated into the recharge calculations. In the classical CMB equation, only averages are used without taking into consideration spatial or temporal variations quantitatively by considering additional statistical variables in the calculations.

## 5.5.1 CLASSICAL CMB METHOD

The application of the classical CMB method is simple with no dependence on sophisticated instruments. It is based on the knowledge of annual precipitation and chloride concentrations in rainfall and groundwater storage. In arid regions, recharge is sporadic, depending on rare rainfall events during the year. Moreover, rainfall occurs at the upstream portion of wadis where, most often due to topographic heights, orographic rainfall occurrences take place over locally concentrated areas. This leads to small recharge areas where the rainfall reaches the water table through the infiltration process. The extent of spatial averaging will depend to some extent on the method of groundwater sampling, particularly the length of screened intervals of the piezometer. Some spatial averaging will occur within the aquifer due to dispersion processes. Recently, Harrington et al. (2002) attempted to quantify spatial averaging inherent in the CMB method. The chloride concentration is homogeneously distributed within the aquifer. Rainfall duration is comparatively very short in arid regions and, therefore, after the rainfall occurrence due to rather intensive solar irradiation, evapotranspiration takes place from the moist and unconfined soil surfaces, which implies an increase in the chloride concentration. Generally, in the application of the CMB method, a set of assumptions should be taken into consideration in interpreting the results. The assumptions in the classical CMB approach for recharge calculations are as follows:

1. In groundwater storage, there is no chloride source prior to the rainfall. This is the reason why the application of the method is valid for the upstream portions of the catchments.
2. There are no additional sources or sinks for chloride concentration in the area of application. This is mostly a valid assumption in the upstream portions of wadis. In thousands of geochemistry studies within the wadis, the chloride is found only in negligible amounts except after a rainfall event (Basmaci and Al-Kabir, 1988; Subyani and Bayumi, 2001; Bazuhair et al., 2002).
3. Rainfall either evaporates or infiltrates in the study region without any runoff. This is rather unrealistic and can be valid just for low-intensity rainfall events. However, most often rainfalls are intense, especially at the upstream portions of the wadi system due to the orographic condition.
4. Long-term rainfall and its chloride concentration amounts are in a balanced situation, i.e., steady-state condition. This implies stable and long-term averages, as the classical CMB method requires. The standard deviations

around the averages are completely ignored in this assumption. A hidden assumption is that the fluctuations around the average rainfall and chloride records must be very small so that they are negligible, and hence do not appear in the classical CMB equation.

On the basis of these assumptions, the fundamental equation applicable for recharge calculations is presented by Wood and Sanford (1995) as

$$q = \frac{RCl_{war}}{Cl_{gw}} \qquad (5.1)$$

in which $q$ is the *recharge flux*, $R$ is the average annual rainfall, $Cl_{war}$ is the weighted average chloride concentration in rainfall, and $Cl_{gw}$ is the average chloride concentration in groundwater. Unfortunately, there is no consistency in the description of the averages in the classical CMB equations. The following points are significant and indicate that there is arbitrariness in the interpretation and use of this equation.

1. Equation 5.1 is valid for short time durations, i.e., the smaller the time duration and the areal coverage of the system, the more valid the equation under steady-state conditions. Otherwise, the equation must be viewed as a gross simplification.
2. More important than the first point is the use of averages in the equation. For instance, it is stated that $q$ is the recharge flux, which may stand for the amount of water reaching the groundwater storage per time per area. It is not specified whether it is average and, if average, which type—arithmetic or weighted.
3. Classical CMB equation does not have homogeneity in "averages," similar to "dimensional" homogeneity in scientific methodologies. The question is, "Is it on an arbitrary basis that one selects arithmetic and weighted average concepts for each variable in the equation?"
4. What about the deviations around the averages? Equation 5.1 does not account for such variations at all.

## 5.5.2 Refined CMB Method

It is certain that during a short time duration, Equation 5.1 is valid without description of averages for its terms. In this form, it is derived from the physical mass conservation principle with no change in the storage, as simple as input is equal to output, i.e., a steady-state condition. In practical applications, it is necessary to consider rather long time intervals, such as a month or year, and the application area will be of at least several square kilometers. These scales of practicality both in time and space give rise to variations in each term, both temporally and spatially. Because there are many sampling measurements, the question is, "Which one of these measurements or what type of averages must be inserted into Equation 5.1?" The simplest view is to consider the arithmetic averages of each term, which leads to a similar expression as in Equation 5.1 with explicit averages (Subyani and Şen, 2006) as

$$\overline{q} = \frac{\overline{RCl_r}}{\overline{Cl_{gw}}} \tag{5.2}$$

where the overbars indicate arithmetic averages. The subscript in the chloride concentration is indicated by $r$ and not by *war* as in Equation 5.1. This indicates that one cannot decide arbitrarily on a weighted-average concept. Equation 5.2 has homogeneity in "averages" and in this sense surpasses that of Equation 5.1, because there are no arbitrary uses of averages. Equation 5.2 is written without a mathematical or statistical basis, but on the basis of common sense, which is not quite valid in practical applications, as will be shown in the following sequel.

In the case of random variations in each term during a long time period, hydrogeologists will have to sample the phenomena by several measurements. Practically, these measurements may not be close to each other, and therefore there are deviations around the averages. Hence, the basic variables can be considered in addition to their averages with their deviations ($q'$, $R'$, $Cl'_r$, $Cl'_{gw}$) as

$$q = \overline{q} + q' \tag{5.3}$$

$$R = \overline{R} + R' \tag{5.4}$$

$$Cl_r = \overline{Cl_r} + Cl'_r \tag{5.5}$$

and

$$Cl_{gw} = \overline{Cl_{gw}} + Cl'_{gw} \tag{5.6}$$

Each one of these equations is referred to as the perturbation of the variable concerned, with deviations added to the arithmetic averages. In order to simplify the mathematical derivations, it is assumed herein that chloride concentration deviations within the aquifer are homogeneous due to long-term mixture, and therefore, its deviations will be ignored ($Cl_{gw} = \overline{Cl_{gw}}$). The substitution of Equations 5.3 to 5.6 with this simplification into Equation 5.1 leads to

$$\overline{q} + q' = \frac{\left(\overline{R} + R'\right)\left(\overline{Cl_r} + Cl'_r\right)}{\overline{Cl_{gw}}} \tag{5.7}$$

The expansion of the parenthesis gives

$$\overline{q} + q' = \frac{\overline{R}\,\overline{Cl_r} + \overline{R}Cl'_r + R'\overline{Cl_r} + R'Cl'_r}{\overline{Cl_{gw}}} \tag{5.8}$$

If both sides of this expression are averaged arithmetically, keeping in mind that by definition the perturbation terms have zero arithmetic averages, one can simplify it to

$$\overline{q} = \frac{\overline{RCl_r} + \overline{R'Cl_r'}}{\overline{Cl}_{gw}} \qquad (5.9)$$

The second term in the numerator indicates the average of rainfall perturbation multiplication by chloride concentration perturbation, which might be thought of as similar to the weighted average but does not lead to the same result. In statistics literature, the arithmetic average of multiplication of two variables is defined as the covariance. By definition, covariance is equal to the multiplication of the correlation coefficient between the two variables by their standard deviations. Hence, it is possible to rewrite the final equation as

$$\overline{q} = \frac{\overline{R}\,\overline{Cl_r} + \hat{\rho}_{RCl_r}\hat{\sigma}_R\hat{\sigma}_{Cl_r}}{\overline{Cl}_{gw}} \qquad (5.10)$$

where $\hat{\rho}_{RCl_r}$ is the correlation coefficient between the rainfall and its chloride concentration, $\hat{\sigma}_R$ and $\hat{\sigma}_{Cl_r}$ are the standard deviations of rainfall and its chloride concentration measurements. This last expression is the refined CMB equation, which reduces to Equation 5.2 under either of the following two circumstances:

1. The correlation coefficient is equal to zero, which implies independence of chloride concentration from the rainfall amount.
2. One of the standard deviations is equal to zero. In this case, the variable with zero standard deviation implies homogeneity (temporal or spatial constancy or both).

In practical applications, the correlation coefficient between the rainfall and its chloride concentration could be a positive value, depending on the sources and seasonal rainfall. Perhaps, this is physically plausible, because the more the rainfall, the bigger the chloride concentration. Table 5.1 indicates the rainfall and corresponding chloride concentrations of four sites during 5 months in Taiwan as presented by Ting et al. (1998).

**TABLE 5.1**
**Rainfall and Chloride Measurements in Taiwan**

| Month | Site I Rainfall (mm) | Site I Cl (mg/l) | Site II Rainfall (mm) | Site II Cl (mg/l) | Site III Rainfall (mm) | Site III Cl (mg/l) | Site IV Rainfall (mm) | Site IV Cl (mg/l) |
|---|---|---|---|---|---|---|---|---|
| May | 393 | 1.88 | 210 | 0.99 | 200 | 1.07 | 175 | 0.82 |
| June | 247 | 0.39 | 292 | 0.47 | 328 | 1.16 | 336 | 1.22 |
| July | 307 | 0.95 | 338 | 1.32 | 372 | 1.04 | 354 | 0.74 |
| August | 915 | 0.57 | 943 | 1.16 | 1,073 | 3.06 | 1,090 | 3.71 |
| September | 186 | 0.71 | 193 | 0.83 | 145 | 1.28 | 138 | 4.46 |
| Correlated coefficient | **−0.092** | | **0.39** | | **0.95** | | **0.33** | |

*Source:* From Ting, C.S. et al. 1998. *Hydrogeol. J.*, 6:282–292. With permission.

It is obvious that some sites have very significant correlation coefficients between the rainfall amounts and their chloride concentrations. It is possible to rewrite the classical CMB equation by considering Equations 5.2 and 5.10 as

$$\overline{q} = \frac{\overline{q}_c + \hat{\rho}_{RCl_r} \hat{\sigma}_R \hat{\sigma}_{Cl_r}}{Cl_{gw}} \tag{5.11}$$

where $\overline{q}_c$ is the *average recharge rate* according to the classical CMB method.

## 5.5.3 APPLICATION

In order to show the performance of the refined CMB method, data from Wadi Yalamlam in the western part of Saudi Arabia is considered (Subyani, 2005). It is one of the important wadis in the western part of Saudi Arabia and lies about 125 km southeast of the city of Jeddah and 70 km south of Makkah (Figure 5.2). It flows into the Red Sea coastal plain called *Tihamah*.

Wadi Yalamlam is bounded by latitudes 20° 30' and 21° 10' N and longitudes 39° 45' and 40° 30' E. This wadi is a part of the Scarp-Hijaz Mountains of the Arabian Shield, which extends from north to south parallel to the Red Sea. This escarpment is one of the outstanding landscape features of Saudi Arabia. Three physiographic units, namely, the Red Sea coastal plain, the hills, and the Scarp-Hijaz Mountains, characterize the area. This basin drains a wide catchment area of about 1,600 km² which starts from the Scarp-Hijaz mountains (*escarpment*) and is characterized by high annual rainfall, exceeding 200 mm. Toward the drainage opening on the plain, the wadi loses its defined course and becomes wide spans of sheet wash, while further downstream, it is integrated as part of the Red Sea coastal plain (El-Khatib, 1980; Noory, 1983). The hydrology and precipitation features of the region are identified extensively by Şen (1983), Alehaideb (1985), and Al-Yamani and Şen (1992).

In order to perform the application of refined CBM as presented by Equation 5.10, mean monthly rainfall records (1981–2000) and rainfall chloride concentration data are collected from Ashafa station, which lies on the Red Sea escarpment, at the upstream of Wadi Yalamlam. This station is the most representative one for rainfall and subsequent runoff event evaluations near the study area. Rainfall in arid regions is highly variable with high evapotranspiration. The noneffective months (June and July) with low rainfall amounts are not considered in this study (Table 5.2). Groundwater samples for average chloride concentration are also taken from the wells that are upstream of the Wadi Yalamlam basin. In addition, there are no irrigation effects in this area. All the needed data for the application of refined CMB method are given in Table 5.2.

It is obvious from this table that the harmonic mean concentration of chlorides in the rainfall is $\overline{Cl}_r = 8.16$ mg/L. Arithmetic, harmonic, and weighted means are not significantly different. The harmonic mean is used due to its value between the two other averages, which makes the harmonic means more representative. The average concentration of chloride during the same period is calculated as $\overline{Cl}_{gw} = 80$ mg/L. The rainfall chloride concentration standard deviation is 0.96 mg/L with the standard deviation of monthly rainfall as 6.2 mm, in addition to the monthly harmonic mean

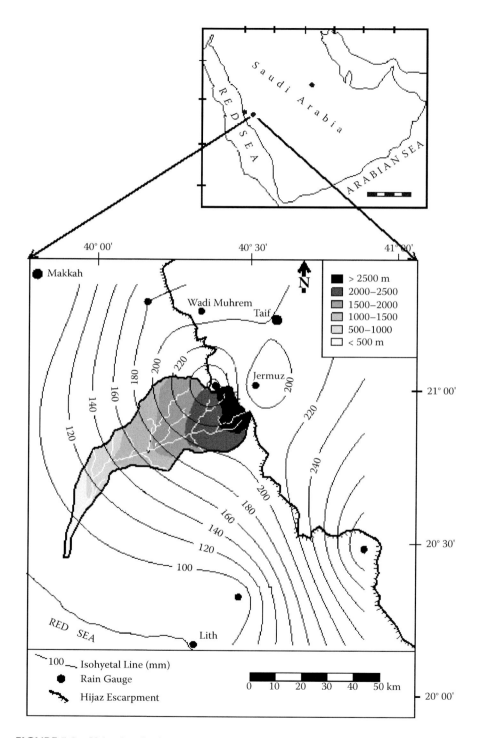

**FIGURE 5.2**  Yalamlam basin topography and annual rainfall distribution.

**TABLE 5.2**
**Statistical Summary of Mean Monthly Rainfall and Rainfall Chloride Concentration**

| Month | Mean Monthly Rainfall (mm) | Chloride Concentration (mg/l) |
|---|---|---|
| January | 19.8 | 7.5 |
| February | 13.4 | 7.0 |
| March | 21.9 | 8.0 |
| April | 30.6 | 9.5 |
| May | 28.0 | 9.9 |
| August | 20.4 | 8.5 |
| September | 12.3 | 7.8 |
| October | 14.0 | 8.1 |
| November | 17.0 | 9.0 |
| December | 15.4 | 7.3 |
| *Mean* | **19.28** | **8.26** |
| *Harmonic mean* | **17.71** | **8.16** |
| *Geometric mean* | **18.45** | **8.21** |
| *Standard deviation* | **6.18** | **0.96** |
| *Correlation coefficient* | | **0.77** |

*Note:* The noneffective months of June and July with low rainfall amounts are not considered in this study.

of rainfall as 17.7 mm. The estimated correlation coefficient is $\rho_{rCl_r} = 0.77$, which indicates that there is a significant dependence between the two variables. Substitution of all the relevant values into Equation 5.10 gives 1.9 mm/month, and it is possible to conclude that the refined approach results in 11% of the recharge rate. The difference between the classical and refined approaches increases with the increase in the standard deviations of monthly rainfall and/or rainfall concentration.

## 5.5.4 RECHARGE OUTCROP RELATION (ROR)

In extremely arid regions of the world, the recharge from surface flow to aquifers is very small, if it exists at all. Therefore, every single drop of water is very precious for groundwater storage. Consequently, the recharge calculations in these regions pose special problems, which should be solved by refined techniques suitable for the prevailing conditions in the environment. Among these conditions is the sporadic areal distribution of a rainfall event, its rare temporal occurrence, high intensity, and a barren earth surface with virtually no plant cover. The central Arabian Peninsula is located in an extremely arid zone belt of the world with the characteristics of very little and unpredictable amounts of irregularly occurring rainfall (Şen, 1983). Rainfall happens usually during a period from November to May. In addition, spatially haphazard variations of infiltration and porosity properties, in extremely arid regions, make the estimation of total recharge to a deep aquifer system rather difficult. Basic

## TABLE 5.3
## Recharge Rates from Different Studies

| Study | Method | Recharge Rate (mm/year) |
|---|---|---|
| Sogreah (1968) | Hydroclimatological data | 3.0–5.0 |
| Dincer et al. (1974) | Isotope | 20.0 |
| S.M.M.P. (1975) | Piezometry and transmissivity | 5.2 |
| B.R.G.M. (1976) | Hydroclimatological data | 6.5 |
| Caro and Eagleson (1981) | Dynamic model | 6.0 |
| This study | Recharge outcrop relation | 4.0 |

evaporation rates are high, which lead to an excess loss of water whenever it is available at the surface (Salih and Şendil, 1984).

There are studies for estimating the recharge contribution to the Wasia and Biyadh (WB) aquifers (Subyani and Şen, 1991). A brief summary of these studies is presented in Table 5.3. Sogreah (1968) estimated very roughly the recharge from the available hydroclimatological data. The average amount of rainfall likely to percolate deeply enough to contribute to groundwater storage is found to be about 3 to 5 mm/year on the average. Sogreah (1968) has studied to some extent piezometric gradient, aquifer thickness, and permeabilities, which suggest that some $11 \times 10^6$ m$^3$/year of recharge takes place within Wadi As-Sahba only.

On the other hand, two distinctive methods are proposed by Dincer et al. (1974) in calculating the recharge amount that goes through sand dunes around the Khurais area near the city of Riyadh. One of these methods depends on temperature, grain size, and sand moisture measurements, and shows that infiltration from rainfall in the sand dunes is a complex phenomenon. Another method is based on the thermonuclear tritium content of the sand moisture, which neglects all of the previous physical properties. On the basis of these methods, one can conclude that the recharge is about 20 mm/year. Furthermore, S.M.M.P. (1975) estimated the recharge through the WB formations based on piezometry as well as transmissivity, which vary areally, and accordingly calculated the average recharge value as 5.2 mm/year. B.R.G.M. (1976) concluded that their study of recharge depends on the hydroclimatological data, which were preprocessed for the purpose of numerical model simulations, leading to the estimation of an annual recharge amount of about 6.5 mm/year. However, Caro and Eagleson (1981) estimated median annual recharge by using a dynamic model of annual water balance with the maximum depth of about 6 mm/year. At the end, they reached the conclusion that the WB aquifer is being mined due to excessive pumping.

These studies missed some common points as the data is not enough to give a clear picture of the recharge phenomenon. For example, the Sogreah (1968) study depends on very short records of hydroclimatological data (about 5 years). Dincer et al. (1974) measured the recharge rate from the isotope technique, which yielded reliable answers about the groundwater mixture, age, and recharge possibility only. However, it fails to provide reliable numerical answers for groundwater recharge amounts. The S.M.M.P. (1975) study is more precise, especially around unconfined portions.

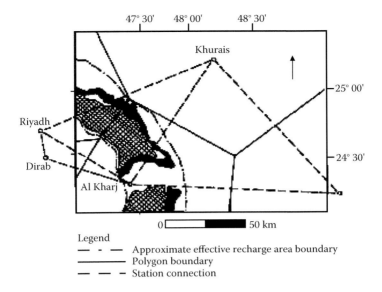

**FIGURE 5.3**    Thiessen polygon subareas.

B.R.G.M. (1976) concluded from long period data and model simulation an opti-
mized value, which did not consider the outcrop of geological formations. Besides, it
requires digital computer usage for calculations, which might not be practical.

Subyani and Şen (1991) proposed a recharge outcrop relation (ROR) technique,
which combines the geological lithology with a water budget. It provides a simple
way of calculating mean monthly recharge from daily rainfall amounts. In the study
area, there are four climatological stations for rainfall measurements, located at
Riyadh, Al-Kharj, Dirab, and Khurais, as shown in Figure 5.3. In order to apply the
ROR method, the study area is divided by using the Thiessen polygon technique into
representative subareas for each station. Hence, each subarea will be referred to by
the name of a station within the subarea.

An outcrop of WB falls under Riyadh and Al-Kharj subareas only. This means
that the recharge of these aquifers is direct, whereas the recharge from Dirab the
subarea is indirect, i.e., prior to recharge there is runoff. Tables 5.4 and 5.5 show the
mean monthly recharge rates (7 months) in Riyadh and Al-Kharj stations.

The second column in these tables is the mean monthly values of actual evapotranspiration for the same daily rainfall period, depending on the equation for estimation of the actual evapotranspiration, $(ETR)_{act}$, as

$$(ETR)_{act} = 1.16 \, (ETR)_{J-H} - 0.37 \tag{5.12}$$

where $(ETR)_{J-H}$ is the evapotranspiration according to the Jensen–Haise method
which was recommended by Salih and Şendil (1984) for central Arabian Peninsula conditions. To calculate the lake (evapotranspiration) ETR, the observed ETR
records are multiplied by a pan coefficient, which is equal to about 0.7 for arid zones
(Linsley et al., 1975). The results are presented in Tables 5.4 and 5.5 in column 4.

**TABLE 5.4**

**Monthly Recharge Rates for Riyadh Area**

| Months with Rainy Days (1) | Mean Rainfall (mm) (2) | Mean Actual ETR (mm) (3) | Lake ETR = 0.7 ETR (mm) (4) | Runoff = 0.1 Rainfall (mm) (5) | Recharge Rate (mm) (6) |
|---|---|---|---|---|---|
| January | 10.2 | 5.6 | 3.6 | 1.0 | 5.6 |
| February | 10.3 | 5.0 | 3.5 | 1.0 | 5.7 |
| March | 12.6 | 19.7 | 22.5 | 15.0 | 0.0 |
| April | 11.5 | 17.3 | 12.1 | 1.1 | 0.0 |
| May | 4.7 | 10.0 | 7.0 | 0.5 | 0.0 |
| November | 2.3 | 2.3 | 1.6 | 0.23 | 0.5 |
| December | 5.0 | 3.7 | 2.6 | 0.5 | 0.8 |
| Total | | | | | 12.7 |

**TABLE 5.5**

**Monthly Recharge Rates for Al-Kharj Area**

| Months with Rainy Days (1) | Mean Rainfall (mm) (2) | Mean Actual ETR (mm) (3) | Lake ETR = 0.7 ETR (mm) (4) | Runoff = 0.1 Rainfall (mm) (5) | Recharge Rate (mm) (6) |
|---|---|---|---|---|---|
| January | 5.0 | 4.1 | 2.9 | 0.5 | 1.6 |
| February | 5.0 | 4.0 | 2.8 | 0.5 | 1.7 |
| March | 17.0 | 23.0 | 16.1 | 1.7 | 0.0 |
| April | 10.0 | 12.0 | 8.4 | 1.0 | 0.0 |
| May | 1.8 | 7.6 | 5.3 | 0.2 | 0.0 |
| November | 1.7 | 2.3 | 1.6 | 0.2 | 0.0 |
| December | 3.4 | 3.8 | 2.6 | 0.3 | 0.5 |
| Total | | | | | 3.8 |

The study area is devoid of runoff measurements, and therefore the runoff percentages are calculated empirically from rainfall for arid zones as 10% (Chow, 1964), which are shown in column 5. Finally, the mean monthly recharge is calculated by applying the following water balance expression:

$$\text{Recharge rate} = \text{rainfall} - (\text{runoff–lake ETR}) \qquad (5.13)$$

which leads to the results presented in column 6 of Tables 5.4 and 5.5. In order to calculate the recharge rates by the ROR method, first of all estimates of particular outcrop formation percentages within each subarea are calculated from respective polygons by using a planimeter. Subsequently, the multiplication of these percentages with the respective recharge gives the amount of water received by the WB outcrop as recharge. Details of the calculations are presented in Table 5.6.

**TABLE 5.6**
**ROR Calculations**

| Subarea | Total Area (km²) | Effective WB[a] Area (km²) | Percentage of WB Area (%) | Recharge (mm/year) | Recharge to WB Area (mm/year) |
|---|---|---|---|---|---|
| Riyadh | 2,710 | 1,924 | 71 | 12.6 | 9.0 |
| Al-Kharj | 9,250 | 4,350 | 49 | 3.8 | 1.8 |
| Weighted average (mm/year) | | | | | 4.0 |

[a] Wasia/Biyadh.

The recharge contribution areas to WB in Riyadh and Al-Kharj are 1,924 k² and 4,530 km², respectively. The weighted average recharge in Table 5.6 is calculated as,

$$r_w = \frac{r_R A_R + r_K A_K}{A_R + A_K} \tag{5.14}$$

where $r_R$ ($r_K$) and $A_A$ ($A_K$) are the recharge rate and area in Riyadh (Al-Kharj). Substitution of the necessary quantities from Table 5.6 into Equation 5.14 leads to

$$r_w = \frac{9 \times 1,920 + 1.8 \times 4,530}{1,920 + 4,530} = 4.0 \text{ mm/year}$$

This weighted average represents the recharge with the use of the ROR method. Comparison of this value with the results in Table 5.3 indicates the obvious overestimation of other techniques which are valid for humid regions only.

## 5.6 ENVIRONMENTAL ISOTOPES

Various researchers indicated that the environmental isotopes *oxygen-18* ($\delta^{18}O$) and *deuterium* ($\delta^2H$) can be used to know the altitude of recharge (Musgrove and Banner, 1993; Scholl et al., 1996). In isotope hydrology, the depletion of $\delta^{18}O$ and $\delta^2H$ with altitude has been used to show the groundwater source areas from different sources as well as the identification of a groundwater flow system. Because $\delta^2H$ is not affected by water–rock interaction, its use has been preferred over $\delta^{18}O$ by many researchers (Lyles and Hess, 1988; Kirk and Campana, 1990; Thomas et al., 1996). As stated by Clark and Fritz (1997) the altitude effect of $\delta^{18}O$ ranges generally between 0.1 and 0.5% for each 100 m gained, which is valid for humid regions.

In arid and semi-arid regions the isotopic composition is affected by other factors such as the rainfall amount, variation seasonality, and air temperature. Al-Yamani (2001) has shown that per each 100 m increase in the Red Sea area, the decrease is −0.08%. Jones et al. (2000) noticed that there are lower $\delta^{18}O$ values than with small rainfall events. On the other hand, Gat (1987) stated that the rainfall during the winter season is lighter in isotopes of $\delta^{18}O$ and $\delta D$ than during the summer season due to evapotranspiration. Hence, an isotopic signature provides evidence of the climatic conditions prevailing at the time of discharge. According to a study in the United States, paleo-groundwaters are characterized by a large depletion in $\delta^{18}O$ and $\delta^2H$ values with respect to modern meteoric water. Gat (1987) reported that the local runoff recharge shows higher $\delta^{18}O$ values than the average rainfall composition, whereas the groundwater recharge associated with heavy floods displays lower $\delta^{18}O$ values.

The isotopic variations in natural waters help to determine the ages, sources, mixtures, and water–rock and water–gas interactions. Analysis of $\delta^2H$ and $\delta^{18}O$ data also helps to know *focused* or *diffused* recharge where, in the former case, there will be a higher *isotopic gradient* away from the recharge area. Hence, samples from beneath the recharge are isotopically heavier than those collected away from it. It is assumed that the difference is a result of recharge temperature differences as a function of latitude and altitude, both of which decrease in the flow direction. In the case of diffuse recharge, the isotopes of the recharged water would mix with those of the regional flow, generating a low gradient in the isotopic ratios over the entire area (Wood and Sanford, 1995). Values of $\delta^2H$ and $\delta^{18}O$ can be used to date waters between 1,000 and 40,000 years old because long-term climatic changes are reflected in the values of these isotopes.

Due to chemical and physical changes in different isotopes of an element, the isotopic fractionations take place, and they are directly proportional to relative differences in their masses. The main environmental isotopes that are used in water studies are stable hydrogen (H), oxygen (O), carbon, (C), and sulphur (S). Their average abundances in percentages and half-lives are presented in Table 5.7.

In this table radioisotopes ($^3H$ and $^{14}C$) have very low abundance. The fractionation process is initiated by the large mass difference between heavy and light isotopes. The typical example of the fractionation is the evaporation process from the

## TABLE 5.7
## Environmental Isotope Features

| Element | Isotope | Atomic Mass | Average Natural Abundance (%) | Half-Life (year) |
|---------|---------|-------------|-------------------------------|------------------|
| H | $^1H$ = Protium | 1.00782 | 99.985 | Stable |
|   | $^2H$ = Deuterium | 2.0141 | 0.015 | Stable |
|   | $^3H$ = Tritium | 3.01603 | 0.00013 | 12.3 |
| O | $^{16}O$ | 15.9949 | 99.759 | Stable |
|   | $^{17}O$ | 16.9991 | 0.037 | Stable |
|   | $^{18}O$ | 17.9991 | 0.204 | Stable |
| C | $^{12}C$ | 12.0000 | 98.892 | Stable |
|   | $^{13}C$ | 13.0033 | 1.108 | Stable |
|   | $^{14}C$ | 14.0032 | $\sim 10^{-10}$ | $5730 \pm 40$ |
| S | $^{32}S$ | 31.9721 | 95.02 | Stable |
|   | $^{33}S$ | 32.9715 | 0.75 | Stable |
|   | $^{34}S$ | 33.9679 | 4.21 | Stable |
|   | $^{35}S$ | 35.9671 | 0.02 | Stable |

surface of oceans, which enriches the heavy isotopes and the diffusion out of isotopically light molecules from seawater. According to Fritz and Fontes (1980), there is a striking isotopic content uniformity ($\delta^{18}O = 0 \pm 1$ and $\delta^2D = 0 \pm 5$ per mil) within the ocean waters relative to the standard mean ocean water (SMOW) measure. Meteoric water is always depleted in $\delta^{18}O$ and $\delta^2D$ relative to sea water. According to Dansgaard (1964), the degree of depletion depends on different factors such as the altitude, latitude, temperature, amount (rainfall), and continental (distance from sea) effects. He found that

$$\delta^{18}O = 0.695T \, (^\circ C) - 13.6 \tag{5.15}$$

On the other hand, Craig (1961) suggested a good relationship between the isotopic composition of $\delta^{18}O$ and $\delta^2D$ as,

$$\delta^2H = 8\delta^{18}O + 0.001 \tag{5.16}$$

This expression is known as the average global meteoric water line (GMWL) or as meteoric water line (MWL). Later, Yurtsever (1975) used worldwide data for the possible relationships of these effects on $\delta^{18}O$. He found that the main effect was due to temperature variations without significant effects of other variations. He found the relationship between $\delta^{18}O$ and mean surface temperature as

$$\delta^{18}O = (0.521 \pm 0.014) \, T \, (^\circ C) - (14.96 \pm 0.21) \tag{5.17}$$

On the other hand, Yurtsever and Gat (1981) found a good linear relationship between the weighted $\delta^{18}O$ and $\delta^2D$ values in the precipitation samples of the International Atomic Energy Association (IAEA) network (website: www.iaea.org/programs/ri/gnip/gnipmain.htm) as

$$\delta^2H = (8.17 \pm 0.28) \, \delta^{18}O + (10.55 \pm 0.64)\% \tag{5.18}$$

with a correlation coefficient of 0.997. Rozanski et al. (1993) have updated the MWL by additional average precipitation data from the World Meteorological Organization (WMO) as

$$\delta^2H = 8.13 \, \delta^{18}O + 10.8\% \tag{5.19}$$

The slope and intercept of any local meteoric water line (LMWL) based on the rainfall samples from a single site or different sites can be significantly different from GMWL corresponding features. In general, most of these local lines have slopes as $8 \pm 0.5$ or as extremes from 5 to 9, depending on geographic and climatic parameters. The most fractionation effect is due to temperature in arid regions. Logically, the evaporated water or water mixed with already evaporated water plots below the GMWL that intersects the LMWL at the composition of the original water (before the evaporation) composition. Commonly, evaporation lines with slopes 2 to 5 are frequent. In the case of evaporated surface water the slopes go below 4, and the slope

**TABLE 5.8**
**Local Meteoric Water Line (LMWL) Expressions**

| Station | Regression Equation | Correlation Coefficient | Source |
|---|---|---|---|
| Vienna, Austria | $\delta^2H = 7.07\ \delta^{18}O - 1.38$ | 0.961 | 1 |
| Ottawa, Canada | $\delta^2H = 7.44\ \delta^{18}O + 5.01$ | 0.973 | 1 |
| Addis Ababa, Ethiopia | $\delta^2H = 6.95\ \delta^{18}O + 11.51$ | 0.918 | 1 |
| Tokyo, Japan | $\delta^2H = 6.87\ \delta^{18}O + 4.70$ | 0.835 | 1 |
| Arctic Ocean | $\delta^2H = 8.00\ \delta^{18}O + 0.00$ | — | 1 |
| Weathership E | $\delta^2H = 5.96\ \delta^{18}O + 2.99$ | 0.940 | 1 |
| Canadian MWL | $\delta^2H = 7.80\ \delta^{18}O + 10.00$ | — | 2 |
| United States MWL | $\delta^2H = 8.11\ \delta^{18}O + 8.99$ | 0.980 | 3 |
| East Mediterranean MWL | $\delta^2H = 8.00\ \delta^{18}O + 22.00$ | — | 4 |
| Jordan MWL | $\delta^2H = 6.53\ \delta^{18}O + 13.65$ | 0.930 | 5 |
| Global Meteoric Water Line (GMWL) | $\delta^2H = 8.13\ \delta^{18}O + 10.80$ | — | 6 |
| Global Meteoric Water Line (GMWL) | $\delta^2H = 8.00\ \delta^{18}O + 10$ | — | 7 |

[1]  Clark and Fritz (1997).
[2]  http://science.uwaterloo.ca/~twdedwar/cnip/cniphome.html.
[3]  http://pups.water.usgs.gov/ofr-00-160.
[4]  Leguy et al. (1983).
[5]  http://www.angelfire.com/wy2/bajjali/proj/jordan/proj5.htm.
[6]  Rozanski et al. (1993).
[7]  Craig (1961).

can be as low as 2 for soil water in the unsaturated zone. Hence, the groundwater subjected to evaporation can be identified on this basis. Some of the LMWL expressions for different parts of the world are given in Table 5.8.

The intercept value in the last equation of this table is 10, and it is referred to as the *deuterium excess* value for the equation, i.e., deuterium excess = $10 = \delta^2H - 8\ \delta^{18}O$. The precise value of deuterium excess depends on climatological and geographical conditions, and therefore each equation has different values. The global average of deuterium excess is equal to 10 with its typically varying values between 0 and 20. Especially, the effects on the sea surface on the deuterium excess are due to relative humidity, temperature, and wind speed variations. Thus, the global atmospheric moisture, which originates at the sea surface, admixes with or without the recycled continental vapor, distribute varying deuterium excess values of different regions. It also shows seasonal variations. Kondoh and Shimada (1997) found the higher values of deuterium excess in winter than summer, which is attributed to different moisture sources.

The application of isotope studies is achieved for the purpose of the groundwater recharge evaluation in the Kingdom of Saudi Arabia, and the results are presented in Table 5.9 (Bazuhair et al., 2002).

**TABLE 5.9**
**Isotope Equations**

| Wadi Name | Sample Type | Regression Equation | Correlation Coefficient |
|---|---|---|---|
| Aqiq | Rainfall | $\delta^2H = 5.64\ \delta^{18}O + 5.87$ | 0.99 |
| | Groundwater | $\delta^2H = 5.20\ \delta^{18}O + 3.32$ | 0.92 |
| **Khulays (major wadi)** | | | |
| Murawani | Rainfall | $\delta^2H = 1.68\ \delta^{18}O + 0.37$ | 0.52 |
| | Groundwater | $\delta^2H = 8.13\ \delta^{18}O + 10.8$ | N/A |
| Ghuran | Rainfall | $\delta^2H = 5.80\ \delta^{18}O + 6.92$ | 0.93 |
| | Groundwater | $\delta^2H = 5.39\ \delta^{18}O + 5.13$ | 0.90 |
| Wajj | Rainfall | $\delta^2H = 6.68\ \delta^{18}O + 7.80$ | 0.98 |
| | Groundwater | $\delta^2H = 6.81\ \delta^{18}O + 8.45$ | 0.92 |
| Turabah | Rainfall | $\delta^2H = 10.0\ \delta^{18}O + 15.67$ | 0.92 |
| | Groundwater | $\delta^2H = 3.96\ \delta^{18}O + 0.83$ | 0.90 |
| Abha | Rainfall | $\delta^2H = 5.85\ \delta^{18}O + 7.14$ | 0.55 |
| | Groundwater | $\delta^2H = 8.13\ \delta^{18}O + 10.8$ | N/A |
| Jizan | Rainfall | $\delta^2H = 7.83\ \delta^{18}O + 7.48$ | 0.99 |
| | Groundwater | $\delta^2H = 6.20\ \delta^{18}O + 5.11$ | 0.99 |
| **Regional** | **Rainfall** | $\delta^2H = 5.85\ \delta^{18}O + 6.43$ | **0.96** |
| | **Groundwater** | $\delta^2H = 5.84\ \delta^{18}O + 4.47$ | **0.94** |
| **Global** | **GMWL** | $\delta^2H = 8.13\ \delta^{18}O + 10.80$ | **0.95** |
| **Eastern Mediterranean** | **GMWL** | $\delta^2H = 8.00\ \delta^{18}O + 22.00$ | N/A |

## 5.7   DEPRESSION CONE

Prior to pumping from any aquifer, the *piezometric* or *phreatic surface* should be almost horizontal, i.e., zero hydraulic gradient (no groundwater movement). Right after the pump start, a decrease in pressure occurs in the vicinity of the well, giving rise to an increase in *hydraulic gradient* toward the well in a radial manner. During radial groundwater flow toward a well, the *hydraulic head* around the well takes the shape of an inverted cone with its apex at the pumping well and horizontal base on the original *piezometric surface*; this is called a *depression cone* (Figure 5.4).

The base area of the cone is called the area of influence, and its radius is the *radius of influence*, R. Within the area of influence, the vertical distance from the base to any hydraulic head position at time, t and radial distance, r, from the well center is *drawdown*, s(r,t). In a *confined aquifer* the depression cone is imaginary but in an unconfined aquifer it is real. If the material of the aquifer is homogeneous, uniform, and isotropic then the base area is circular. The depression cone goes on expanding indefinitely with the abstraction of water from the well if the withdrawal is not compensated by any recharge.

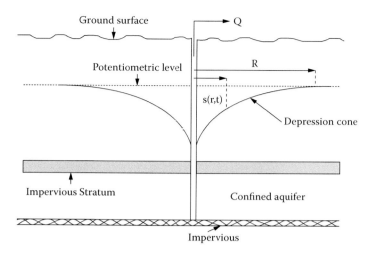

**FIGURE 5.4**   Depression cone.

The evolution of the depression cone depends upon a number of factors such as the aquifer type (confined, *unconfined*, or *leaky*), nature of the geological materials (homogeneous or heterogeneous, isotropic or anisotropic), and their geometrical relationships (occurrence of different hydraulic conductivity layers, impervious or recharge boundaries, presence of clay lenses in alluvial deposits, etc.). A depression cone possesses directional as well as temporal dynamicity. With time it expands or remains in steady state with constant discharge from the main pumping well and passes through different phases, depending upon the aforementioned reasons. The appropriate prediction of depression cone behavior in a pumping test leads one to decide whether the aquifer is confined, unconfined, leaky, *bounded*, or *layered*, and this identification helps in the determination of truly representative hydrological parameters such as *transmissivity*, hydraulic conductivity, *storativity*, leakage factor, etc.

On the other hand, in homogeneous, isotropic, and saturated medium, the hydraulic conductivity is defined as the volume of the groundwater that will move in unit time under unit hydraulic gradient through a unit cross-sectional area perpendicular to the streamlines. On the basis of such a definition, the hydraulic conductivity has a unit of volume per time per area.

## 5.8   DISCHARGE CALCULATION IN LARGE DIAMETER WELLS

During field studies in arid regions, a common difficulty in discharge measurement is that the leading pipe from the well either branches into many small pipes supplying water to different agricultural lands or to a common reservoir from which the water is distributed by means of many uncontrollable outputs. It is well known from experience that as the water is abstracted from a well, initially all the water comes from the well storage. During this period the relationship between the drawdown and time appears as a straight line. The physical consequences of this straight line

**TABLE 5.10**
**Five Different Large-Diameter Wells**

| Time Since Pump Start (min) | W1 $r_w = 1.3$ m | W2 $r_w = 0.98$ m | W3 $r_w = 1.08$ m | W4 $r_w = 1.19$ m | W5 $r_w = 0.95$ m |
|---|---|---|---|---|---|
| 1 | 0.08 | 0.08 | 0.12 | 0.05 | 0.12 |
| 2 | — | 0.21 | 0.19 | 0.10 | 0.28 |
| 3 | 0.24 | 0.32 | 0.29 | 0.15 | 0.44 |
| 4 | 0.31 | 0.45 | — | 0.20 | 0.60 |
| 5 | 0.40 | 0.55 | 0.51 | 0.26 | 0.72 |
| 6 | — | 0.70 | 0.62 | — | 0.86 |
| 7 | 0.56 | — | — | — | 0.99 |
| 8 | 0.66 | 0.89 | 0.81 | — | 1.10 |
| 9 | — | — | — | — | — |
| 10 | 0.79 | — | 0.96 | — | — |

are that the pumping discharge comes from the well storage contribution only. This relationship is presented by Şen (1986a) as

$$Q = \pi r_w^2 \frac{ds_w(t)}{dt} \qquad (5.20)$$

where t is the time since pump start and $r_w$ is the radius of the well, $ds_w(t)/dt$ is equal to the initial slope of the time drawdown curve. The initial drawdown measurements during a pumping test can be plotted against the corresponding times. The slope can be read off from this plot and, after the measurement of the well diameter, Equation 5.20 gives the discharge value. In the absence of direct discharge measurements, this method gives practical, reliable results.

This method is applied for five large-diameter wells (W1, W2, W3, W4, and W5) in one of the wadis. Quaternary alluvial deposits constitute the only potential aquifer in the wadi. The early drawdown measurements in these wells with their radii are given in Table 5.10 for 10 min. The plots of time versus drawdown for each well are shown in Figure 5.5, where the best straight lines are drawn to match the scatter points.

The calculated discharge and direct measurements in the field are presented in Table 5.11. Also, the relative errors are given in the same table, and each individual relative error is less than 10% with an average error of 3%.

## 5.9   AQUIFER TESTS

An aquifer test is the record of drawdown at a set of predetermined times in the main or preferably observation wells. The drawdown is created by pumping a constant (*aquifer test*) or a sequence of constants (*step drawdown test*) discharge in a controlled manner. The pump discharge and drawdown can be measured simultaneously at the surface. Aquifer test data cannot be regarded as a basis for mechanical and

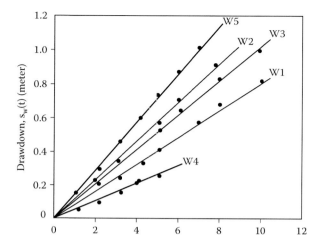

**FIGURE 5.5**   Linear time–drawdown relation for small times.

numerical hydrogeological parameter calculations through available strict methodologies only. The evaluation of an aquifer test data is more an art and needs philosophical thinking for proper interpretations. It is advised that the mechanical type curve matching in aquifer parameter evaluations should have a preliminary study by looking on the scatter diagram of time versus drawdown or distance versus drawdown data so as to depict the verbal (linguistic) features that can indicate the type of flow medium (Şen, 1995). Due to mechanical and numerical appreciation models, the majority of the groundwater specialists overlook the verbal interpretations. It also helps to identify which assumptions are valid from a bundle of basic assumptions in determining the aquifer parameters according to a convenient mathematical model. In order to have a philosophical evaluation, one should look at the scatter diagrams without any restrictive assumptions. In practice, aquifer data assessments can be achieved through the scatter diagrams on ordinary, semi-logarithmic, and double logarithmic papers. It is necessary that the analyst also know the distinction between these papers. For philosophical evaluation of the aquifer test, it is necessary that the researcher have logical rule bases, not equations, concerning the basic features of the aquifer composition. One should always keep in mind that the mathematical

## TABLE 5.11
## Comparison of Measured and Calculated Discharges

| Well No. | Discharge (m³/min) | | Relative Error (%) | Slope |
|---|---|---|---|---|
| | Measured | Calculated | | |
| W1 | 0.38 | 0.41 | 8 | 0.078 |
| W2 | 0.31 | 0.33 | 6 | 0.113 |
| W3 | 0.33 | 0.35 | 10 | 0.100 |
| W4 | 0.24 | 0.22 | −9 | 0.050 |
| W5 | 0.44 | 0.44 | 0 | 0.240 |

formulations and models require a set of simplifying assumptions, which should be questioned prior to any aquifer test calculations. A set of restrictive assumptions are given in many textbooks (Kruseman and De Ridder, 1970; Freeze and Cherry, 1989; Fetter, 1994; Şen, 1995).

Hydrogeologic parameter determination of the aquifers depends on various factors including the field setup of the well–aquifer configuration and groundwater flow regime. Different and complementary solutions are suggested for *steady state* (Theim, 1906; De Glee, 1930) and a set of *unsteady groundwater flow* conditions. One of the major factors is the material property of the aquifer as porous, fractured, or karstic media. Especially, porous medium solutions are abundant (Theis, 1935; Hantush, 1956; Boulton, 1954; Neuman and Witherspoon, 1968, 1969a,b, 1972; Witherspoon et al., 1968; Şen, 1985, 1989, 1990, 1996) for confined, unconfined, leaky and fractured aquifers. All these methodologies require a theoretical type curve and its matching to field data through a matching point; then, the classical aquifer parameter calculation equations are used for the numerical estimations. Except for leaky aquifers, many of these types of curves are thought to reach the same asymptote, which appears as a straight line on semi-logarithmic paper, with a horizontal logarithmic axis representing time and vertical ordinary axis drawdown.

Although the assumptions are necessary for rational analytical approaches, they are restrictions to philosophical thinking, which should take into account the heterogeneity, unisotropy, nonuniformity, and radial nonsymmetry. Hence, the analytical solutions are approximations to the real situation, and they provide the general trend solutions in the averages only.

It is almost a fixed rule that in porous and many fractured media, the late time drawdown data appears as a straight line on a semi-logarithmic paper, and therefore, the Jacob straight-line (Cooper and Jacob, 1946) method is applied without any further consideration. Such an operation implies philosophically that all straight-line appearance of late time data matches the late time portion of the Theis (1935) type curve, whereas this type curve has only one trace, and consequently it should represent a corresponding single straight line (Section 5.11). The question is then how can one use the Jacob approach for any straight-line occurrence on semi-logarithmic paper? Most often late time–drawdown data come along a straight line on a semi-logarithmic paper. The more the slope of this line, the smaller is the transmissivity. Hence, one can conclude verbally that the slope is inversely related to transmissivity. Other information that can be obtained from a straight line is the intercept on one of the two axes, and it is related to the storage coefficient of the aquifer.

Good test data are unavailable for the vast majority of the world's wells. Even where time-drawdown data are available, the nonequilibrium methods involving type curve matching or fitting straight lines are often misapplied. It is also possible that the wide availability of software for interpreting pumping test data, enabling automatic fitting of straight lines or type curve matching hinder proper thought in analysis. Most often engineering solutions are obtained from a set of analytically derived formulations.

Even where good test data are available, there is a tendency among many hydrogeologists to adopt a ready solution with classical methods, almost without any regard to the assumptions of the formulations. Thus, in the Jacob analysis, it is not uncommon to encounter a straight line drawn as a best fit through a semi-logarithmic data scatter, with no discussion of the reasons for possible variances between theory and the conditions in the field. Such approaches may lead to large errors in aquifer parameter estimations. It is the duty of the aquifer test data analyst to correlate all pertinent information and provide a realistic picture of the system. It is safe to say that aquifer test data must be analyzed with careful considerations of the geological composition around the well.

Prior to type curve matching or straight-line (Jacob) method application to available data, is it possible to deduce from the scatter of time–drawdown or distance–drawdown data some clues as to the type of aquifer medium (porous, fractured, or karstic)? If porous, is it fine or coarse material? If fractured, is it double porosity or a single fracture? Horizontal or vertical? Extensive or limited, with big aperture or small? Is the fracture surface smooth or rough?

It is advised that after the collection of field data, time–drawdown and distance–drawdown scatter diagrams must be examined with a critical view so that the aquifer features around the well can be mimicked. This will help groundwater hydrologists to augment their expertise about the aquifer tests with an accumulation of experience that may not be easily found elsewhere.

### 5.9.1 AQUIFER TEST PITFALLS

Aquifer parameter determination in practical hydrogeological studies exposes one of the most significant steps in groundwater resources assessment and *management* studies. Successful parameter estimation necessitates both a reliable methodology and field data. There are different analytical solutions in the forms of *type curves*, depending on well and aquifer configuration such as porous or fractured media; confined, unconfined, and leaky aquifers; small and large diameter wells, etc.

A pitfall may be defined as the taking of a false logical path that may lead to absurd conclusions, a hidden mistake capable of destroying the validity of an entire argument. It is, in fact, a conceptual error into which analysts frequently and easily fall, if they do not have sound plausible physical reasoning about the concerned phenomenon. Unfortunately, in many parts of the world, aquifer test analysts may overlook any geological, physical, or configuration prospects of the test environment, but rather rely on mechanically matched type curves or straight lines for the parameter estimations. Such a mathematical treatment of time–drawdown data might lead easily to unrealistic and unrepresentative aquifer parameter estimations and consequently to over- or underestimations of groundwater potentiality in an area. Many of the analytical models may not even have examples of actual data to fit. On the other hand, the multiplicity of such models gives rise to similarities that may exist among the performances of completely different systems.

In order to reach a reliable identification of aquifer parameter estimations, it is useful to be aware of pitfall types that might occur either before the beginning of actual field data collection, during the data collection, or after the type curve matching. The following points are among the pitfalls in an aquifer test:

1. Unfamiliarity not only with different type curves, but with the validity of underlying assumptions about the data at hand and the field environment.
2. Data collection started by pumping before the confirmation of a steady-state flow condition.
3. Other plausible explanations not considered regarding possible outcomes.
4. Overdependence on statistical significance, particularly at the expense of practical significance.
5. Focusing only on the overall (average) results and not considering the local changes.
6. Not indicating the assumptions, uncertainties, and other limitations of the evaluation when presenting the findings.
7. Poor presentation of findings so that potential users cannot really understand them.
8. Late time–drawdown and distance–drawdown plots may not necessarily mean that the Jacob method is applicable.
9. Overconfidence in the literature. It may be stated that, actually, the data lie along straight lines only for sufficiently large time or sufficiently small distances (Bear, 1972). The second part of this statement is not plausible physically because the smaller the distance, the greater the hydraulic gradient and, consequently, there is a possibility of nonlinear flow for which the definition of transmissivity is not valid (Şen, 2000). On the other hand, most of the line sinks or source solutions assume that the well diameter is equal to zero! Consequently, theoretical drawdown in such a well is extremely big. However, in physical reality the drawdowns are finite, hence theoretical drawdown cannot be compared with actual ones in the well vicinity.
10. Not distinguishing the physical reality, for example, if leaky aquifer curves are similar to exponentially decreasing discharge type curves (Şen et al., 2003).

## 5.10   AQUIFER HETEROGENEITY AND METHODS

Among the classical Theis (1935) assumptions is the large-scale homogeneity, which is used in most aquifer test analyses and interpretations with various corrections, because it cannot account for the behavior of time–drawdown variation plots (Meier et al. 1998). In order to alleviate the assumption of homogeneity in the Theis-type curve and Jacob straight line methods, other analytical solutions assume that the aquifer can be subdivided into, at most, two or three regions with uniform parameters in each (Streltsova, 1988; Butler, 1988, 1990; Butler and Liu, 1991, 1993). When the aquifer parameters depend on radial distance (axial symmetry), changes in the aquifer parameters reflect aquifer properties only within a ring of depression cone influence through which the front of the pressure depression passes within the considered time interval. Therefore, estimated transmissivity, T, and storativity, S, coefficients are dependent on the aquifer material properties (geological composition) between the inner radius of the ring and the pumping well. Furthermore, S estimates are dependent on the variations in T between the pumping well and the front of the cone of depression (Butler, 1988). Butler and Liu (1993) conceptualized the non-uniform aquifer as a uniform matrix into which a disk of anomalous properties has been placed. Among other things, they concluded that Jacob's method can be used in any laterally nonuniform system to estimate matrix transmissivity if the flow to the pumping well is approximately radial during the period of analysis (Sanchez-Vila et al., 1999).

It is possible to observe in practice that most often the Jacob straight-line method applications lead to constant T estimates, whereas the S estimates expose great spatial variability (Schad and Teutsch, 1994; Herweijer and Young, 1991). Suggests that the actual aquifer material T is strongly heterogeneous. S depends on the formation porosity in addition to rock and water compressibility. There is a paradoxical point in the statement that long-term aquifer (time–drawdown) tests lead to the small variability of T, and large variability of S, whereas the opposite should be expected for point values of T and S. Meier et al. (1998) conjectured that this paradox could be attributed to the fact that methods developed for homogeneous media are being used for interpretation of tests performed in heterogeneous formations.

The time–drawdown field records from a well include spatial variations within the geological formation around the well due to depression cone expansion up to about 400 to 500 m distance. Hydrogeologists and groundwater engineers apply the classical Theis or any other type curve or Jacob straight-line methods through matching of field data to available type curves for estimating the aquifer parameters. Unfortunately, the local deviations from the type curve or straight-line fittings are overlooked entirely, and hence there are unique T and S values. This complies with the basic assumption of the type curve models that the aquifer is homogeneous and extensively large with uniform thickness. In fact, each deviation from the type curve implies a local heterogeneity, and this is the point taken in this section for the aquifer parameters evaluation. The Şen (1986b) slope matching procedure (SMP) is revised and applied to arid-region aquifer test measurements (time–drawdown records) for estimating series of T and S values. Such an approach provides T and S estimates right from the early time drawdown measurements, and on the average these estimates yield the same results with the classical methodologies, provided that the aquifer is homogeneous and isotropic. There are other methods of analyzing an aquifer test in radially symmetrical but heterogeneous aquifers, namely, the use of a radial flow numerical model, such as presented by Rushton and Redshaw (1979).

### 5.10.1 Slope Matching Procedure

One of the ways to know any possible heterogeneity within an aquifer is through the SMP application (Şen, 1986b). In nature, aquifers are not homogeneous, isotropic, uniform, and extensive. However, these are the fundamental assumptions in any mathematical modeling of groundwater movement toward the wells. Unfortunately, under these assumptions the aquifer is idealized and, consequently, there are single estimation values for each aquifer parameter. The expansion of the depression cone is not considered at all in the calculations and interpretations. The SNMP provides a sequence of aquifer parameters that reflect the possible heterogeneity within the aquifer. In this manner, it is possible to make statistical assessment of the aquifer parameters with confidence, and likewise risk assessments can also be suggested.

The aquifer parameters are dependent on the geological characteristics, which are controlled by the evolution of rocks (sedimentation, volcanic eruptions, etc.) and by subsequent secondary geological events such as folding, faulting, fracturing, and fissuring, in addition to chemical changes, especially in the limestone aquifers. Although the conventional aquifer tests tend to average these conditions in the aquifer response to pumping, the field test data will still have some local deviations from any analytically derived type curve. In the conventional type curve matching procedures, the aquifer is assumed to represent an equivalent homogeneous medium, and consequently the parameters are expected to be temporally and spatially constants. However, due to deviations, one should physically expect some variations in these parameters. The main objective of the SMP is to identify the likely variations in aquifer parameters by matching slopes of the type curve and field data. Şen (1986b) gave the analytical slope, $\alpha$, expression for the Theis (1935) curve on a double logarithmic paper as,

$$\alpha = -\frac{e^{-u}}{W(u)} \tag{5.21}$$

This expression helps to convert the Theis type curve table into slope values as presented in Table 5.12.

By considering this table, the processing of the aquifer test data can be achieved through the following steps without type curve matching.

## TABLE 5.12
## Slope Matching Values

| u | 1 | 2 | 3 | 4 | 5 | 6 | 7 | 8 | 9 |
|---|---|---|---|---|---|---|---|---|---|
| x10-0 | -1.6798 | -2.7619 | -3.8298 | -4.8199 | -6.1254 | -6.8854 | -7.5990 | -8.8279 | -10.2841 |
| x10-1 | -0.4971 | 0.6711 | -0.8141 | -0.9576 | -1.0831 | -1.2196 | -1.3421 | -1.4494 | -1.5637 |
| x10-2 | -0.2451 | -0.2929 | -0.3278 | -0.3585 | -0.3851 | -0.4095 | -0.4337 | -0.4547 | -0.4760 |
| x10-3 | -0.1578 | -0.1769 | -0.1906 | -0.2014 | -0.2104 | -0.2189 | -0.2262 | -0.2329 | -0.2394 |
| x10-4 | -0.1159 | -0.1259 | -0.1322 | -0.1379 | -0.1424 | -0.1461 | -0.1494 | -0.1525 | -0.1551 |
| x10-5 | -0.0914 | -0.0976 | -0.1016 | -0.1047 | -0.1072 | -0.1094 | -0.1112 | -0.1128 | -0.1144 |
| x10-6 | -0.0755 | -0.0797 | -0.0824 | -0.0844 | -0.0859 | -0.0873 | -0.0886 | -0.0896 | -0.0906 |
| x10-7 | -0.0643 | -0.0673 | -0.0692 | -0.0707 | -0.0728 | -0.0727 | -0.0735 | -0.0743 | -0.0745 |
| x10-8 | -0.0560 | -0.0683 | -0.0597 | -0.0607 | -0.0616 | -0.0623 | -0.0629 | -0.0634 | -0.0640 |
| x10-9 | -0.0496 | -0.0514 | -0.0524 | -0.0533 | -0.0539 | -0.0545 | -0.0549 | -0.0553 | -0.0557 |
| x10-10 | -0.0445 | -0.0459 | -0.0468 | -0.0475 | 0.0480 | -0.0484 | -0.0488 | -0.0491 | -0.0494 |

1. After the second time–drawdown measurement, calculate the slope between the two successive points in the double-logarithmic scale as

$$\alpha_i = Ln(s_i/s_{i-1})/Ln(t_{i-1}/t_i) \qquad (5.22)$$

where $i = 2,3, ..., n$ and $n$ is the number of drawdown records.

2. Find the $u_i$ value corresponding to this slope from Table 5.12, and, if needed, interpolate.

3. Knowing $\alpha_i$ and $u_i$ values, find the well function from Equation 5.21 as

$$W_i(u) = \frac{e^{-u_i}}{\alpha_i} \qquad (5.23)$$

4. Calculate local $T_i$ and $S_i$ values from the well function and dimensionless time factor definitions (Theis, 1935) as follows:

$$T_i = \frac{QW_i(u_i)}{4\pi s_i} \qquad (5.24)$$

and

$$S_i = \frac{4t_i T_i u_i}{r^2} \qquad (5.25)$$

where $t_i$ and $s_i$ are the time and drawdown measurements from the field.

5. Repeat the previous steps as the new time–drawdown measurements are available. Finally, sequences of estimation are obtained for each aquifer parameter.

## 5.10.2 FIELD APPLICATIONS

The application of this method is performed for aquifer tests in Wadi Na'man, Kingdom of Saudi Arabia, by Şen and Wagdani (2007) as shown in Figure 5.6. It extends eastward from the Red Sea coast at longitude 40° 30' E to the summit of the Hijaz escarpment at longitude 39° 00' E. It thus embraces mainly the western slopes of the central part of the escarpment highlands (about 2,000 m elevation above mean sea level [msl]) with an area of about 2,080 km².

The natural alluvial deposits are composed of gravelly coarse sand and contain boulders of significant sizes. Sieve analyses have shown a composition of the alluvial with 50 to 70% sand, 20 to 30% gravel, and 5% silt. The material is derived from quartz-diorite, migmatite, amphibolite, diorite, dolerite, and hornblendite, i.e., mostly local basement origin.

The groundwater from Wadi Na'man through wells comes from both alluvium and fractured bedrock. The hydraulic behavior of the wells is a combined effect of these formations, namely, porous alluvium at the top, fractured layer at the bottom and weathered zone in between. The classical calculations (Theis type curve matching) indicate that the combination of the alluvium aquifer has hydraulic conductivity equal to $5.0 \times 10^{-4}$ m/sec, whereas the same for the fractured media is $1.6 \times 10^{-4}$

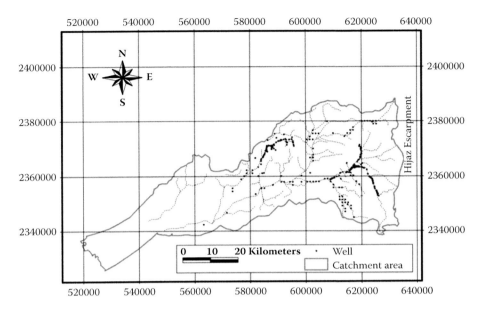

**FIGURE 5.6**   Wadi Na'man and well locations (Courtesy of the Saudi Geological Survey).

m/sec. It is understood from the extensive studies by RSMC (1986) that the rock conductivity decreases with increasing T in the horizontal direction and becomes impervious when T approaches $3.3 \times 10^{-3}$ m$^2$/sec. The alluvium and the underlying fractured zone have a much higher hydraulic conductivity than the underlying basement rock. When the fractured zone is considered in the analysis, the hydraulic conductivity of the rock mass decreases and the effective conductivity of the rock will be in the same order of magnitude as recorded from the hydraulic tests of the boreholes in the Wadi Na'man area.

The RSMC (1986) aquifer test data are reevaluated with the hope of obtaining more detailed numerical and linguistic interpretations about the hydrogeological setup in Wadi Na'man. The following points are the most significant steps in calculations and interpretations by the SMP.

1. Neither the porous aquifers nor the fractured aquifers are assumed as homogeneous and isotropic during the whole aquifer test evaluation.
2. The parameters are assumed to change from one reading to the next. This means to say that the aquifer is considered as a heterogeneous medium, which affects the time–drawdown behavior of the water level change due to pumping.
3. The statistical parameters of each hydrogeological parameter (storativity and transmissivity) are presented in addition to the relative frequency diagrams (histograms).
4. The percentages of different parameter values are calculated so as to reflect the aquifer composition.

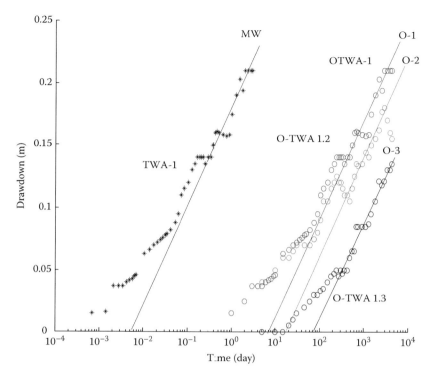

**FIGURE 5.7** Aquifer test data on semi-logarithmic paper.

There are many well bores with aquifer tests in Wadi Na'man but only one of them is selected for the detailed application and presentation of the SMP. The selected main (pumping) well, MW, lies at the downstream part of Wadi Na'man with three observation wells, O-1, O-2, and O-3. The plot of all the data from the main and observation wells are presented in Figures 5.7 and 5.8 on semi-logarithmic and double-logarithmic papers, respectively. In the same figures, Jacob straight lines and Theis type curve are also presented for comparison purposes.

RSMC (1986) used Theis type curve and Jacob straight-line methods for the aquifer parameter estimation as shown in Figure 5.7, but without any visual inspection concerning time–drawdown scatter. Prior to the application of any method, the following linguistic information can be deduced from Figure 5.7:

1. The aquifer around the well location has a general tendency toward homogeneity with significant local deviations along each test data. For instance, prior to about 0.01 days, the drawdown in the main well has two jumps, which indicate that with the expansion of depression cone, the groundwater flow toward the well enters a comparatively lower permeability region, which has either a very fine grained patch or silty zone. Such a situation does not occur with observation well data, and hence the high permeability zone is local and not extensive.

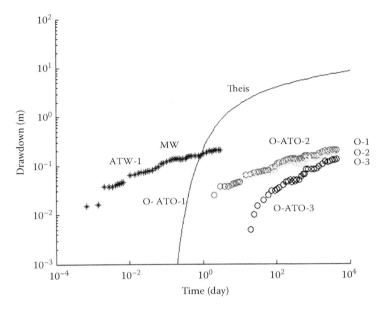

**FIGURE 5.8** Aquifer test data on double-logarithmic paper.

2. The aquifer material is not isotropic because the behavior of drawdown in each observation well at different directions is different from each other. For instance, in O-2 there are significant drawdown reductions during the aquifer test. Such incidences are as a result of internal groundwater recharge due to the local water bodies within comparatively high permeability regions of the same aquifer, which might be either coarse grained or fractured zones.

3. In spite of the fact that the distance of observation well O-2 is more than twice the distance from the main well compared to O-1, their responses fall almost over each other especially at the early and moderate times. This is the indication that the aquifer portion between these observation wells has more or less the same properties up to a certain distance from the main well.

4. Sudden drawdown decrease in O-2 data at large times is due to some additional groundwater flow into the well.

5. In each time–drawdown data there are local horizontal portions, which are evidences of leakage from different limited extended layers. This is the evidence that there is local horizontal layering within the aquifer.

On the other hand, according to the double-logarithmic plot in Figure 5.8 none of the data give the impression of the Theis type curve tendency, except in O-3. Others are more or less as a straight line at moderate and large time durations. Sudden drawdown increment appears in the main well, and the first two readings are almost at the same level, which imply the possibility of high rate groundwater entrance into the well. This is a good indication that the transmissivity of this well location is relatively high.

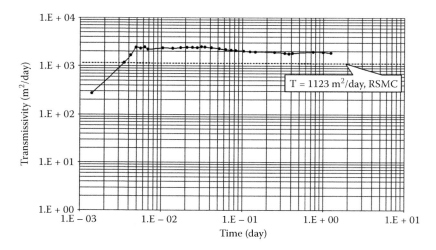

**FIGURE 5.9**   Successive transmissivity values for O-1.

Although Jacob straight-line and Theis type curve methodologies are applied for the aquifer data assessment (RSMC, 1986), both approaches do not take into consideration local deviations, but consider the data along a line (late times) or curve (early and moderate times) as homogeneous and isotropic media, which are against what the field data present. After the application of the SMP, Figures 5.9 to 5.11 indicate the successive averages of T and S variations, and the T–S relationship for O-1, respectively. On the relevant figures, the classical method results as obtained by RSMC (1986) are also shown for comparison purposes. It is obvious that the aquifer parameters vary around the constant classical estimations. As the depression cone expands by time, it covers successively bigger portions of the aquifer around the main well where the geological composition is not homogeneous.

Figures 5.9 and 5.10 indicate the successive advancement of the depression cone and its average cumulative successive transmissivity and storage coefficient values by time, respectively.

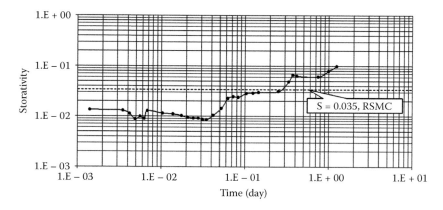

**FIGURE 5.10**   Successive storativity values for O-1.

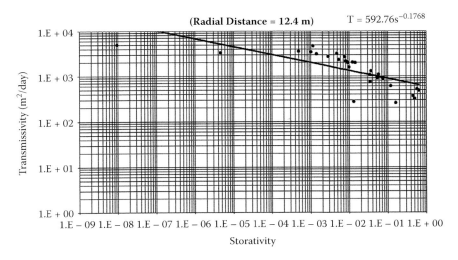

**FIGURE 5.11** Transmissivity–storativity relationship for O-1.

The successive transmissivity values become more stable. Theoretically, if the aquifer material is homogeneous and isotropic as the RSMC (1986) assumes, then the late time (long distance) successive transmissivity average should be expected to approach the classical methodology result. However, in Figure 5.9 this is not the case, but after a long time duration, the transmissivity value stabilizes near 2000 m$^2$/d, which is higher than what RSMC calculated as 1123 m$^2$/d.

The successive S variation in Figure 5.10 shows initially a more or less constant part, but then there is a trend that appears toward an increased size. This implies that up to a certain distance around the well the aquifer material is more or less homogeneous. The increase in the storage coefficient is due to possible coarse grain or fracture contributions. The RSMC (1986) storage coefficient value cannot represent most of the aquifer. At early times, it yields overestimations, whereas for late times there are underestimations.

The SMP also provide a quantitative way of relating the aquifer parameters to each other. Although in the classical techniques they are assumed as completely independent and constant, Figure 5.11 shows that there is an inverse relationship between the two parameters at O-1 location. This general trend may be because of refinement of aquifer material away from the well.

The collective results of parameter calculation based on the SMP and the classical approaches are presented in Table 5.13. The SMP yields the variation domain (maximum and minimum) of each aquifer parameter in addition to the average values.

SMP results appear in the form of aquifer parameter estimation sequences, which provide a common basis for the preparation of aquifer parameter frequency diagrams. These diagrams give a foundation for the following interpretations.

**TABLE 5.13**

**Aquifer Parameter Estimates**

| Slope Matching Procedure (SMP) | Main Well | | O-1 | | O-2 | | O-3 | |
|---|---|---|---|---|---|---|---|---|
| | T (m²/d) | S | T (m²/d) | S | T (m²/d) | S | T (m²/d) | S |
| Minimum | 33.74 | 5.62E-08 | 255.0435 | 9.59E-09 | 189.1664 | 1.86E-06 | 24.16410 | 5.96E-03 |
| Average | 104.90 | 1.10E-02 | 1837.377 | 9.80E-02 | 1089.239 | 6.70E-02 | 1054.685 | 3.52E-02 |
| Median | 120.71 | 2.63E-06 | 1764.839 | 1.29E-02 | 918.4315 | 1.67E-02 | 775.1704 | 1.98E-02 |
| Maximum | 139.43 | 1.70E-01 | 4915.463 | 6.48E-01 | 3865.560 | 3.67E-01 | 2657.099 | 1.11E-01 |
| RSMC[a] | 319.68 | — | 1123.200 | 3.35E-02 | 1425.600 | 7.00E-03 | 1166.400 | 1.75E-02 |

[a] RSMC = Red Sea Mining Company.

1. Generally, in a porous medium the transmissivity histogram is expected to have a logarithmic normal distribution shape, where the frequency diagram has a single mode close to zero value with an extensive tail toward the right (Freeze and Cherry, 1989).
2. The histogram provides a visual inspection of whether the hydrogeological parameter has one type of features or the mixture of more than two different types.
3. It is possible to group, on the basis of the aquifer parameter, frequencies identified as having "low," "medium," and "high" values.
4. The validity of the arithmetic average parameter value can also be assessed. Only in the case of symmetrical histogram graphs is the arithmetic average advised for use. Otherwise, the most frequently occurring value, i.e., mode, must be adopted for groundwater assessment calculations.
5. Histograms are key elements in making the risk assessment of groundwater volume availability and exploitability.
6. It is possible to decide on the number of different aquifer materials or fracture sets.

Figures 5.12 and 5.13 present T and S frequency diagrams for O-1, respectively. In Figure 5.12, T does not have a log-normal distribution but rather an exponential form where "low" ("high") T values occur at "high" ("low") frequencies. Additionally, there are two transmissivity groups where the high T group varies between 4,500 to 5,000 m²/d. Likewise, the S histogram in Figure 5.13 has also two groups with peak values at very low frequencies. The right-hand part of this frequency diagram can be ignored in further calculations because the S values are beyond practically acceptable levels.

Aquifers in arid-region alluvial-filled wadis (drainage basins) are not homogeneous due to lateral facies change and horizontal layering of alluvium underlain by weathered and fractured zones. Classical aquifer tests that are often employed in practical applications, such as the Theis and Jacob approaches, require among the basic assumptions homogeneity of aquifer material and uniform thickness, which are not naturally the case.

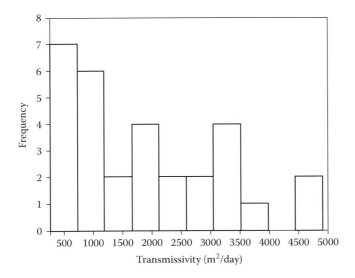

**FIGURE 5.12**    Transmissivity histogram of O-1.

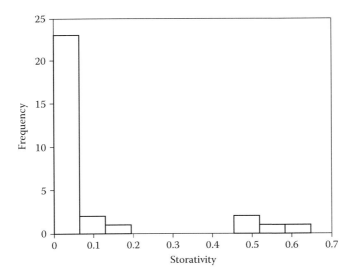

**FIGURE 5.13**    Storativity histogram of O-1.

## 5.11   TYPE STRAIGHT-LINE METHOD

There is a wrong impression that any straight-line appearance in a time–drawdown plot on a semi-logarithmic paper warrants the application of the Jacob straight-line method. The warranty of Jacob method application should depend on the fact that the late time–drawdown measurements fall along the final portion of the Theis type curve. Otherwise, there are misuses in the Jacob straight-line method application and, therefore, improvements are necessary for its reliable application.

A large number of analytic solutions are available for determining aquifer properties from pumping test data. Different methods are presented to cope with a wide variety of well and aquifer configurations, ranging from the simple homogeneous and isotropic aquifers to more complex situations involving anisotropy, barrier boundaries, leaky aquifers, *fractured* or *porous* medium, partial well penetration, and so forth. The application of these analytical methods requires both good test data and understanding of the assumptions inherent in the methods themselves as explained in Section 5.9.

A general simplified theory of groundwater movement toward an infinitesimally small diameter fully penetrating well in a leaky aquifer system has been developed (Şen, 1996). The basis of the theory is the depression cone volume concept, which separates the leaky aquifer into a conventionally confined aquifer after the isolation of the pumped aquifer elastic storage contribution to the pump discharge. The solution of the groundwater movement equation for the pumped aquifer with variable discharge leads to the desired type curve expression, where the unpumped aquifer elastic storage and the aquitard hydraulic parameters are taken into consideration. This general solution reduces to different special cases available in the literature such as the Theis (1935), De Glee (1930), and Hantush (1959) solutions.

As already mentioned in Section 5.9, the multiplicity and similarity of type curves make application very difficult in searching for the best and unique matching curve for the field data plot. In order to alleviate this situation *type straight lines* are suggested. The straight lines render such a difficulty into a manageable form leading to unique aquifer parameter estimations and useful interpretations. In general, Şen (1996, 2007) studies led to the following type straight-line expression between the dimensionless time factor, $u$, and the well function, $W(u,\eta,r/L)$ for $u < 0.01$, which is the Jacob straight-line applicability condition.

$$W(u,\eta,r/L) = \left(1 - \frac{1}{\eta}\right) K_0\left(\frac{r}{L}\sqrt{\frac{\eta}{\eta-1}}\right) - \frac{0.5772}{\eta} - \frac{2.3}{\eta}\log(u) \qquad (5.26)$$

where $\eta$ is a dimensionless factor greater than 1; $r$ is the radial distance between the main and observation wells; $L$ is the *leakage factor*; and $K_0(.)$ is a modified Bessel function of second kind with zero order. Equation 5.26 shows a straight-line relationship between the well function, $W(u,\eta,r/L)$ and u on a semi-logarithmic paper for any given value of $\eta$, and it reduces to two extreme cases, as follows:

1. $\eta = 1$ implies physically that the aquifer is confined and therefore Equation 5.26 yields to the classical Jacob type straight-line method as

$$W(u) = -0.5772 - 2.3\,log(u) \qquad (5.27)$$

2. for $\eta \to \infty$ the aquifer is of leaky type as studied by Hantush (1956) and, therefore, Equation 5.26 gives

$$W(u,r/L) = K_0\left(\frac{r}{L}\right) \qquad (5.28)$$

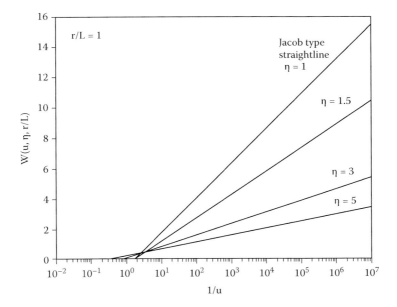

**FIGURE 5.14**  Various type straight lines.

This is the De Glee (1930) solution for steady-state flow, which corresponds to a horizontal line. The graphical representation of Equation 5.26, together with the Jacob straight line, is given for a set of parameters in Figure 5.14.

It is possible to deduce the following points, which combine the type straight-line method with the classical Jacob approach.

1. The set of straight lines confirms the statement that not every straight line on semi-logarithmic paper implies the use of the Jacob method (Şen, 1990).

2. Invariably the leaky aquifer straight lines have smaller slopes than the Jacob line. The smaller the $\eta$ value, the closer become the slopes of these lines to the Jacob line. For very big $\eta$ values, the straight line becomes horizontal, implying that the aquifer is equivalent to the classical leaky aquifer. It is well known in the Jacob method that the aquifer transmissivity is inversely related to straight-line slopes (Cooper and Jacob, 1946).

$$T = \frac{2.3Q}{4\pi \, s} \tag{5.29}$$

This implies that without distinguishing whether the aquifer is leaky or not, direct use of slope value in the Jacob method leads to the overestimation of aquifer transmissivity.

3. The intercept of leaky aquifer straight lines is always bigger than the Jacob line intercept. The storage coefficient is in direct relationship with the intercept value in the Jacob method as

$$S = \frac{2.25t_0 T}{r^2} \tag{5.30}$$

Therefore, the substitution of intercept value in the Jacob formulation leads to underestimation of the storativity value.

If any type of straight line is extended until the intercepts on the time axis where drawdown is equal to zero, the intercept point coordinates are $W(u,\eta,r/L) = 0$ and $u = u0$. These coordinates correspond to $s = 0$ and $t = t0$, respectively, in a field data plot on semi-logarithmic paper. Substitution of these values into Equation 5.26 gives, after some algebraic manipulations,

$$u_0 = 0.56 e^{(\eta-1)K_0\left[\left(\frac{r}{L}\right)\left(\frac{\eta}{\eta-1}\right)\right]} \tag{5.31}$$

By considering Equation 5.25, this expression provides a practical formula for the storage coefficient estimation as

$$S_L = \frac{2.25 T t_0}{r^2} e^{(\eta-1)K_0\left(\frac{r}{L}\sqrt{\frac{\eta}{\eta-1}}\right)} \tag{5.32}$$

For $\eta = 1$, it reduces to the classical Jacob formula, which gives the storage coefficient, $S$, the estimate of a confined aquifer according to Equation 5.30, and its comparison with Equation 5.32 leads to

$$S_L = S e^{(\eta-1)K_0\left(\frac{r}{L}\sqrt{\frac{\eta}{\eta-1}}\right)} \tag{5.33}$$

Because the exponential term is always greater than one, always $S_L > S_J$. This is tantamount to saying that the use of the Jacob method without aquifer type identification leads to underestimation of storage coefficient.

The slope, $\Delta W(u,\eta,r/L)$, of the leaky aquifer type straight line becomes, from Equation 5.26,

$$W(u,\eta,r/L) = \frac{2.3}{\eta} \tag{5.34}$$

As already mentioned by Şen (1989, 1990, 1995, 2007), the slope of the Jacob type straight line on a dimensionless plot is equal to 2.3. Consideration of explicit well function expression renders Equation 5.34 as a useful formulation for transmissivity estimation in leaky aquifers in general, as

$$T_L = \frac{2.3Q}{4\pi\eta \ s} \tag{5.35}$$

in which $\Delta s$ is the slope of a straight line fitted to field data on semi-logarithmic paper. Comparison with the Jacob formulation indicates that

$$T_L = \frac{T}{\eta} \qquad (5.36)$$

It is clear from this expression that always $T_L < T$. Furthermore, from Equations 5.33 and 5.36 it appears that the key factor for accurate calculations of aquifer parameters is the storage coefficient ratio parameters, $\eta$.

### 5.11.1 APPLICATION

The application of methodology suggested in the previous section is performed for aquifer test data obtained from the Arabian Shield. In one of the wadis in quaternary deposits, a well is pumped with a discharge of 2,000 m³/d, and the time–drawdown measurements are collected from an observation well which is 102 m away from the pumped well. The late time–drawdown data is given in Table 5.14.

The following steps are necessary for an effective application of the presented methodology.

1. Plot the large time–drawdown data on a semi-logarithmic paper with time on the logarithmic axis (see Figure 5.15).
2. Find the aquifer parameters, namely S and T, of the pumped aquifer by using the conventional Jacob method. For this purpose, match the most suitable straight line through late time–drawdown points, and find the intercept, $t_0 = 3.9 \times 10^{-3}$/d, on the time axis and the slope, $\Delta s = 0.35$. Substitutions of these values into Jacob expression (Equations 5.29 and 5.30) yield $T = 104.64$ m²/d and $S = 9.18 \times 10^{-4}$. An implied assumption in this step is that the whole aquifer configuration is considered as a confined aquifer. Such an assumption is regarded as a null hypothesis with an alternative hypothesis that the aquifer is not of confined type.
3. In order to confirm the validity of this null hypothesis, calculate the dimensionless time, $U_J$, and drawdown, $W_J$, from the following two equations:

$$U_J = \frac{r^2 S}{4tT} \qquad (5.37)$$

and

$$W_J = \frac{4\pi T}{Q} s \qquad (5.38)$$

Application of these equations to data in Table 5.14 gives the results in Table 5.15.
4. Plot these dimensionless values on the same scale semi-logarithmic paper as for the type straight lines (see Figure 5.16).
5. Draw the best straight line matching the dimensionless data and calculate its slope value, $(\Delta W_J)_L = 0.24$.

**TABLE 5.14**
**Aquifer Test Data**

| Time (day) | Drawdown (m) | Time (day) | Drawdown (m) |
| --- | --- | --- | --- |
| 0.0056 | 0.0601 | 0.5556 | 0.8097 |
| 0.0058 | 0.0684 | 0.5848 | 0.8180 |
| 0.0062 | 0.0772 | 0.6173 | 0.8268 |
| 0.0065 | 0.0866 | 0.6536 | 0.8361 |
| 0.0069 | 0.0964 | 0.6944 | 0.8460 |
| 0.0074 | 0.1069 | 0.7407 | 0.8565 |
| 0.0079 | 0.1182 | 0.7937 | 0.8677 |
| 0.0085 | 0.1302 | 0.8547 | 0.8798 |
| 0.0093 | 0.1432 | 0.9259 | 0.8929 |
| 0.0101 | 0.1574 | 1.0101 | 0.9070 |
| 0.0111 | 0.1729 | 1.1111 | 0.9225 |
| 0.0123 | 0.1901 | 1.2346 | 0.9397 |
| 0.0139 | 0.2092 | 1.3889 | 0.9588 |
| 0.0159 | 0.2310 | 1.5873 | 0.9806 |
| 0.0185 | 0.2561 | 1.8519 | 1.0057 |
| 0.0222 | 0.2857 | 2.2222 | 1.0354 |
| 0.0278 | 0.3220 | 2.7778 | 1.0717 |
| 0.0370 | 0.3689 | 3.7037 | 1.1185 |
| 0.0585 | 0.4432 | 5.5556 | 1.1929 |
| 0.0617 | 0.4520 | 6.1728 | 1.2016 |
| 0.0654 | 0.4613 | 6.1728 | 1.2109 |
| 0.0694 | 0.4712 | 6.9444 | 1.2208 |
| 0.0741 | 0.4817 | 6.9444 | 1.2313 |
| 0.0794 | 0.4930 | 7.9365 | 1.2425 |
| 0.0855 | 0.5050 | 9.2593 | 1.2546 |
| 0.0926 | 0.5180 | 9.2593 | 1.2676 |
| 0.1010 | 0.5322 | 11.1111 | 1.2818 |
| 0.1389 | 0.5840 | 13.8889 | 1.3337 |
| 0.1852 | 0.6309 | 18.5185 | 1.3805 |
| 0.2222 | 0.6605 | 18.5185 | 1.4102 |
| 0.2778 | 0.6969 | 27.7778 | 1.4465 |
| 0.3704 | 0.7437 | 55.5556 | 1.4933 |

## TABLE 5.15
### Dimensionless Field Data

| $U_J$ | $W_J$ | $U_J$ | $W_J$ |
|---|---|---|---|
| 0.3948 | 0.0395 | 0.0039 | 0.5321 |
| 0.3750 | 0.0450 | 0.0038 | 0.5376 |
| 0.3553 | 0.0508 | 0.0036 | 0.5433 |
| 0.3356 | 0.0569 | 0.0034 | 0.5495 |
| 0.3158 | 0.0633 | 0.0032 | 0.5559 |
| 0.2961 | 0.0702 | 0.0030 | 0.5628 |
| 0.2763 | 0.0776 | 0.0028 | 0.5702 |
| 0.2566 | 0.0856 | 0.0026 | 0.5781 |
| 0.2369 | 0.0941 | 0.0024 | 0.5867 |
| 0.2171 | 0.1034 | 0.0022 | 0.5960 |
| 0.1974 | 0.1136 | 0.0020 | 0.6062 |
| 0.1777 | 0.1249 | 0.0018 | 0.6175 |
| 0.1579 | 0.1375 | 0.0016 | 0.6301 |
| 0.1382 | 0.1518 | 0.0014 | 0.6444 |
| 0.1184 | 0.1683 | 0.0012 | 0.6609 |
| 0.0987 | 0.1878 | 0.0010 | 0.6804 |
| 0.0790 | 0.2116 | 0.0008 | 0.7042 |
| 0.0592 | 0.2424 | 0.0006 | 0.7350 |
| 0.0375 | 0.2913 | 0.0004 | 0.7839 |
| 0.0355 | 0.2970 | 0.0004 | 0.7896 |
| 0.0336 | 0.3032 | 0.0004 | 0.7958 |
| 0.0316 | 0.3096 | 0.0003 | 0.8022 |
| 0.0296 | 0.3165 | 0.0003 | 0.8091 |
| 0.0276 | 0.3239 | 0.0003 | 0.8165 |
| 0.0257 | 0.3318 | 0.0002 | 0.8244 |
| 0.0237 | 0.3404 | 0.0002 | 0.8330 |
| 0.0217 | 0.3497 | 0.0002 | 0.8423 |
| 0.0197 | 0.3599 | 0.0002 | 0.8525 |
| 0.0178 | 0.3712 | 0.0002 | 0.8638 |
| 0.0099 | 0.4341 | 0.0001 | 0.9267 |
| 0.0079 | 0.4579 | 0.0001 | 0.9505 |
| 0.0059 | 0.4887 | 0.0000 | 0.9813 |

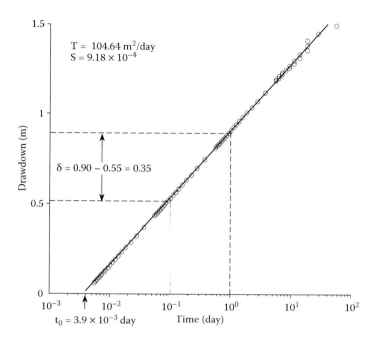

**FIGURE 5.15**  Semi-logarithmic plot.

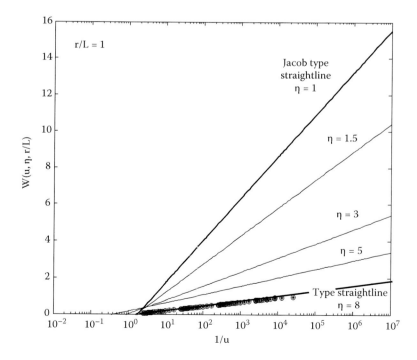

**FIGURE 5.16**  Type straight lines and dimensionless field data.

6. Compare the slope of this line with the dimensionless Jacob slope, $\Delta W_J$, which is equal to 2.3. If they both have the same slopes, then the null hypothesis is acceptable and the Jacob method application is correct and does not need any correction. Otherwise, when $\Delta W_J > (\Delta W_J)_L$ the aquifer is of the leaky type. The more the deviation of slope from 2.3, the more is the leakage contribution to the overall pump discharge. However, for slopes bigger than 2.3, Şen (1990) has already shown that the flow is of non-Darcian type.

7. As an alternative hypothesis the aquifer is considered as leaky. Hence, find the best matching type straight line to the plotted dimensionless time–drawdown data (Figure 5.16). Read off the values of $\eta = 8$ and $r/L = 1$ for the problem at hand. The plot of dimensionless field data values in Figure 5.16 indicates that the slope of the data points is far away from the Jacob straight line.

8. With the pair of $\eta$ and $r/L$ determine the corresponding value of Bessel function, $K0\{[r/L]\sqrt{\eta/(\eta-1)}\}$ which turns out to be 0.7341.

9. Calculate the aquifer transmissivity value from Equation 5.35, which yields for the data at hand that $T_L = 191$ m$^2$/d.

10. Calculate the storage coefficient value from Equation 5.3. Hence, $S_L = 1.5 \times 10^{-1}$.

The dimensionless time factor and dimensionless well function plots on double logarithmic paper do not overlap with Theis type curve, which indicates that the given data cannot be assessed by the classical Jacob straight-line method (Figure 5.17).

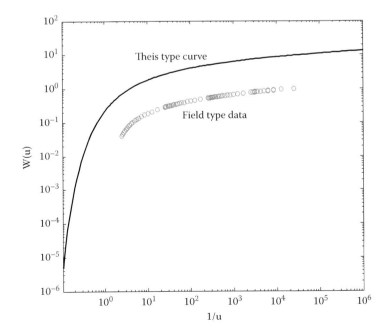

**FIGURE 5.17**    Theis type curve and dimensionless field data plot.

## 5.12    STANDARD ION INDEX FOR GROUNDWATER
           QUALITY EVOLUTION

Groundwater in any part of the world is not homogenous in quality. Its occurrence, distribution, and movement lead to natural mixture of different water types. The natural mixture determines the groundwater quality in nature, depending on the geological setup and the hydrologic cycle activities (Şen et al., 2003). For instance, during metamorphism, the hydrous clay minerals are converted to less hydrous forms and water is expelled from the rock. It appears that as this metamorphic water migrates toward the earth's surface, it generally mixes with meteoric water. Consequently, the groundwater quality might vary spatially to a significant extent within the same aquifer at different well locations and depths. In addition, due to different storm events and subsequent direct and indirect recharge, fluctuations in groundwater levels cause temporal and spatial changes. Depending on the well location, penetration depth, pumping discharges, and periods, groundwater quality shows significant differences from one well to another or in the same well from time to time. In order to protect or to improve water quality, many authors presented different suggestions (Fair et al., 1971; Ayers and Wescott, 1976). In addition, a comprehensive review and interpretation of the chemical characteristics of natural water are presented by Hem (1970).

    Groundwater quality fluctuates as a response to hydrological, hydrogeological, and other relevant environment effects in arid-region wadis. For instance, after long durations of dry periods its salinity increases because there is not enough recharge, but after rainfall and its subsequent recharge through infiltration, quality improves and its taste might change slightly. Such fluctuations are evidence that there cannot be a rigid characterization of groundwater quality, and that it varies within natural fluctuation limits. The records provide an objective and experimental basis for the empirical establishment of some characteristic range of groundwater quality. There are many hydrochemical assessment methods in the literature, but unfortunately they are more suitable for complete groundwater quality classification, and each one has its own set of assumptions and drawbacks, in addition to the methodological restrictions (Piper, 1953; Wilcox, 1955). They are dependent on arithmetic means or percentages or some of them have dimensions, and therefore they cannot provide a complete common basis for the comparison and identification of any significant characteristic in the spatio-temporal groundwater quality evolution. The classical statistical techniques such as multiple regression, principle component, and other methods all have restrictive assumptions about the data behavior. *Groundwater quality* measurements are scarce and therefore their applications cannot be successful with scarce data.

    Groundwater quality variables help to identify water mixture regions and suitability of water for various purposes. There are many extensive aquifers in arid regions of the world such as the Umm Er Radhuma aquifer in the eastern part of the Arabian Peninsula, which lies within the Arabian Shelf sedimentary geological configuration, in addition to many others in African Sahara, and Central Asia and South America. Deep groundwater resources have been replenished since time immemorial, and therefore they provide invaluable fossil groundwater in confined or leaky aquifers. Especially, Umm Er Radhuma aquifer groundwater chemistry features are examined by Şen and Dakhil (1986) through the use of simple and classical statistical and conventional diagrams.

### 5.12.1 STANDARD ION INDEX

In order to achieve a reliable, robust, simple, objective, effective, and all-ion-dependent measure, the *standard ion index* (SII) concept is suggested. Similar to Equation 3.51 for the standard precipitation index, standardization is a statistical procedure whereby the arithmetic average of a given sequence, say, $X_1, X_2, ..., Xn$ is rendered to zero and standard deviation to one, through the following simple procedure:

$$x_i = \frac{X_i - \overline{X}}{S_x} \qquad (5.39)$$

where $x_i$ is the standardized value, $\overline{X}$ is the arithmetic average, and $S_x$ is the standard deviation of the original series. Such a simple transformation provides the following properties:

1. Standard series has zero mean.
2. Standard series has unit variance or standard deviation.
3. Standard series is dimensionless.
4. Standard series provides equal footing for comparison of different samples.

After standardization, the remaining significant statistical parameters are skewness and kurtosis coefficients for distinctive representation of different samples. In this case, the application of Equation 5.39 to temporal or spatial groundwater samples leads to a distinctive pattern in the form of a parabola as a relationship between the skewness and kurtosis. In Figure 5.18 different symbols refer to different samples, either from a set of wells or from a single well at a set of different times. It implies a systematic relationship between the skewness and kurtosis parameters.

In the application of the SII to each single water sample, two different approaches are considered, depending on the purpose:

1. *Individual* SII approach, which provides change of each ion SII value separately on the same graph.
2. *Successive* SII approach provides temporal evolution of groundwater quality change by looking at the successive sample concentrations in a systematic manner.

### 5.12.2 INDIVIDUAL SII METHOD

The purpose is to search for a variation pattern based on the ions (cations and anions) in different samples by considering the SII values. It helps to make visual inspections to see whether the current sample fits with the general pattern of the previous samples. If there are differences, then the reasons must be sought accordingly.

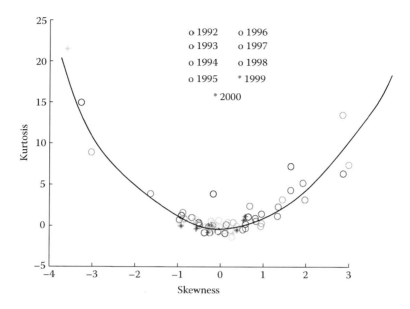

**FIGURE 5.18**    Skewness–kurtosis relationship.

The groundwater samples can be collected either systematically or irregularly in time from a given set of wells in a study area. Individual SII graphs consist of eight-corner broken lines for each sample including major cations (Ca, Mg, Na, and K) and anions (Cl, $HCO_3$, $NO_3$, and $SO_4$). Hence, in Figure 5.19 there are many eight-corner broken lines, each corresponding to a different sampling. If needed, it is possible to increase the number of ions. Before further explanations, one should keep in mind that, in the preparation of an individual SII graph, the following fundamental points are important for consideration:

1. The laboratory analyses of ions in ppm should be converted into epm values so as to see the equilibrium condition. All groundwater data are checked through this basic balance procedure.
2. It is known that in epm the summation of ions and cations must be equal to each other. This implies a requirement that the cations and anions groups contribute 50% to the overall equilibrium condition.
3. The epm values for reliable samples are standardized so as to have zero mean and unit variance according to Equation 5.39.
4. The individual SII graph is the plot of these standard values against each ion as in Figure 5.19.

There is no significance in the ion sequence order, but in any study the same sequence must be kept for comparison purposes. Although different ion sequences can be selected on the horizontal axis, it is recommended to have a sequence of cations first, which is followed by anions. The most convenient way is to group cations on

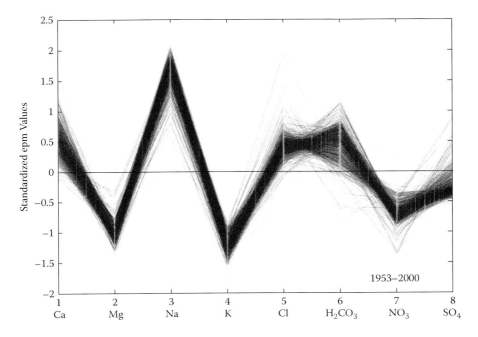

**FIGURE 5.19**   Individual SII graph for all groundwater sample ion concentrations.

the right-hand side and anions on the left-hand side of the horizontal axis as shown in Figure 5.19. The following points are important in order to better understand the individual SII graphs:

1. Some of the standard ion index values appear as positive and others as negative because by definition their summation is equal to zero.
2. Positive (negative) SII dimensionless means that, in the sample, this ion concentration exists more than (less than) the average value.
3. Horizontal zero line (perfect uniformity of each ion) represents the zero SII level. Hence, the closer the SII values to this line, the less active the ion in the overall quality balance.
4. Some of the ions may assume positive (negative) SII values whatever the time-wise sample number is, these ions are considered as stable ions. However, if there is crossover around the horizontal line for different samples, then the concerned ion is rather unstable in the chemical composition evolution of the groundwater quality.
5. It is possible to see the variation range for each ion on the vertical axis. The smaller the range, the less variable is the concentration of this ion in the temporal (or spatial) evolution of the ion.
6. It is not possible for any water sample to have all the cations positive and the anions negative. This is against the concept of the chemical (epm) balance and the SII concept.

It is possible to see in Figure 5.19 that there appears to be a general pattern with fluctuations in water quality. The general features can be depicted as follows:

1. Mg and K assume negatives, whereas Na has positive values only. In regard to cations, although Ca has positive values in a majority of samples, there are a few cases with negative values.
2. Regarding anions, the Cl SII values are all positive, but the $NO_3$ takes negative values. However, most of the time both $HCO_3$ and $SO_4$ assume positive values along with several negative values.
3. In general, the individual SII values assume alternative positive and negative SII values.

### 5.12.3 Successive SII Method

This is a procedure that helps to control the association between two samples and possible deviations that may occur. If the two samples have exactly the same ion concentrations, then their plot on an ordinary paper will result along a straight line with 45° (Figure 5.20). The basic straight line is referred to as the *ideal similarity line*; the closer the scatter points are to this line, the more the similarity between the samples or successive samples in time. The template in Figure 5.20 is referred to as the successive standard ion index (SII) graph. It has four major parts with different interpretations.

The following points can be stated about the features of the successive SII paper:

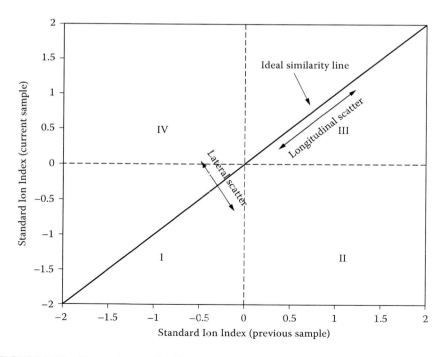

**FIGURE 5.20** Successive standard ion index (SII) graph.

1. The variables for the plot do not have any dimensionality, and therefore this graph provides a universal approach for any study.

2. If the purpose is to look for temporal groundwater quality evolution, it provides a dimensionless basis for the successive sample SII values in such a manner that there is no human interference as to the sequence of ions.

3. The scatter points can be inspected visually and preliminary interpretations can be deduced from the closeness of the scatter points to the ideal similarity line.

4. The first quadrant, I, includes "low" SII values, which implies similarity or closeness of the sample values. Similarly, the third quadrant, (III), is also of close similarity in the "high" ion concentrations.

5. It is possible to interpret quadrants I and III such that "low" ("high") ion concentrations follow "low" ("high") ion concentrations. This implies that for the two samples to become similar to each other, the first condition is the "low" ("high") concentrations in the previous sample must be followed by the "low" ("high") concentrations. Otherwise, they cannot be similar; in other words, they cannot have the same characteristics.

6. The second quadrant, II, includes the "high" ion concentration values of the previous sample with "low" concentrations of the current sample. Likewise, the fourth quadrant, IV, implies that "low" ion concentrations of the previous sample are associated with the "high" concentrations in the current sample.

7. It can be concluded from the aforementioned discussions that quadrants I and III are for the similarity of the overall water quality, whereas quadrants II and IV are for dissimilarity.

8. Another feature of this graph is that many ions can be plotted on the horizontal and vertical axes, so it is not restricted to the number of ions as in the Piper diagram where, at the maximum, three ions can be represented.

9. In such graphs two perpendicular deviations or scatters can be considered for meaningful interpretations. These are along the 45° ideal straight line and perpendicular to it. They are referred to as the longitudinal and lateral dispersions, deviations or scatters, respectively (see Figure 5.19). This concept is very similar to the principle component analysis (Davis, 2002), but without assumptions such as the normal (Gaussian) distribution of the data.

10. The longitudinal dispersion is significant for the identification of any different ion groups, whereas the lateral scatter is meaningful for the deviations from the similarity within each group.

11. In an ideal characteristic situation, it is expected that along the ideal similarity line, there are distinct ion groups with small lateral deviations. Hence, rather than the longitudinal dispersion, the lateral one is more significant in deciding whether the current sample transgresses allowable error as quality limits.

In Figure 5.21 a set of the groundwater quality SII values are plotted on successive SII graphs, and the relevant comparative interpretations are deduced. For instance, the plot of eight major ions on the successive SII graph appears as scatter points with dominant points in quadrants I and III, which implies that overall the samples are

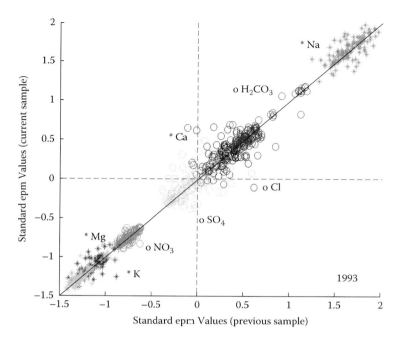

**FIGURE 5.21** Successive SII graph for 2003.

similar to each other. This figure indicates that sodium appears as a very distinctive group at the high ion concentration levels.

Some other ions also started to appear as individual groupings such as nitrate and bicarbonate. The longitudinal scatter of nitrate decreases with almost the same lateral deviations. Magnesium ions penetrate the potassium group at the lower concentration levels.

## 5.13   CLIMATE CHANGE AND GROUNDWATER

Groundwater is the major freshwater source especially for arid and semi-arid regions. Little attention is given on the potential *climate change* effects on these freshwater resources. Aquifers in arid and semi-arid regions are replenished by floods at possible recharge outcrop areas through fractured and fissured rocks, and solution cavities in dolomite or limestone geological setups, as well as through main stream channels of Quaternary alluvium deposits (Section 5.5.4). At convenient places along the main channel, engineering infrastructures such as levees, dikes, and successive small-scale groundwater recharge dams may be constructed for groundwater recharge augmentation. The groundwater recharge areas must be cared for in isolation from fine silt accumulation after each flood occurrence or at periodical intervals. Furthermore, flood inundation areas are among the most significant groundwater recharge locations in arid and semi-arid regions. Accordingly, their extents must be delimited by considering future climate change effects.

Understanding the relative importance of climate, vegetation, and soils in controlling groundwater recharge is critical for estimating recharge rates and for assessing the importance of these factors in controlling aquifer vulnerability to contamination. The role of climate and vegetation in controlling recharge is valuable in determining impacts of climate change and *land use* change on recharge.

The aquifers that are in contact with the present-day hydrological cycle will be affected by climate change. These are referred to as *unconfined* or *shallow aquifers*. On the other hand, deep aquifers are not directly in contact with the present-day hydrological cycle, and consequently their response to climate change is virtually negligible. They are fossil groundwater storage areas only.

An unsaturated zone has the unique capability of helping to assess impacts of climate change on groundwater resources. The potential impacts of climate change can be assessed by focusing on porous, fractured, and karstic (carbonate rock, dolomite, limestone) aquifer systems. Fractured and karstic aquifers are the most responsive to changes in recharge, as typically they have low specific yields (i.e., they have drainable porosities) in comparison with porous flow systems. Karstic rocks are soluble, and the aquifers might show exacerbated water table lowering if predicted increases in atmospheric $CO_2$ contents, along with temperature rise, induce rapid enlargement of fracture apertures and enlargement in the solution cavities. Dissolution of carbonate rocks (karstic media) might become more vigorous in time and, accordingly, the hardness of groundwater sources is expected to increase, leading to possibly unacceptable water quality.

The direct groundwater recharge mode is more sensitive to climate change than indirect natural recharge. Direct recharge can be defined as water added to the aquifer through the unsaturated zone by direct percolation of rainfall at the spot where it falls. Indirect recharge occurs where water fulfills the soil moisture deficits and evapotranspiration process before reaching a groundwater reservoir.

Coupled with the changes in the hydrological cycle and probable inducement of climate change basic elements, the groundwater recharge is also interactively affected due to the following events:

1. Changes in precipitation, evapotranspiration, and runoff are expected to influence recharge. It is possible that increased rainfall intensity may lead to more runoff and less recharge.
2. Sea-level rise may lead to increased saline intrusion of coastal and island aquifers, depending on the relative position of the sea level to the groundwater table level.
3. Changing in precipitation imply changes in $CO_2$ concentrations, which may influence carbonate rocks dissolution and hence formation and development of karstic groundwater aquifers.
4. Natural vegetation and crops changes reflection of climate change may influence recharge.
5. Increased flood events contribute to unconfined aquifers in arid and semi-arid zones and, hence, they affect groundwater quality in alluvial aquifers of wadis.

6. Changes in soil organic carbon may affect the infiltration properties above aquifers and consequently the groundwater recharge.

The above mentioned factors indicate that organizations focused on groundwater problems should take an interest in global climate change issues in order to protect groundwater resources from the implications of their effect.

Another great advantage of groundwater is that as water slowly percolates down into the aquifer it is usually purified of biological pollutants. Thus, groundwater is usually the best source of drinking water, especially in arid, semi-arid, and rural areas of developing countries where water treatment facilities or desalination plants are not available. In water-resources-poor regions, such as the Arabian Peninsula countries, desalination plants are used to maintain groundwater resources as strategic planning assets for future generations or emergency situations (Al-Sefry, et al., 2004). Groundwater resources assessment for any purpose in arid region wadi drainage basins requires that uncertainties be taken into all calculations. Among these uncertainties the impact of climate change may well need to be accounted for in future planning, operations, and management. Any mechanistic or deterministic approach applied to risk assessment involving fluctuating climate, recharge, and aquifer and drainage basin. Hence, rather than hoping for useful regional average parameter values, a risk level, say, of 10% may need to be calculated when trying to determine any parameter value properties will not produce the most useful results.

In arid regions, wadis have Quaternary depositions of different facies and different grain size distributions, depending on the paleogeologic environmental processes they have undergone. Consequently, any measurement at a particular point may be significantly different from other points within the same wadi. For instance, several aquifer tests have led to different hydrogeologic parameter estimations, and the rainfall variability has temporal and spatial trends. It is necessary to manage the groundwater resources under such risk levels. The application of the methodology is presented for Wadi Fatimah that lies in the central western part of the Kingdom of Saudi Arabia (Al-Sefry et al., 2004).

The rainfall and infiltration elements of the hydrological cycle in arid regions indicate temporal and spatial variations in a random and sporadic manner. Such variations may be exacerbated groundwater replenishment facilities.

Groundwater recharge estimation has become a priority issue for both developed and underdeveloped countries, especially in dry areas like central Africa, where rainfall is both temporally and spatially irregular. The rapid agriculture and industrial growth in such areas has dramatically changed the groundwater resources withdrawal pattern. Groundwater withdrawal in excess of recharge has lowered the hydraulic heads in the aquifers, and resulted in increased pumping energy costs and reduced the rate of removal. Continual withdrawal in excess of recharge and possible climate change effects will in time ultimately remove all of the recoverable water.

Rainfall-based annual recharge maps should be prepared for active aquifer monitoring. The quantity of recharge to an aquifer must be considered in any integrated management program as equivalent to the *safe yield* or quantity of groundwater that could be withdrawn form an aquifer on a sustainable basis. This type of map will be very helpful in evaluating the effects of climate variability on groundwater recharge and exploitable groundwater resources. Many researchers believe that climatic conditions mainly govern the recharge rates, although vegetation and soils also interactively exert control.

The unsaturated zone has a unique capability in helping to assess impacts of climate change on groundwater resources. The potential impacts of climate change can be assessed focusing on fractured carbonate aquifer systems (having variable degrees of fracture development, i.e., karstification) for these reasons:

1. Fracture flow aquifers are the most responsive to changes in recharge as typically they have low specific yields (i.e., they have drainable porosities) in comparison with intergranular flow systems.
2. Carbonate rocks are soluble, and the aquifers might show exacerbated water table lowering if predicted increases in atmospheric $CO_2$ contents, along with temperature rise, induce rapid enlargement of fracture apertures.
3. If dissolution of carbonate rocks does become more vigorous, then potentially the hardness of groundwater could be expected to increase, possibly leading to unacceptable water quality.

## REFERENCES

Abdulrazzak, M. J., Şorman, A .U., and Alhames, A. S., 1989. Water balance approaches under extreme arid conditions—a case study of Tabalah basin, Saudi Arabia, *Hydrol. Proc.*, Vol. 3, 107–122.

Alehaideb, I., 1985. Precipitation Distribution in the Southwest of Saudi Arabia, Ph.D. thesis, Arizona State University, p. 215.

Al-Kabir, M., 1985. Recharge characteristics of groundwater aquifers in Jeddah-Makkah-Taif area, M.Sc. thesis, King Abdulaziz University, Jeddah, Saudi Arabia.

Allison, G. B., 1988. A review of some of the physical, chemical and isotopic techniques available for estimating groundwater recharge. In: Simmers, I. (Ed.), *Estimation of Natural Groundwater Recharge*. D. Reidel Publishing Co., Dardrecht. 49–72.

Allison, G. and Hughes, M., 1983. The use of natural tracers as indicators of soil-water movement in a temperate semi-arid region, *J. Hydrol.* 60, 157–173.

Al-Yamani, M.S., 2001. Isotopic composition of rainfall and groundwater recharge in the western province of Saudi Arabia, *J. Arid Environ.* 49, 751–760.

Al-Yamani, M. S. and Şen, Z., 1992. Regional variation of monthly rainfall amounts in the Kingdom of Saudi Arabia. *J. King Abdulaziz University*: Faculty of Earth Sciences, 6, 113–133.

Al-Sefry, S., Şen, Z., Al-Ghamdi, S. A., Al-Ashi, W., and Al-Baradi, W., 2004. Strategic ground water storage of Wadi Fatimah – Makkah region Saudi Arabia. Saudi Geological Survey, Hydrogeology Project Team, Final Report.

Andersen, N. J., Wheater, H. S., Timmis, A. J. H., and Gaongalelwe, D., 1998. Sustainable development of alluvial groundwater in sand rivers of Botswana. In *Sustainability of Water Resources under Increasing Uncertainty*, IAHS Pubn. No. 240, pp. 367–376.

Athavale, R. N. and Rangarajan, R., 1990. Natural recharge measurements in the hard rock regions of semi-arid India using tritium injection- a review. In: Lerner, D. N., Issar, A. S., and Simmers I., (Eds.) *Groundwater Recharge: A Guide to Understanding and Estimating Natural Recharge*. International Association of Hydrogeologists (IAH) Publication, International Contribution to Hydrogeology, Vol. 8, 235–256.

Ayers, R. S. and Wescot, D. W., 1976. Water Quality for Agriculture. Food and Agriculture Organization of the United Nations, Irrigation and Drainage Paper 29, Rome, Italy.

Basmaci, Y. and Al-Kabir, M., 1988. Groundwater recharge over Western Saudi Arabia, In *Estimation of Natural Groundwater Recharge*, Simmers, I. (Ed). D. Reidel Publishing Co., Dardrecht; 395–403.

Basmaci, Y. and Hussein, J. A. A., 1988. Groundwater recharge over Western Saudi Arabia, In: *Estimation of Natural Groundwater Recharge,* Simmers, I. (Ed.), D. Reidel Publishing Co., Dardrecht, 395–403.

Bazuhair, A. and Wood, W. W., 1996. Chloride-Mass-Balance method for estimating groundwater recharge in arid areas: example from western Saudi Arabia. *J. Hydrol.* 186: 153–159.

Bazuhair, A., S., Nassief, M. O., Al-Yamani, M. S., Sharaf, M. A., Bayumi, T. H., and Ali, S., 2002. Groundwater recharge estimation in some wadi aquifers of the western Saudi Arabia. King Abdulaziz City for Science and Technology; Project No. AT-17–63, Riyadh, Saudi Arabia., p. 389.

Bear, J., 1972. Dynamics of fluid in porous media. Elsevier Applied Science, New York.

Boulton, N. A., 1954. The drawdown of the water table under non-steady conditions near a pumped well in an unconfined formation. *Proc. Inst. Civil Eng.* Vol. 3, No. 3, London, England, 564–579.

B.R.G.M., 1976. Hydrogeological investigation of Al-Wasia in the eastern province of Saudi Arabia. Regional study. Final Report, Ministry of Agriculture and Water, Riyadh, Saudi Arabia.

Butler, J. J., Jr., 1988. Pumping test in nonuniform aquifers: The radially symmetrical case, *J. Hydrol.*, 101: 15–30.

Butler, J. J., Jr., 1990. The role of pumping tests in site characterization: Some theoretical considerations. *Ground Water*, 28(3): 394–402.

Butler, J. J., Jr. and Liu, W. Z., 1991. Pumping tests in nonuniform aquifers: The linear strip case. *J. Hydrol.*, 128: 69–99.

Butler, J. J., Jr. and Liu, W. Z., 1993. Pumping tests in nonuniform aquifers: The radial asymmetric case. *Water Resour. Res.*, 29(2): 259–269.

Caro, R. and Eagleson, P. S., 1981. Estimation of aquifer recharge due to rainfall, *J. Hydrol.*, Vol. 53, 185–211.

Chow, V. T., 1964. *Handbook of Applied Hydrology,* McGraw-Hill Co., New York, pp. 11–38.

Clark, I. and Fritz, P., 1997. *Environ. Isotopes Hydrogeol.*, CRC Press-Lewis Publishers, p. 328.

Cooper, H. H., Jr. and Jacob, C. E., 1946. A generalized graphical method for evaluating formation constants and summarizing well-field history, *Eos Trans.*, AGU, 27(4): 526–534.

Craig, H., 1961. Isotopic variation in meteoric water. *Science*, 133: 1702–1703.

Dansgaard, W., 1964. Stable isotopes in precipitation: *Tellus*, Vol. 16: 436–467.

Davis, J. C., 2002. *Statistics and Data Analysis in Geology.* John Wiley & Sons, New York, p. 638.

Deardorff, J. W., 1977. A parameterization of ground-surface moisture content for use in atmospheric prediction models. *J. Applied. Meteorol.*, 16, 1182–1185.

De Glee, G. J., 1930. Over Grondwaterstroomingen bij wateronttrekking door middel van Putten, Waltman, T. Jr., (Ed.), Delft.

Dincer, T., 1980. Use of environmental isotopes in arid zone hydrology. In: *Arid Zone Hydrology, Investigation with Isotope Techniques.* IAEA, Vienna, 6–9 Nov. 1978, 23–30.

Dincer, T., Al-Mugrin, A., and Zimmermann, U., 1974. Study of the infiltration of recharge through sand dunes in arid zones with special reference to stable isotopes in thermonuclear tritium. *J. Hydrol.*, Vol. 23: 79–87.

Edmunds, W. M., Darling, W.G., and Kinniburgh, D. G., 1990. Solute profile techniques for recharge estimation in semi-arid and arid terrain. In: Lerner, D. N., Issar, A. S., and Simmers I., (Eds.), *Groundwater Recharge: A Guide to Understanding and Estimating Natural Recharge*. International Association of Hydrogeologists (IAH) Publication, International Contribution to Hydrogeology, Vol. 8: 257–270.

El-Khatib, A., 1980. *Seven Green Spikes*, Ministry of Agriculture and Water, Riyadh, Saudi Arabia, p. 362.

Eriksson, E. and Khunaksem, V., 1969. Chloride concentration in groundwater, recharge rate and rate of deposition of chloride in the Israel coastal plain, *J. Hydrol.*, Vol. 7: 178–197.

Eriksson, E. 1976. The distribution of salinity in groundwater of the Delhi region and the recharge rates of groundwater. In *Interpretation of Environmental Isotopes and Hydrochemical Data in Groundwater Hydrology*, IAEA, Vienna, 171–177.

Fair, G. M., Geyer, C. J., and Okun, D.A., 1971. *Elements of Water Supply and Waste Water disposal*, 2nd ed. John Wiley & Sons, Inc., New York, p. 752.

Fetter, C. W., 1994. *Applied Hydrogeology*. 3rd ed., Merrill Publishing Company, Columbus, OH.

Flint, A. L., Flint, L. E., Kwicklis, E. M., Fabryka-Martin, J. T., and Bodvarsson, G. S., 2002. Estimation recharge at Yucca Mountain, Nevada, USA, comparison methods. *Hydrogeol. J.* 10:180–204.

Foster, S. S. D., 1988. Quantification of groundwater recharge in arid regions: a practical view for resource development and management. In: Simmers, I. (Ed.), *Estimation of Natural Groundwater Recharge*. D. Reidel Publishing Co. Dardrecht, 323–338.

Freeze, R. A. and Banner, J., 1970. The mechanism of natural groundwater recharge and discharge, 2. Laboratory column experiments and field measurements. *Water Resour. Res.* 6: 138–155.

Freeze, R. A. and Cherry, J. A., 1989. *Groundwater*. Prentice-Hall International, Inc., London, p. 605.

Fritz, P. and Fontes, J., Ch., 1980. *Handbook of Environmental Isotope Geochemistry*, Amsterdam, Elsevier Scientific Pub. Co., Vol. 1: p. 545.

Gat, J. R., 1987. Variability (in time) of isotopic composition of precipitation: Consequences regarding the isotopic composition of hydrologic systems. In: *Isotopic Techniques in Water Resources Development*, Rep. IAEA-SM-299, pp. 551–563, IAEA, Vienna.

Gee, G. W. and Hillel, D., 1988. Groundwater recharge in arid regions: review and critique of estimation methods. *Hydrol. Process,* 2: 255–266.

Ghurm, A. and Basmaci, Y., 1983. Hydrogeology of Wadi Wajj: In *Seminar on Water Resources in the Kingdom of Saudi Arabia*, 17–20 April, King Saud University, Riyadh, Saudi Arabia.

Gleick, P., 1993. *Water in Crisis: A Guide to the World's Fresh Water Resources*. Oxford University Press, Oxford, England.

Grismer, M. E., Bachman, S., and Powers, T., 2000. A comparison of groundwater recharge estimation methods in semi-arid, coastal avocado and citrus orchard, Ventura County, California. *Hydrol. Processes,* 14: 2527–2543.

Hantush, M. S., 1956. Analysis of data from pumping test in leaky aquifers. *Trans. Amer. Geophys. Union.*, 37, 702.

Hantush, M. S., 1959. Nonsteady flow to flowing wells in leaky aquifers. *J. Geophys. Res.*, Vol. 64, 1043.

Harhash, I., 1980. Runoff rainfall and recharge evaluation in Qatar 1972–79. FAO, Rome, Tech. Note 7, *Water Res. Agric. Devel. Project*, Govt. Qatar, p. 84.

Harrington, G. A., Cook, P. G., and Herczeg, A. L., 2002. Spatial and temporal variability of groundwater recharge in central Australia: a tracer approach. *Ground Water* 40(5): 518–528.

Hellwig, D. H. R., 1973. Evaporation of water from sand: The loss of water into the atmosphere from a sandy river bed under arid climatic conditions. *J. Hydrol.*, 18, 305–316.

Hem, J. D., 1970. *Study and Interpretation of the Chemical Characteristics of Natural Water*, 2nd Ed. U.S. Geological Survey Water Supply Paper 1473, U. S. Department of Interior, Washington, D.C., p. 363.

Herweijer, J. C. and Young, S. C., 1991. Use of detailed sedimentological information for the assessment of aquifer tests and tracer tests in a shallow fluvial aquifer, In *Proc. 5th Annual Canadian/American Conf. Hydrogeol.* Parameter Identification and Estimation for Aquifer and Reservoir Characterization, Natl. Water Well Assoc., Dublin, OH, pp. 101–115.

Houston, J., 1990. Rainfall-runoff-recharge relationships in the basement rocks of Zimbabwe. In: Lerner, D. N., Issar, A. S., and Simmers I., (Eds.) *Groundwater Recharge: A Guide to Understanding and Estimating Natural Recharge.* International Association of Hydrogeologists (IAH) Publication, International Contribution to Hydrogeology, Vol. 8: 271–283.

Jones, I. C., Banner, J. L., and Humphrey, J. D., 2000. Estimating recharge in a tropical Karst aquifer. *Water Resour. Res.,* Vol. 36, No. 5: 1289–1299.

Kirk, S. T. and Campana, M. E., 1990. A deuterium-calibrated groundwater flow model of carbonate-alluvial system, *J. Hydrol.,* Vol. 119: 357–388.

Kondoh, A. and Shimada, J., 1997. The origin of precipitation in Eastern Asia by deuterium excess, *J. Japan Soc. Hydrol. Water Resour.,* 10 (6): 627–629.

Kruseman, G. P. and De Ridder, N. A., 1970. *Analysis and Evaluation of Pumping Test Data.* Int. Inst. For land reclamation and Improvement, Bull. 11, Wageningen.

Leguy, C., Rindsberger, M., Zangwil, A., Issar, A., and Gat, J. R., 1983. The relation between the oxygen-18 and deuterium contents of rainwater in the Negev Desert and air mass trajectories. *Isotope Geosciences,* 1: 205–218.

Lerner, D., 1990. Techniques: precipitation recharge. In: Lerner, D. N., Issar, A. S., and Simmers I., (Eds.) *Groundwater Recharge: A Guide to Understanding and Estimating Natural Recharge.* International Association of Hydrogeologists (IAH) Publication, International Contribution to Hydrogeology, Vol. 8: 11–147.

Linsley, J. K., Kohler, M. A., and Paulhus, J. L., 1975. *Hydrology for Engineers*, McGraw-Hill Co., New York, p. 482.

Lloyd, J. W., 1986. A review of aridity and groundwater. *Hydrol. Process,* 1: 63–78.

Lyles, B. F. and Hess, J. W., 1988. Isotope and iron geochemistry in the vicinity of the Las Vegas Valley shear zone, University of Nevada, *Desert Research Institute Publication* 41111, p. 78.

Meier, P. M., Carrera, J., and Sanchez-Vila, X., 1998. An evaluation of Jacob's method for the interpretation of pumping tests in heterogeneous formations. *Water Resour. Res.,* Vol. 34, No. 5, 1011–1025.

Munich, K. O., 1968a. Moisture movement measured by isotope tagging, *Guidebook on nuclear techniques in Hydrology,* IAEA, Vienna, 112–117.

Munich, K. O., 1968b. Use of nuclear techniques for the determination of groundwater recharge rates, *Guidebook on Nuclear Techniques in Hydrology,* IAEA, Vienna, 191–197.

Musgrove, M. and Banner, J. L., 1993. Regional ground-water mixing and the origin of saline fluids: Mid-continent, United States, *Science*, p. 259.

Neuman, S. P. and Witherspoon, P. A., 1968. Theory of flow in aquicludes adjacent to slightly leaky aquifers. *Water Resour. Res.,* 4(1): 103.

Neuman, S. P. and Witherspoon, P. A., 1969a. Transient flow of groundwater to wells in multiple-aquifer systems, Geotechnical Engineering Report, University of California, Berkeley.

Neuman, S. P. and Witherspoon, P. A., 1969b. Theory of flow in a two aquifer system. *Water Resour. Res.,* 5(4): 817.

Neuman, S. P. and Witherspoon, P. A., 1972. Field determination of the hydraulic properties of leaky multiple aquifer systems. *Water Resour. Res.*, 8: 1284–1298.

Noory, M., 1983. *Water and the Progress of Development in the Kingdom of Saudi Arabia* (in Arabic), Jeddah, Saudi Arabia, Tihama Publisher, p. 302.

Piper, A. M., 1953. A graphical procedure in the geochemical interpretation of water analysis. *USGS Groundwater Note*, 12.

Rehm, B. W., Moran, S. R., and Greenwood, G. H., 1982. Natural groundwater recharge in an upland area of central North Dakota, *U.S. J. Hydrol.* 59, 293–314.

Rozanski, K, Araguias, L., and Gonfiantini, R., 1993. Isotopic pattern in modern global precipitation. In: *Climate Change in Continental Isotopic Record, Geophysical Monograph* 78, Swart, P. K., Lohman, K. C., McKenzie, J., and Savin, S., (Ed.), American Geophysical Union, Washington, D.C., 1–36.

RSMC, Red Sea Mining Company, 1986. Hydrogeological Investigations Makkah Project, Compiled Report, Wadi Ibrahim.

Rushton, K. R., 1988. Numerical and conceptual models for recharge estimation in arid and semi-arid zones: In: Simmers, I. (Ed.), *Estimation of Natural Groundwater Recharge*, D. Reidel Publishing Co., Dardrecht, 223–238.

Rushton, K. R., 1990. Recharge in the Mehsana alluvial aquifer, India. In: Lerner, D. N., Issar, A. S., and Simmers I., (Eds.) *Groundwater Recharge: A Guide to Understanding and Estimating Natural Recharge.* International Association of Hydrogeologists (IAH) Publication, International Contribution to Hydrogeology, Vol. 8: 297–312.

Russo, S. L., Zavattaro, L., Acuits, M., and Zuppi, G. M., 2003. Chloride profile technique to estimate water movement through unsaturated zone in a cropped area in sub-humid climate, Po Valley, NW Italy, *J. Hydrol.* 270: 65–74.

Sakthivadivel, R., Nihal, F., and Jeffrey, B., 1997. *Rehabilitation Planning for Small Tanks in Cascades: A Methodology Based on Rapid Assessment.* IWMI Research Report 13. Colombo, Sri Lanka, International Water Management Institute.

Salih, A. M. and Şendil, U., 1984. Evapotranspiration under extremely arid climates. *J. Irrigation and Drainage Eng.*, Vol. 110, No. 3, pp. 289–303.

Sanchez-Vila, X., Meier, P. M., and Carrera, J., 1999. Pumping test in heterogeneous aquifers: An analytical study of what can be obtained from their interpretation using Jacob's method, *Water Resour. Res.*, Vol. 35, No. 4, 943–952.

Scanlon, B. R., Healy, R. W., and Cook, P. G., 2002. Choosing appropriate techniques for quantifying groundwater recharge, *Hydrogeol. J.,* 10: 18–39.

Schad, H. and Teutsch, G., 1994. Effects of investigation scale on pumping test results in heterogeneous porous aquifers, *J. Hydrol.*, 159: 61–77.

Scholl, M. A., Ingebritsen, S. E., Janik, C. J., and Kauahikaua, J. P., 1996. Use of precipitation and groundwater isotopes to interpret regional hydrology on a tropical volcanic island, Kilaue volcano area, Hawaii, *Water Resour. Res.*, 32(12), 3525–3537.

Simmers, I., Hendricks, J. M. H., Kruseman, G. P., and Rushton, K. R., 1997. Recharge of phreatic aquifers in semi-arid regions, *IAH Inrl. Contrbsn. Hydrogeol.,* 19.

S.M.M.P., 1975. Riyadh Additional Water Resources Study Report, Ministry of Agriculture and Water, Riyadh, Saudi Arabia.

Sogreah, 1968. Water and Agricultural Development Survey for Area IV, Final Report, Ministry of Agriculture and Water, Riyadh, Saudi Arabia.

Sorey, M. L. and Matlock, W. G., 1969. Evaporation from an ephemeral streambed, *J. Hydraul. Div. Am. Soc. Civ. Eng.*, 95, 423–438.

Stone, W. J., 1984. Preliminary estimates of the Ogallala aquifer recharge using chloride in the unsaturated zone, Curry country, New Mexico, In: George A. Whetstone (Ed.), *Proc. Ogallala Aquifer Symposium II*, Water Resources Center, Texas Tech. Univ., Lubbock, p. 593.

Streltsova, T. D., 1988. *Well Testing in Heterogeneous Formations*, John Wiley, New York, p. 413.

Subyani, A. M. and Bayumi, T. H., 2001. Evaluation of groundwater resources in wadi Yalamlam basin, Makkah area. King Abdulaziz University, Project No. 203/420, Jeddah, Saudi Arabia.

Subyani, A. M., 2005. Hydrochemical identification and salinity problem of ground-water in Wadi Yalamlam basin, Western Saudi Arabia, *J. Arid Environments,* 60: 53–66.

Subyani, A. M. and Şen, Z., 1991. Study of recharge outcrop relationship of the Wasia Aquifer in central Saudi Arabia. King. Abdulaziz Univ., *Earth Sci.*, Vol. 4, p. 137–147.

Subyani, A. M. and Şen, Z., 2006. Refined chloride mass-balance method and its application in Saudi Arabia, *Hydrol. Process,* 20, 4373–4380.

Şen, Z., 1983. Hydrology of Saudi Arabia, water resources in the Kingdom of Saudi Arabia, management, treatment and utilization. Vol. 1, College of Eng., King Saud University, Riyadh, A68–A94.

Şen, Z. 1985. Volumetric approach to type curves in leaky aquifers, *J. Hydraulic Eng.-ASCE* 111 (3): 467–484.

Şen, Z., 1986a. Discharge calculation from early drawdown data in large-diameter wells, *J. Hydrol.* 83 (1–2): 45–48.

Şen, Z., 1986b. Determination of aquifer parameters by the slope matching method, *Ground Water*, Vol. 24, No. 2, pp. 217–223.

Şen, Z., 1989. Volumetric approach to multiaquifer and horizontal fracture wells, *J. Hydraulic Eng.-ASCE* 115 (12): 1646–1666.

Şen, Z., 1990. Nonlinear radial flow in confined aquifers toward large-diameter wells, *Water Resources Research*, 26 (5): 1103–1109.

Şen, Z., 1995. *Applied Hydrogeology For Scientists and Engineers*, CRC Press, Inc., Boca Raton, FL, p. 444.

Şen, Z., 1996. A graphical method for storage coefficient determination from quasi steady state flow data. *Nordic Hydrol.,* 27 (4): 247–254.

Şen, Z., 2000. Non-Darcian groundwater flow in leaky aquifers. *Hydrol. Sci. J.*, 54(4), 1–14.

Şen, Z., Saud, A. A., Altunkaynak, A., and Özger, M., 2003. Increasing water supply by mixing of fresh and saline ground waters, *J. Am. Water Resour. Assoc.*, Vol. 39, No. 5., 1209–1215.

Şen, Z., 2007. Groundwater flow from fractured layer to porous-media blind variable large diameter well. ASCE, *Hydraul. Eng. Div.*, (in print).

Şen, Z. and Dakheel, A. 1985. Hydrochemical Facies Evaluation in Umm Er Radhuma Limestone Eastern Saudi Arabia. *Ground Water* 24(5): 626–635.

Şen, Z. and Wagdani, A, 2007. Aquifer heterogeneity determination through slope method. *Hydrol. Process.*, (in print).

Şorman, A. U. and Abdulrazzak, M. J., 1993. Infiltration — recharge through wadi beds in arid regions, *Hydrol. Sci. J.*, 38, 3, 173–186.

Theim, G., 1906. *Hydrologische Methoden*, Gebhart, J.M., Leipzig, p. 56.

Theis, C. V., 1935. The relationship between the lowering of piezometric surface and the rate and duration of discharge of a well using ground water storage, *Trans. Am. Geophys. Union*, Vol. 16, pp. 519–529.

Thomas, J. M., Wolch, A. H., and Dettinger, M. D., 1996. Geochemistry and Isotope hydrology of representative aquifers in the great basin region of Nevada, Utah, and adjacent states, U. S. Geological Survey Professional Paper 1409–C, p. 100.

Ting, C. S., Kerh, T., and Liao, C. J., 1998. Estimation of groundwater recharge using the chloride mass-balance method, Pingtung Plain, Taiwan. *Hydrogeol. J.* 6: 282–292.

Wilcox, L. V., 1955. *Classification and Use of Irrigation Waters*, U. S. Dept. Agric. Circ. 969, Washington, D.C., p. 19.

Witherspoon, P. A., Javandel, I., and Neuman, S. P., 1968. Use of the finite element method in solving transient flow problems in aquifer systems, In *The Use Of Analog and Digital Computers in Hydrology*, AIHS Publication 81, 2:687.

Wood, W. W. and Sanford, W. E., 1995. Chemical and isotopic methods for quantifying groundwater recharge in a regional, semiarid environment, *Groundwater*, Vol. 33, No. 3, 458–468.

Yurtsever, Y., 1975. Worldwide survey of stable isotopes in precipitation. *Rep. Sect. Isotope Hydrol, IAEA*, November 1975, p. 40.

Yurtsever, Y. and Gat, J. R., 1981. In: *Stable Isotope Hydrol.* (Eds.) Gat, J. R. and Gonfiantini, R. (IAEA Tech Rep Ser 210) 103–142.

Zimmerman, U., Munninch, K. O., and Roether, W., 1967. Downward movement of soil moisture traced by hydrogen isotopes, *Geophy.* Monograph 11, American Geophysical Union, Washington, 28–36.

# 6 Groundwater Management Methods for Arid Regions

## 6.1 GENERAL OVERVIEW

In addition to severe water scarcity in arid regions, water resources in wadi hydrologic systems are characterized by significant spatial and temporal variability. Such variations lead to either drought periods that last for several years or to short-duration heavy rainfall that causes floods with loss of life and property damage. Recently, the developing countries, especially in arid and semi-arid regions, have experienced important socioeconomic developments with an increase in water demand. Water resources mobilization for water supply purposes, mainly for agriculture and industry, has increased almost to 80% in these regions. This rate exceeds 100% in some countries, and in order to meet the expected socioeconomic development and to ensure food security, there are countries that have had to take recourse in unconventional water resources, such as desalination units, despite the high cost.

In general, assessment and management are the key factors in any *integrated development* strategy for water resources. Without a correct and detailed assessment, it is almost impossible to plan, design, realize, and manage any water resources development project. The results of the assessment and subsequent management are the bases for any decision-making process, as they can lead to large investments and serious consequences to the environment. Proper management and utilization of wadi flow depend essentially on the availability of the data, as well as analysis techniques. Unfortunately, in many arid and semi-arid countries water resources data are not always available, which has negative effects on the development process. Water resources assessment consists of determining their quantity, quality, and variability for sustainable development and rational management.

This chapter provides simple groundwater management rules in a single wadi or among multiple wadis by considering a set of wells or sub-basins with rational and logical approaches for arid regions.

## 6.2 SAFE YIELD AND GROUNDWATER MANAGEMENT

Groundwater management deals with the complex interaction between human societal activities and the physical environment, which poses an extremely complex and difficult problem to solve for the benefit of all parties involved. Aquifers are common groundwater storage resources for society that cannot be regulated. Those using them are little motivated toward conservation as they are aware anyone at any time can exploit them without restriction. Consequently, there is free usage by well owners with no thought of recharge issues or how society at large may be affected by any resultant water shortages. Among management strategies the following actions need consideration:

1. Adjustment of the annual rates of pumping, especially in arid and semi-arid regions; monthly rates are important.
2. Adjustment of well configurations.
3. Augmentation of the water supply from other sources or groundwater recharge enhancement.
4. Increasing awareness of groundwater beneficiaries.
5. Limiting use according to data from monitoring.

Meinzer (1920) defined the *safe yield* of an aquifer as the water that can be abstracted permanently from an aquifer without producing any undesirable results. On the other hand, Todd (1976) defined the *perennial yield* as the flow of water that can be abstracted from a given aquifer without producing results that lead to an adverse situation. Each aquifer will have a different safe yield, depending on the way it is exploited, on changes in natural recharge due to land-use modifications, and on the introduction of engineered changes in recharge, such as different forms of artificial recharge. The key point in each of these definitions is the consensus on the "undesirable" or "adverse" results. This is due to the fact that the assessment of groundwater resources differs with the consumer and the administrator, scientists, government officers, farmers, environmentalists, etc. In arid regions, the deeper the well, the more the groundwater potentiality, which causes friction between rich and poor people. All the above mentioned points are from the quantity point of view; however, the groundwater quality is more of a concern in many cases than the quantity. The groundwater quality variations are more difficult to control, manage, and solve, and they are not well perceived and understood by the public, mass media, and even decision makers, engineers, and hydrologists. Once an aquifer is depleted, especially in arid regions, it will take generations to recover. The issue is not conservation for conservation's sake, but maintaining a vital resource for use in perpetuity. Strategic planning does not mean either aquifer overexploitation or storage of groundwater more than necessary.

In any strategic groundwater resources planning the following points must be taken into consideration for a successful management policy.

1. Physical dimensions are a determining factor in the exploitable water quantity and possible groundwater recharge amounts. It is necessary that a hydrodynamical equilibrium is held between the natural and artificial recharge and groundwater abstraction rates. It is always better to try to reestablish a depleted resource.
2. Chemical dimensions are concerned with the groundwater quality variations during recharge or exploitation processes. Degradation of groundwater quality must not be allowed.
3. Economic consequences in the short and long term must be considered.

4. Social impact is also important. Groundwater use between competing users on unequal terms with detrimental effects suffered by third-party users must be avoided.
5. Environmental sustainability is also important. Damage to the natural environment, especially of sensitive aquatic ecosystems, must be avoided.

In any strategic planning of groundwater resources, the following questions are the preliminary requirements for efficient planning:

1. From what point of view are we considering exploitation of groundwater resources? Physical, hydraulic, social, engineering?
2. Is it necessary to maintain a minimum level for the groundwater table, especially in unconfined aquifers?
3. Is it necessary to maximize the annual benefits and overturn from an economic point of view?
4. Do the groundwater quality variations play a significant, or perhaps the only role, in groundwater allocation in emergency cases?
5. Is it necessary to store a certain amount of groundwater volume in the aquifer so as to cover the needs of a certain size of population?

There are two main questions facing the decision makers and experts concerning the possibility of aquifer overexploitation:

1. How far might the aquifer be over exploited from its long-term equilibrium situation? There must be a definite guideline for the allowable overexploitation of the aquifer.
2. What are the ways of using aquifer reserves? The variety of hydrogeological conditions must be considered with priorities.

In nonequilibrium conditions and especially for overexploitation, it is not necessary only that the total pumping discharge, $Q_p$, is greater than the average recharge, $Q_r$, as

$$Q_p > Q_r$$

but, more specifically,

$$Q_p > \Delta Q_a + \Delta Q_r$$

where, $\Delta Q_a$ is the induced decrease of the average outflow, which is equivalent to the natural yield of the aquifer, and $\Delta Q_r$ is the induced increase in the recharge flow.

In any effective groundwater management, it is necessary to link flow management with storage management in variable proportions so the exploitation will continue in a harmonious way. All groundwater exploitations begin with a phase of nonequilibrium during which some part of the water volume produced is removed from the reserve. The length of the nonequilibrium phase depends on these two factors:

1. The rate of groundwater abstraction and, hence, on the aquifer parameters.
2. Boundary conditions and the rapidity of compensatory reactions, e.g., reduction in the natural outflow and increase in the induced inflow.

According to the exploitation strategies, three methods of use are conceivable and practical to carry out management practices of the aquifer reserve, but under certain constraints (Margat and Saad, 1983):

1. A strategy of maximum and lasting exploitation of the renewable resources should be devised in a regime of dynamic equilibrium, with average abstraction which is approximately equal to average recharge (possibly enhanced by boundary effects), but not without taking into account seasonal or even possible annual variations (increased abstraction in periods of drought). Thus, after a decrease in the initial phase of nonequilibrium, the stabilized reserve is used, usually, as a regulatory factor, annually or multiannually. Its natural regulatory function will be amplified under the constraint of preserving a minimum flow rate at the outflow boundaries of the aquifer (springs, low water level in draining streams, etc.) or preserving freshwater/seawater equilibrium in a coastal aquifer.
2. In a strategy of repeated exploitation of the storage and then through flow only, in a prolonged unbalanced regime, that may be "guided" or unintentional in the initial phase, and in which abstraction (increasing or stabilized) is greater than recharge (even when enhanced by boundary effects), a second phase involves reducing abstraction to restore the equilibrium. In this case, the depletion of the reserve contributes largely, and sometimes predominantly, to the production of water, during an initial phase of limited medium to long-term duration, which is limited either by external constraints (see strategy 1 in later text) or by a reduction in the productivity of wells (drawdown by deteriorating aquifer conditions and limited by the base of the aquifer). In the later phase of possible reequilibration, the reserve may be:

   a. Stabilized, on average, provided that it has reached an equilibrium bringing abstraction close to recharge.
   b. Restored, in part, by reducing the abstraction the below average (repayment of the loan) and sometimes by artificially increasing recharge, and then at a new equilibrium.

3. A strategy of unrestricted mining or exhaustion exploitation, with abstraction increasing or otherwise, greater than the average recharge. In this case, depletion of the reserve provides most of the water produced, and the exploitation is in the long term more or less limited when drawdown becomes excessive without later returning to a regime of reequilibrium. The recovery of the reserve may be too slow and sometimes hindered by irreversible degradation of the original capacity of the reservoir due to subsidence.

All aquifers do not lend themselves indifferently to these methods of management. A detailed account is given as follows.

1. Strategy 1: This is appropriate for unconfined aquifers of small and medium capacity and of limited thickness, with a high rate of recharge (reserves of the order of one to ten times the average annual volume of discharge), and includes the case where a confined aquifer is exploited near the point where it becomes unconfined. This strategy is used particularly in cases where constraints limit the possible drawdown (to preserve open water bodies or watercourses, conserve hydraulic links with surface water supply sources, prevent the displacement of the freshwater/seawater interface, etc.), whatever the rate of recharge of the aquifer.

2. Strategy 2: This is more convenient for unconfined and semiconfined aquifers with high capacity, and a small to medium recharge rate (reserves of the order of ten to a hundred times greater than the average volume of annual recharge), without appreciable constraints for the conservation of water levels. For example, this strategy is appropriate in an aquifer with outlets that are independent of watercourses (areas of evaporation in arid regions).

3. Strategy 3: This is the only one possible for aquifers of high capacity with little recharge (reserves on the order of one hundred to thousands of times greater than the average annual volume of recharge). Development can take place in the unconfined zone and also in the confined zone where the diffusivity is high. This is the case for most of the deep aquifers in sedimentary basins, the most transmissive layers of which drain the less permeable but higher-storage layers (aquitards).

## 6.3  MANAGEMENT ENVIRONMENTS IN ARID REGIONS

Water resources management can be defined as a set of actions to be realized for medium- and long-term planning. These actions have to consider all recommendations and measures resulting from the planning process. *Water resources management* is a daily or monthly task that has to be carried out with the participation of all parties related to the water sector. Quantitative and qualitative aspects must be taken into consideration as:

1. Equitable distribution of water according to the allocation plan.
2. Rational and efficient use of water.
3. Protection of water quality and aquatic ecosystems.
4. Management of exceptional phenomena, such as drought, floods, etc.

Groundwater reservoirs are the most abundant and dependable freshwater sources in many parts of the world. Their mineralogical composition makes the taste and chemical composition very suitable for various human activities including domestic, agricultural, and industrial uses. Depending on the type of geological formations, this suitability has the best quality, especially in quartz and sandstone formations where soft and filtered groundwater appears as potential supply resources. Groundwater is almost the only water source in arid and semi-arid regions, but it is not available in desired quantities temporally and spatially. Sporadic, rare, and haphazard rainfall occurrences cause rather unpredictable natural groundwater recharge possibilities (Chapter 5). It is, therefore, necessary to lay down the strategic groundwater reservoir assessment basic variables and their subsequent scientific and technical processing for future development and management. The general features in any groundwater future planning should include the following points:

1. Spatial environment—In general, groundwater resources may cover more than 4 km in depth within the earth's crust. Its exploitation cannot be done practically at more than 1 km depth. However, most often the groundwater wells and boreholes have depths varying between a couple of meters and 300 to 500 m. According to the depth, the groundwater resources can be thought of as *shallow* or *deep*. With the advancement of future technology, greater depths may be reached, but this is not advisable strategically because there may be a general decrease in the groundwater levels (tables or piezometric levels) all over the world. Besides, in the wadi alluviums of arid regions, the groundwater is frequently in the form of unconfined aquifers where the saturation thickness is rarely over 100 m. There is no possibility of drilling deeper wells to obtain more groundwater volume. Hence, depth is one of the restrictive factors in groundwater resources planning. Apart from thickness, the areal extent of the groundwater reservoir plays a role as one of the spatial variables. It is necessary to know the surface area of recharge exposure in any strategic groundwater resources planning.

2. Temporal environment—In any groundwater exploitation, the connection of groundwater reservoirs with the present-day hydrological cycle must be considered, and according to replenishment rates, groundwater withdrawals should be planned and managed. Groundwater reservoirs may be classified as *renewable* and *nonrenewable* according to their connections with the hydrological cycle. In arid regions, deep formations are meagerly in contact with the present-day hydrological cycle. However, occasional floods, flash floods, or large runoff volumes provide some natural and artificial recharge possibilities. For instance, wadi alluviums are direct recharge exposure areas, but temporal recharge source as rainfall is scarce. It is, therefore, necessary to take into account the haphazard variability of rainfall in any groundwater recharge study.

3. Hydrogeological environment—Geological features play a dominant role in any groundwater resources evaluation, assessment, and exploitation. These features help to connect the surface runoff to the groundwater reservoir during a recharge phenomenon. Such a connection is dependent on the composition of the geological formation through voids, fissures, fractures, crevices, solution cavities, faults, and other structural features. The amount of *porous, fractured,* and *karstic* media should be considered in effective groundwater planning. Wadis are composed of mainly quaternary deposits, in addition to a weathered and fractured basement. The hydrogeological

parameters such as the storativity and transmissivity are essential quantities in any groundwater planning and management.

4. Hydrochemical environment—Groundwater quality, ionic exchange, salt-water intrusion, and proximity to the seashore are additional factors that must be taken into consideration in any planning and *management* of an *aquifer*. In general, quality deteriorations occur from the upper part of a wadi toward the downstream. Comparatively worse quality water zones within the study area cannot be overlooked, as artificial mixture of different groundwater qualities may give rise to acceptable water quality (Şen et al., 2005). On the other hand, it is also possible to mix desalination plant water with natural groundwater so as to obtain better quality.

5. Alternative strategies—Planners or decision makers should have different alternatives for the same problem, so that they can decide on the most suitable solution under local circumstances. There are alternative water resources such as runoff, desalinization, etc. In dry regions, *desalination plants* provide an additional water source to current water supply system. It is, therefore, very essential in arid regions to exploit this source as much as possible, so as to preserve groundwater reservoirs for emergency uses in the future. If there are two or more wadis as groundwater reservoir alternatives, then the question is which groundwater reservoir should have the priority of exploitation? How long should it be exploited for before the second alternative enters the circuit of planning?

6. Emergency situations—For water resources strategic planning in general, and groundwater reservoirs in particular, the planner must keep in mind that in cases of emergency involving diseases, terrorist attacks, wars, and natural destructions such as earthquakes, landslides, etc., the infrastructure for other water resource supply systems may malfunction or not function at all, and in such cases the groundwater supplements have the highest precious levels.

## 6.4   ARID REGIONS AND WADI CONDITIONS

Apart from the above-mentioned general features, the groundwater resources are also location dependent, including geological, geomorphologic, tectonic, climatologic, social, economical and practical factors. Generally, the world is divided into two broad categories by considering the water availability in arid or humid regions. It is rather interesting that such a division sometimes may not conform to groundwater resources, which are subsurface geology dependent. In arid regions, the groundwater resources play a major role in various sectors, and therefore, their planning, operation, management, and maintenance have key importance in exploitation, especially in emergency stages. For groundwater resources planning in arid regions, the following specific points must be considered:

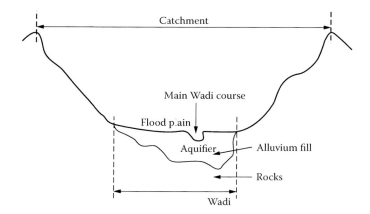

**FIGURE 6.1**   Wadi cross-section.

1. Arid land catchments are almost without vegetation cover but their surface may have volcanic-origin necked rocks, such as *basalts*. In these environments, initially, the rainfall transforms to runoff from the surface in a short time almost without infiltration. It is necessary to know the surface geological and morphological features of any catchments in arid regions so as to define the term *wadi fills*, which is considered herein as the alluvium deposits at lower elevations across any lateral cross-section along the main channel (see Figure 6.1). The plain areas at these lower lands are the main courses of surface flow from relatively higher sides. Infiltration with consequent groundwater recharge takes place in these regions. It is important to know the areal extend and infiltration properties of these plains.

2. The rainfall–runoff relationships must be taken into consideration by simple techniques, in addition to flood as well as flash flood occurrences (Chapter 4). Such simple considerations may lead to further extensive evaluations that may furnish a simple strategy rather than the use of complicated software models, which are developed most often for humid regions. Risk attachment to various variables provides different alternatives to planners who can then make their decisions according to present and local conditions.

3. Across the wadis, beneath the plain areas, the subsurface geological compositions must be prospected by using different approaches such as geophysics, direct drilling, and observations in cuts (road construction, quarrying, etc.). These will help to identify the geometric dimensions of saturated and unsaturated areas and volumes which are necessary in any groundwater resources evaluation.

4. The geometric dimensions of subsurface cross-sections must be determined through extensive geophysical prospecting, especially by use of vertical electrical survey (VES). The surface and subsurface basement longitudinal slopes are also important for surface and subsurface groundwater flow movement evaluations. Here, even more than the groundwater movement, its present and future possible augmentable volumes, i.e., storativity features, are important. The groundwater movement and abstraction through a

set of optimum number of wells with suitable locations is the topic, which may be considered under the title of Groundwater Management and Well Optimization studies (see Section 6.7).

5. Groundwater storage calculations require determination of different hydro-geological parameters such as porosity, specific yield (storage coefficient or effective porosity) and hydraulic conductivity for actual water volume calculations. It is preferable to obtain parameter estimations by field techniques and, especially, through aquifer tests and their proper interpretations in different wells (Chapter 5).

6. Especially in the cases of renewable groundwater reservoir planning, the groundwater quality is significant not only in management, but also regarding excessive exploitation. Naturally, the groundwater reservoir does not have uniformity, and different types of water may exist. In general, the bottom layer is comparatively more saline, and therefore its intrusion to abstractable water must be avoided.

7. Safe yield on exploitation must be considered for the natural replenishment and movement of groundwater within the reservoir. A certain amount of water must be left within the aquifer for uses in cases of possible future emergencies.

8. The population and its future growth must be considered for groundwater resource allocation. The minimum requirements should be taken into account for per capita use, which might be equal at least to 50 L/day/person.

9. Private well owners must be convinced to allow strategic planning of groundwater reservoirs by avoiding haphazard exploitation. This requires a political decision, which should apply to everybody without privilege. Besides, experienced personnel should observe the whole groundwater storage situation both temporally and spatially, in addition to maintaining well-established monitoring networks.

## 6.5  STRATEGIC GROUNDWATER STORAGE PLANNING

Groundwater resources are precious in arid regions where recharge occurrences are rare with sporadically distributed storm events. These storms are the main sources of groundwater replenishment, and recharge capabilities should be planned and managed in such a way that the loss of source should not be allowed, even partially. In arid regions, the groundwater resources must be planned according to some strategic rules, which are marginal calculations at both extremes (dry and wet periods; see Chapter 3). During wet periods, the maximum possible benefit should be rendered for recharge facilities. On the other hand, extensive dry periods should be managed in a rational, restrictive, and systematic manner. Strategic planning is essential, especially where there exist rather random and extreme variations in hydrologic features and demand. These variations should be incorporated economically or socially within any strategic planning program according to a set of sequential rules for maximum benefit achievements.

In general, there are two strategic planning procedures in a region, depending on a number of wadi groundwater sources. Here, the meaning of a region is the area lying within at least two wadis, so that they can complement each other for joint strategic exploitation. Hence, in regional planning, two complementary studies are needed:

1. *Within-wadi* (microplanning)
2. *Inter-wadi* (macroplanning)

### 6.5.1   WITHIN-WADI STRATEGIC PLANNING

The wadi is considered as a whole regarding rainfall occurrences and the wadi course and fills along the main channel for strategic groundwater recharges and withdrawals. Different parts of the wadi must be identified with its special features that are meaningful in the strategic planning and management. For instance, different branches and parts of the main wadi may provide competing alternatives. In a within-wadi strategic planning study, the following points are significant for consideration.

1. The drainage basin is considered as composed of either sub-basins or sub-areas with different strategic parameter variations. Generally, any wadi area is considered as having three parts (upstream, midstream, and downstream), which are rather arbitrarily divided according to expert views. On the other hand, each branch of stream (or sub-basin) can be taken as a potential strategic planning unit. This is a useful division procedure for surface water resources planning.
2. Drainage area may also be considered as composed of different hydro-geo-stratigraphic units as aquifers and nonaquifers. Such a subdivision is useful from groundwater resources point of view, but it needs detailed information about the subsurface geological and, especially, hydrogeological information.
3. Division of any drainage basin can be based on the groundwater quality zones, and such a division favors strategic groundwater resources planning for water quality satisfactions (Section 6.8).
4. The drainage area can be divided according to social, industrial, military, and agricultural activities. This is tantamount to saying that subdivision achievement can be obtained on the basis of land use priorities.
5. It is also possible to consider the subdivision of a drainage basin by recharge (infiltration and its augmentation) possibilities, because the recharge areas have major significance for groundwater storage increments.
6. The strategic significance of a wadi can be based also on human activities, water shortages for certain periods, floods, and other natural and social impacts. For instance, if a drainage basin is far away from human activities, whatever the groundwater resources availability, it will not be included in strategic planning purposes in the short-term projections.

The rational and methodological comparisons of subdivisions will lead to the best management strategy for wadi water resources. Prior to a regional study, strategic sub-areas of each wadi must be depicted with various types of information and knowledge.

## 6.5.2 INTER-WADI STRATEGIC PLANNING

This involves large-scale (macroplanning) strategic groundwater resources planning, including two or more wadis with their competitive properties. In large-scale strategic planning, the following points must be taken into consideration:

1. It is necessary that there is an actual or expected future possible cause for groundwater overexploitation. For instance, expansion of a city might lead to ongoing strategic planning of surface water and, especially, groundwater resources in an arid area. Population growth, land use, scarcity of rainfall occurrences, water quality deterioration due to mismanagement, and uncontrolled pollution activities are among strategic groundwater planning and management ingredients.
2. It is preferable to interchange and to exploit groundwater resources jointly in nearby rather than far-away wadis, especially for efficiency at the time of an emergency. If the regional hydrologic and subsurface groundwater balances require it, it is possible to transport groundwater between adjacent wadis.
3. The competitive divisions on a small scale (within-basin) are considered first as to how they compete with each other. Then, other small-scale features of different wadis in the same region must be allowed to compete with the previous considerations. It may be feasible to transport and consume groundwater resource from some nearby wadis.
4. It is possible that each division from different wadis may have different strategic planning directions, and in such a situation the priorities must be given according to local administration or central government views. Inter-wadi planning should be effective after the microplanning strategies are completed in wadis. Furthermore, the completion of macroplanning is necessary to conduct any study on groundwater management and well optimization, which must be based on all the strategic (micro- or macro-) planning studies.

## 6.6 PROBABILISTIC RISK MANAGEMENT IN AN AQUIFER

Stochastic hydrogeology as suggested by Delhomme (1978) deals with uncertainty in groundwater resources evaluation and contaminant transport problems. Prevailing hydrogeological conditions in an area are controlled by the distribution and occurrence of lithofacies. Although various geometrical (extent, areal coverage, thickness, slope, etc.), structural (fault, folds, fractures, fissures, joints), textural (porosity, specific surface, grain size composition, sorting, etc.), hydraulic (hydraulic conductivity, storage coefficient, etc.) and geological properties are random in various degrees, hydrogeologists generally use deterministic approaches to problem solving. Deterministic approaches give satisfactory single point results on the local scale, but on a regional scale (because of the heterogeneities, discontinuities, and anisotropies), use of such a point value fails to provide a reliable solution. For instance, pumping tests provide representative and reliable aquifer properties within the *radius of influence* that is generally less than 300 to 400 m in confined aquifers and approximately 100 m in unconfined aquifers (Şen, 1995). Deterministic results cannot represent some or larger portions of the aquifer in wadis. Hence, the simplest way of regionalizing a variable or a parameter is through probabilistic and statistical techniques.

Any measured data either in the field or in the laboratory represent quantities that are numerically different from each other. Randomness does not mean that the hydrogeological parameter values are spatially independent, but they are not predictable at any particular site even though their values might be available at a set of neighboring sites. For instance, the storage coefficient for an aquifer will be different at different locations. In order to obtain a representative and reliable estimate for the whole aquifer, the hydrogeologist often uses a single deterministic value such as the mean or the median or an expert value. However, such a simplification brings with it the following disadvantages:

1. It is not possible to make risk assessments of the groundwater problem.
2. The results cannot be related with the real surface or subsurface geology.
3. Theoretical PDFs cannot be established.
4. Correlation analysis cannot be performed between lithological and hydro-geological parameters.

Uncertainties are differences between the true and estimated models or parameters. One way of decreasing uncertainty is by selecting the model that best represents the physical reality of the system. Once the model is judged as adequate, it serves to evaluate the corresponding uncertainties. Hence, two kinds of uncertainties are usually encountered in groundwater studies, namely, model and parameter uncertainties. Model uncertainty results from events that are not known, and at best the model is only at close approximation of reality. On the other hand, parameter uncertainty results because model parameters are estimated from a scarce amount of data. Parameter uncertainty may be estimated by finding the PDF of parameter estimates and by using models with parameters sampled from this distribution (Salas et al., 1985).

The probabilistic approach is but one technique used by geostatisticians to characterize spatial variability and to express a very simple criterion for goodness of estimation (Journel, 1985). Such an approach needs at most a graphical representation of data without the need for computers and hence proves to be more practical.

## 6.6.1 Hydrogeologic Risk

Groundwater occurrence, distribution, and movement are inherently random and, therefore, their measurement and assessment includes elements of uncertainty. Hydrogeologic uncertainty arises as a result of ignorance about hydrogeological and geological phenomena. Although uncertainty and risk might have the same meaning, there exists an important difference between them (Şen, 1978).

A characteristic of *risk* is that it has a definable PDF which, in practice, is presented in the form of a histogram. It can be obtained from a group of observed data values either in the field or in the laboratory (Koch and Link, 1971). Figure 6.2 shows schematically the PDF of, say, groundwater storage capacity, $G_s$.

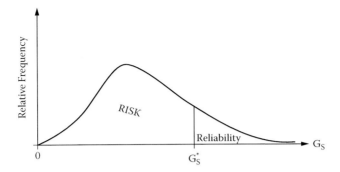

**FIGURE 6.2** Schematic risk–reliability quantification.

However, for a given groundwater storage capacity $G_s^*$, the PDF is divided into two parts, and one of these areas is the quantitative definition of the *risk* according to the problem at hand. For example, if the *groundwater storage capacity* is not desired to fall below $G_s^*$, then the area to the left is considered as risk. Hence, the probabilistic formulation of risk, $R$, is

$$R = P(G_s \geq G_s^*) \tag{6.1}$$

where $P$ (.) is the probability of the argument expression. The definition of risk may, for example, correspond to *safe yield* in a groundwater reservoir. On the other hand, contrary to risk, uncertainties are characterized by unknown and measurable chance caused by sparse data, measurement errors, etc.

## 6.6.2 Risk Assessment of Groundwater

Hydrogeologists may be interested in assessing two important basic composite variables from field measurements. These are:

1. The available volume of a groundwater reservoir within known boundaries
2. The volume of groundwater that the same aquifer can transmit within a given period

Both of these variables are independent physically and give the ability to the hydrogeologist to make decisions about water availability and risk. The volume of the groundwater storage can be found by means of the following simple equation (Davis, 1982)

$$G_s = AhS \tag{6.2}$$

where $G_s$ is the groundwater storage capacity [L³]; $A$ is the aquifer surface area [L²]; $h$ is the aquifer thickness [L]; and $S$ is the storage coefficient. Among the factors on the r.h.s., the area is defined by the aquifer boundary. The larger the area, the more complex the subsurface geology. Therefore, for refining the estimation procedure, it is possible to define the *specific groundwater storage* capacity, $g_s$ as the volume of groundwater per unit area of the aquifer surface, i.e., $g_s = G_s/A$. Hence, from Equation 6.2

$$g_s = hS \qquad (6.3)$$

In this new definition, $h$ represents the thickness of the aquifer per unit area, and $S$ is the storage coefficient for the same unit area. In practical studies, such a unit area may be thought of as a single well where the aquifer thickness and the storage coefficient are both assumed to be spatially and temporally constant. The unit of specific groundwater storage capacity is [L] and it must not be confused with specific storage, which has units of [L$^{-1}$]. On the other hand, the *specific subsurface flow rate, q,* is equal to the volume of groundwater that passes through the whole section of the aquifer per unit width. Hence,

$$q = Ki \qquad (6.4)$$

$K$ is the aquifer *hydraulic conductivity,* and $i$ is the *hydraulic gradient.* In classical groundwater resources assessment, it suffices to substitute into Equation 6.4 a single hydraulic conductivity value obtained from a suitable type curve. Because the basic variables are considered to be random, the composite variables will also be random. There are either probabilistic or statistical ways in order to take into account any over- or underestimation.

### 6.6.2.1   Probabilistic Treatment

In the probabilistic way of assessment, the PDF of a composite variable, say, specific storage capacity, is evaluated from the basic hydrogeological parameter (i.e., the thickness and storage coefficient). The fundamental parameters in hydrogeology for groundwater resources evaluation are the storage coefficient, hydraulic conductivity, hydraulic head and aquifer geometry. Each one of these varies spatially according to a certain PDF. The storage coefficient will have different values at different well locations within the same study area. It is empirically confirmed from several thousand well logs that the specific yield is log normally distributed in aquifers (Seabear and Hollyday, 1966).

On the other hand, Şen (1986) proposed the slope matching procedure for aquifer test data for a single well, whereby the inherent random character in the aquifer parameters can be deduced (see Chapter 5). A plot of storage coefficients obtained by the slope matching procedure from data of Schultz (1973) is shown on the logarithmic probability paper in Figure 6.3. In this plot, the probability, $P_m$, to each storage coefficient is attached by the classical empirical formula as

$$P_m = \frac{m}{n+1} \qquad (6.5)$$

where $n$ is the number of data and $m$ is the rank of the data value considered in the form of ascending data sequence.

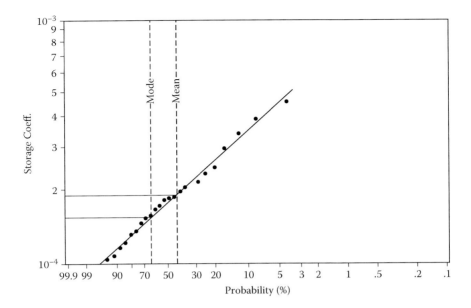

**FIGURE 6.3** Storage coefficient data.

Similarly, Davis (1969) concluded that for unconsolidated water-bearing formations, the hydraulic conductivity also has a log-normal PDF. In addition, during a pumping test, heterogeneities in the transmissivity values have been quantified by the slope matching method (Şen, 1986). The plot of these transmissivity values on a logarithmic probability paper gives straight-line indicating that the transmissivities are also log-normally distributed (see Figure 6.4). Krumbein and Graybill (1965) and Way (1968) observed that the thickness of sedimentary beds is log-normally distributed. Similar observations are confirmed by Gheorghe (1978).

As a result of all the aforementioned discussions, it is reasonable to assume that all the basic hydrogeological variables are generally distributed according to the logarithmic PDF. Detailed information about the logarithmic PDF properties is presented by Aitcheson and Brown (1957).

Fortunately, the composite variables in Equations 6.3 and 6.4 are expressed as simple multiplication of the relevant basic parameters. The multiplication property for the log-normally distributed variables also gives the composite variable a similar PDF. In other words, theoretically, the logarithmic normal distribution of variables on the r.h.s. of Equation 6.3 also gives the composite variable on the l.h.s. normal logarithmic distribution. Furthermore, taking the logarithms of both sides in Equation 6.3 leads to

$$Lg(g_s) = Ln(h) + Ln(S) \qquad (6.6)$$

The logarithmic transformation of logarithmically distributed random variables renders these variables into normally (Gaussian) distributed random variables. Hence, the arithmetic mean and standard deviation of the logs of the composite

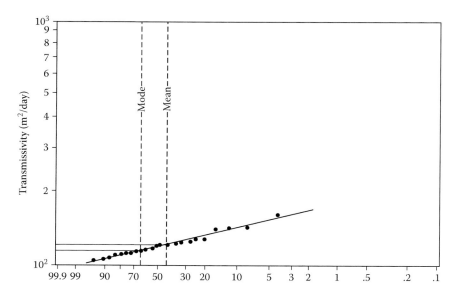

**FIGURE 6.4**   Transmissivity (m²/min) data.

variable are sufficient for a complete definition of the Gaussian PDF. For this purpose, the arithmetic mean, $\mu_{Lngs}$, and the standard deviation of the composite variable
becomes, after some algebraic manipulations,

$$\mu_{Lngs} = \mu_{Lnh} + \mu_{LnS} \tag{6.7}$$

and

$$\sigma^2_{Lngs} = \sigma^2_{Lnh} + \sigma^2_{LrS} + 2\overline{(Lnh)(LnS)} \tag{6.8}$$

respectively. In these expressions, $\mu_{Lnh}$ and $\mu_{Lns}$ are the arithmetic averages and $\sigma^2_{Lnh}$
and $\sigma^2_{LnS}$ are the variances. Finally, $\overline{(Lnh)(LnS)}$ indicates the average of the cross
multiplication of $Lnh$ and $LnS$ values.

### 6.6.2.2   Statistical Treatment

Apart from the probabilistic approach, statistical methodologies consider errors
involved in each variable and their mutual interactions in forming the composite
variable. The simplest treatment is possible by the *perturbation theory* where each
basic variable is composed of random fluctuations around their respective arithmetic averages (see Chapter 5, Section 5.5.2). For instance, a random variable, say $x$,
can be written in two parts, namely, its average value $\bar{x}$ and fluctuations $e_x$ corresponding to errors as $x = \bar{x} + e_x$. In such a definition, the error term $e_x$ has zero
arithmetic average with standard deviation equivalent to the standard deviation of

the original random variable $x$. Similarly, according to the perturbation theory, each basic hydrogeologic variable such as the thickness, $h$, storativity, $S$, transmissivity, $T$, and hydraulic gradient, $i$, can be written as $h = \bar{h} + e_h$, $S = \bar{S} + e_s$, $T = \bar{T} + e_T$, and $i = \bar{i} + e_i$, respectively. Substitution of these expressions into Equations 6.3 and 6.4 leads to the following averages of the composite variables,

$$\overline{g_s} = \overline{hS} + \overline{e_h e_s} \tag{6.9}$$

and

$$\bar{q} = \overline{Ti} + \overline{e_T e_i} \tag{6.10}$$

where $\overline{e_h e_s}$ and $\overline{e_T e_i}$ show the cross-covariance between the two basic hydrogeologic parameters considered. These terms may assume positive or negative values corresponding to overestimation and underestimation, respectively.

It is obvious from these two expressions that errors in terms of deviations from the arithmetic mean play significant roles when they are dependent on each other. These expressions imply that over- or underestimations appear in specific groundwater storage capacity and the subsurface flow rates due to the $\overline{e_h e_s}$ and $\overline{e_T e_i}$ terms, respectively. These error terms indicate the cross-covariances between the parameters as stated earlier. However, their contribution might be equal to zero only under the following circumstances, namely:

1. If the storativity is independent of the aquifer thickness. Although this is valid at a single well, when multitude of wells are considered in a region, then the storage coefficient may be dependent upon the thickness, which may be different at different well locations. Hence, in an area of rough bedrock topography the thicknesses vary significantly in the region. In an ideal case, the thickness of the aquifer is constant over the aquifer area and $\overline{e_h e_s} = 0$. It is well known that a variable is not a constant value and consequently $\overline{e_h e_s} = 0$.
2. If the aquifer reservoir material composition is homogeneous and isotropic, then $e_s = 0$. Contrary to the previous case, the aquifer storage coefficient is constant due to the material composition, but the aquifer thickness may vary due to the surface and subsurface bedrock topography. Areal constancy of the storage coefficient causes $\overline{e_h e_s} = 0$.
3. If both the aquifer saturation thickness and the storage coefficient values are areally constant, then $\overline{e_h e_s} = 0$.

Unfortunately, none of the aforementioned three cases exist in actual field studies. Similar arguments may be stated for $\overline{e_T e_i}$ being equal to zero. Because neither the aquifer material is homogeneous and isotropic nor the saturation thickness uniform, there will always be additional error terms in the form of cross-covariance as in

**TABLE 6.1**

**Pumping Test Results**

| Well No. | Storativity S | Error (%) $e_S$ | Transmissivity (m²/min) T | Error (%) $e_T$ | Thickness h (m) | Error (%) $e_h$ |
|---|---|---|---|---|---|---|
| 1 | $3.3 \times 10^{-2}$ | $-3.6 \times 10^{-2}$ | $1.8 \times 10^{-1}$ | $-7.1 \times 10^{-1}$ | 13.40 | 0.34 |
| 2 | $3.5 \times 10^{-2}$ | $-3.4 \times 10^{-2}$ | $1.2 \times 10^{-1}$ | $-7.7 \times 10^{-1}$ | 12.10 | -0.96 |
| 3 | $2.0 \times 10^{-2}$ | $-4.9 \times 10^{-2}$ | $4.1 \times 10^{-1}$ | $-4.8 \times 10^{-1}$ | 10.50 | -2.56 |
| 4 | $8.5 \times 10^{-2}$ | $1.6 \times 10^{-2}$ | $2.3 \times 10^{-1}$ | $-6.6 \times 10^{-1}$ | 12.20 | -0.86 |
| 5 | $1.9 \times 10^{-1}$ | $1.2 \times 10^{-1}$ | $4.2 \times 10^{-0}$ | $3.3 \times 10^{0}$ | 10.30 | -2.76 |
| 6 | $5.2 \times 10^{-2}$ | $-1.7 \times 10^{-2}$ | $1.2 \times 10^{-1}$ | $-7.7 \times 10^{-1}$ | 14.95 | 1.89 |
| 7 | $7.0 \times 10^{-2}$ | $1.0 \times 10^{-2}$ | $9.6 \times 10^{-1}$ | $-7.0 \times 10^{-2}$ | 18.00 | 4.94 |
| Averages | $6.9 \times 10^{-2}$ | ≈0.0 | $8.9 \times 10^{-1}$ | ≈0.0 | 13.06 | ≈0.0 |
| Variances | $13.2 \times 10^{-2}$ | $7.4 \times 10^{-2}$ | $10.98 \times 10^{-1}$ | $16.9 \times 10^{-1}$ | 35.19 | 6.6 |
| Cross-variances | $\overline{e_S e_T} = 7.45$ | | $\overline{e_S e_a} = 1.66 \times 10^{-2}$ | | $\overline{e_T e_h} = -1.23$ | |

Equations 6.9 and 6.10. In fact, positive cross-covariance leads to overestimation, whereas negative values give rise to underestimations.

The application of the *probabilistic risk* and statistical formulations developed in previous sections is demonstrated for the Wadi Qudaid unconfined aquifer near the city of Jeddah, Kingdom of Saudi Arabia (Şen, 1999). The groundwater resources of this aquifer have been extensively studied by Al-Hajeri (1977). The wadi alluvium is composed of coarse to medium quaternary deposits and surficial sand dunes. The bedrock topography and the alluvial thickness are very irregular, in addition to the heterogeneity of the aquifer material. These features are very obvious from Table 6.1 where the basic hydrogeological parameters, as found in each of the seven wells in the area, exhibit regional variations.

The parameters are obtained from a large-diameter well aquifer test according to Papadopulos and Cooper (1967) type curves. Comparison of each parameter value at individual wells with arithmetic averages indicates rather random behavior. The error terms are also presented in the same table. For the probabilistic risk calculations, first storativity, transmissivity, and thickness values are plotted separately on logarithmic probability paper with empirical probability distributions according to Equation 6.5. These plots are presented in Figures 6.5 to 6.7 where the scatter of points appears along straight lines. This confirms the assumption that the parameters abide by a logarithmic normal PDF.

Fitting a straight line on semi-logarithmic paper is equivalent to obtaining the population of the variable, and hence the probability statements, including risk amounts corresponding to any value, can be made. For instance, the probability of the storage coefficient being equal to or greater than, say, $5 \times 10^{-2}$, is 0.48 (Figure 6.4) and for the transmissivity to be equal to or greater than, say, $6 \times 10^{-1}$, m²/min

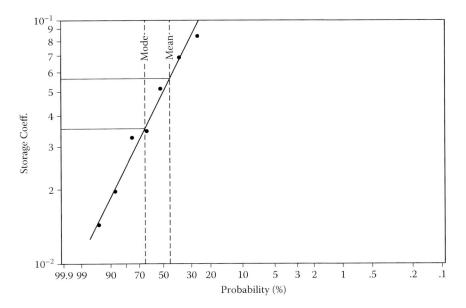

**FIGURE 6.5**   Storage coefficient data from Wadi Qudaid.

is 0.28 (Figure 6.5). On the other hand, in order to be able to find the PDF of a composite variable graphically, the following steps should be followed:

1. Find the mean value of the composite variable from Equation 6.7.
2. Find the variance value of the composite variable from Equation 6.8.
3. Find the median value of the composite variable from the following equation (Aitcheson and Brown, 1957):

$$m_{Lngs} = \mu_{Lngs} \exp\left(-\frac{1}{2}\sigma_{Lngs}\right) \qquad (6.11)$$

4. Plot the two points, namely, the mean and the median on a logarithmic paper.
5. Connect the two points by a straight line.

On the other hand, analytical expressions for the PDF of composite variables can be obtained from Equation 6.5 after finding the parameters in steps 1 and 2 from the following expression of the logarithmic normal distribution for a random variable (Aitcheson and Brown, 1957).

$$f(x) = \frac{1}{\sqrt{2\pi}\sigma_x} \exp\left\{-\frac{1}{2}\left[\frac{1}{\sigma_x} Ln\left(x/m_x\right)\right]^2\right\} \qquad (0 \le x \le \infty) \qquad (6.12)$$

The plots of the resulting PDFs are shown in Figures 6.7 and 6.8 for the specific storage and subsurface flow rate, respectively, for the Wadi Qudaid aquifer.

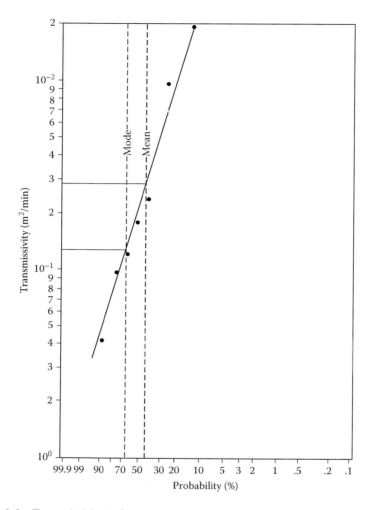

**FIGURE 6.6**  Transmissivity (m²/min) data from Wadi Qudaid.

Deterministically, the total volume of available groundwater within the saturated zone of the Quaternary deposits of Wadi Qudaid is about $3.1 \times 10^6$ m³, provided that the average thickness and the storage coefficient are used with a 3.4 km² area. This volume corresponds to 0.91 m of specific groundwater storage. However, the most frequently occurring specific groundwater storage and the total groundwater storage volume are 0.54 m (see Figure 6.8) and $1.8 \times 10^6$ m³, respectively. Hence, the error of overestimation through the deterministic model is about 73%. This is an extremely large overestimation that may give rise to a multitude of problems such as water shortages for agricultural activities or urban water supply, overpumping and mining of the aquifer, economic losses due to extra wells, pipes, pumps, etc.

In Figure 6.9 the risk is the vertical distance between the cumulative probability curve and the horizontal axis, whereas the reliability is a complementary value (risk plus the reliability is equal to one). For instance, the risk associated with the classical

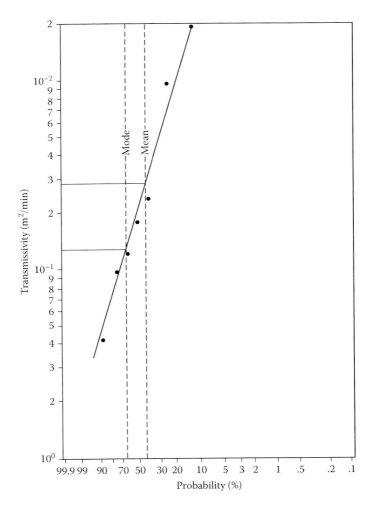

**FIGURE 6.7**  Alluvium thickness in Wadi Qudaid.

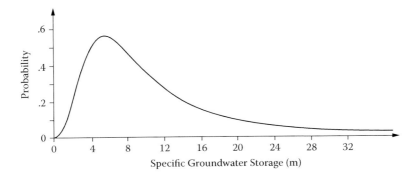

**FIGURE 6.8**  PDF of specific groundwater storage.

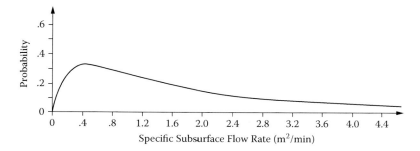

Specific Subsurface Flow Rate (m²/min)

**FIGURE 6.9**  PDF of specific subsurface flow rate.

estimation of the groundwater storage volume is 0.41. However, the most frequently occurring estimate has a risk value of 0.23. As a result of using the latter estimate, the risk decreases as the cost of reliability increases.

In general, the increase in the groundwater storage volume gives rise to an increase in the risk component. Practical questions then arise, including how to estimate the groundwater resources volume in a given area, and what risk level should be accepted in such a determination. Once the estimation of the available groundwater storage volume is determined, all of the subsequent main activities of exploitation, distribution, management, well number and location, development, and overall money to be invested for such activities are dependent primarily on this estimate. In the absence of any further information, such as economic and political conditions, the best estimate corresponds to the maximum frequency as explained above (Figure 6.8).

If the risk level is predetermined, then Figure 6.10 provides an estimate of the groundwater storage volume. Determination of the risk is rather subjective and depends on expert professional as well as personal judgment. For an expert who is ready to work under 20% risk, the groundwater storage volume in Wadi Qudaid will be estimated as $0.5 \times 3.4 \times 10^6$ m³. For the sake of argument, if an inexperienced

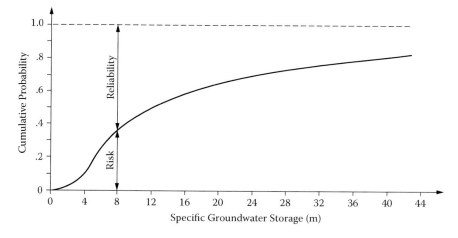

Specific Groundwater Storage (m)

**FIGURE 6.10**  Cumulative PDF of the specific groundwater storage.

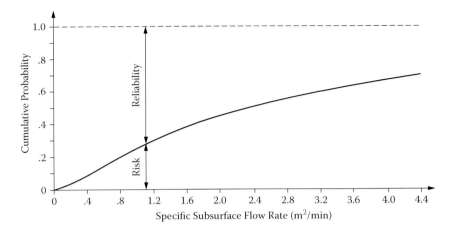

**FIGURE 6.11** Cumulative PDF of the specific subsurface flow rate.

hydrogeologist is ready to accept 50% risk, then his or her estimates will be $1.15 \times 3.4 \times 10^6$ m³, which is too large an overestimation. However, as a hydrologist's experience increases, this risk level will become smaller, and accordingly, he or she will arrive at more conservative estimates.

There is always risk associated with any estimate, and for zero risk the groundwater storage volume must also be equal to zero, which means that the hydrogeologist does not have any idea about the groundwater resources in the study area. As for the specific subsurface flow, the most frequently occurring rate under unit hydraulic gradient is 0.44 m²/min with a risk level equal to 0.09 (see Figure 6.11) and 0.85 m²/min from the classical approach, which has risk value as 0.22. Hence, the relative error percentage is 91, which is too large a difference. Such a large error in the deterministic method affects the water balance equation for the study area.

It implies more output than is likely. It is obvious that risk levels are comparatively small relative to the case of specific groundwater storage. The aforementioned methodology has the following important implications and advantages over commonly used classical deterministic techniques.

1. A quantitative way of assessing hydrogeological risk for groundwater resources evaluation.
2. It provides a flexible procedure to have various alternative estimates of composite decision variables.
3. It provides a probabilistic approach to determine the best estimate in the sense that it is the most likely value to occur.
4. Finding the PDF of the composite variables provides meaningful information in regard to the population behavior of these variables, hence reducing the data necessary to have meaningful results.
5. It provides a way of incorporating the risk concept in groundwater development and management studies.

## 6.7   GROUNDWATER MANAGEMENT AND WELL OPTIMIZATION IN ARID REGIONS

Any management study should indicate how water quantity may change in the aquifers with allowable groundwater flow rate toward wells and drawdown. For this purpose, the transmissivity and hydraulic conductivity are important because the response of an aquifer to pumping depends on these parameters, and especially on their values at well sites where water is being withdrawn from the system. Most often in real aquifers, there is a large number of irregularly spaced wells, and the groundwater level in any well is due to either regional flows in the aquifer, and the pumping within the individual wells or due to their combined effects. A good example of such a situation is the Saq aquifer and tapping wells thereof in the northwestern part of Saudi Arabia (Şen and Somayien, 1991). At the outset of any management program, the interferences must be avoided by considering the optimum values of hydrogeological variables such as the discharge, velocity, drawdown, etc.

Maximum allowable discharge (MAD) is the quantity of water that can be pumped out from a well with no damage either to the aquifer or the well itself. Therefore, pumping must be limited such that the discharge is less than or equal to this optimum yield, which is the quantity of water that can be pumped out from a well with no damage either to the aquifer or the well itself. Therefore, pumping must be limited so that the discharge is less than or equal to this optimum yield. Furthermore, in order to avoid extra costs and unnecessary drawdowns, the interference between the wells must be avoided in an optimum management program. In most of the cases, the wells are already drilled, and hence the distances between them are known exactly. These distances are considered as constraints in the evaluation of optimum discharge so that no chance is left for interference between adjacent wells. In arid regions, optimum discharge is neither unique nor constant, but its value at any time depends on the spacing between wells and the aquifer properties in the vicinity of each well location. Therefore, in any groundwater management for arid regions, the question is what amount of optimum discharges are to be withdrawn from each well so that

1. There is no damage to the well–aquifer interactions in an individual well location.
2. No interference should occur among multiple well locations.

These two conditions should be satisfied simultaneously during any simple management program. It is apparent from the radial flow concept that as the groundwater approaches a well, its velocity increases. Exceedance of velocity over a critical limit gives rise to the transportation of fine particles from the aquifer into the gravel pack. Subsequently, these fine particles cause plugs in the filters that might lead to extra head losses and, consequently, drawdown increments. Approach velocity attains its maximum value at the interface between the aquifer and the filter where the transport of fine materials is most likely to occur. The groundwater velocity at any point within the aquifer is equivalent to the specific discharge, $q$, (Darcy velocity) which is defined as the ratio of pump discharge, $Q$, to the total circumferential area, $A$, perpendicular to the flow lines.

$$q = \frac{Q}{A} \qquad (6.13)$$

This velocity should not be confused with the real flow velocity for which the flow area is the intergranular pore space. The real velocity, $v_r$, is estimated by the dividing of specific discharge, by the porosity, $p$, as

$$v_r = \frac{q}{p} \qquad (6.14)$$

Physically, the approach velocity is rather dependent on the grain size distribution of aquifer material and, by observing a number of operating wells, Sichard (1927) noticed an empirical relationship between the hydraulic conductivity of aquifer, $K$, and drawdown increments as groundwater approaches the wells. His relationship is expressed empirically in terms of approach velocity, $v_a$, with the introduction of the safety factor by Huisman (1972) as

$$v_a = \frac{\sqrt{K}}{60} \qquad (6.15)$$

This relationship provides a valid design criteria where conductivity is determined preferably by *in situ* pumping tests or from the specific discharge measurements within each well. By considering the approach velocity in Equation 6.15, it is possible to rewrite the optimum discharge, $Q_{oi}$, for the i-th well succinctly as,

$$Q_{oi} = 2\pi r_{wi} m_i \frac{\sqrt{K_i}}{60} \qquad (i = 1, 2, 3, \ldots, n) \qquad (6.16)$$

where $r_{wi}$ and $K_i$ are the well radii, and hydraulic conductivity, respectively, for the i-th well, $m$ is the aquifer thickness, and $n$ is the number of wells in the study area.

The second stage in a simple management program is the condition that there will not appear any interference between the wells. In fact, two or more wells pumping from the same aquifer may interfere with each other, causing lower water levels than if only one well is pumping. The effect of several interfering wells at the same point in the aquifer is additive. For confined aquifers the steady-state flow equation for i-th well can be written approximately as (Theim, 1906),

$$Q_{oi} = \frac{2\pi T_i s_{wi}}{\log\left(\dfrac{R_i}{r_{wi}}\right)} \qquad (i = 1, 2, 3, \ldots, n) \qquad (6.17)$$

in which $s_{wi}$ is the drawdown in the well itself, $R_i$ is the radius of influence for the same well, and, $T_i$ is the transmissivity defined as $T_i = mK_i$. This expression represents also the head distribution in the wells without receiving any recharge from the surface sources. In practical studies one can adopt the radius of influence from Holting (1980) as

$$R_i = 3000 s_{wi} \sqrt{K_i} \qquad (6.18)$$

The substitution of which into Equation 6.17 leads to,

$$Q_{oi} = \frac{2\pi T_i s_{wi}}{\log\left(\dfrac{3000 s_{wi}\sqrt{K_i}}{r_{wi}}\right)} \qquad (i = 1, 2, 3, \ldots, n) \qquad (6.19)$$

In order to satisfy simultaneously the aforementioned two stages in the same management model, Equations 6.16 and 6.19 are set equal to each other, and after some algebra one can obtain the optimum drawdown expression:

$$s_{wi} = \frac{r_w \dfrac{\sqrt{K_i}}{60} \log\left(\dfrac{3000 s_{wi}\sqrt{K_i}}{r_w}\right)}{K_i} \qquad (i = 1, 2, 3, \ldots, n) \qquad (6.20)$$

All the quantities are known except the drawdowns, $s_{wi}$. For each well site the optimum drawdown can be calculated from Equation 6.20 by means of the successive approximation technique through trial and error method on digital computers. These solutions do not take into consideration the distances between the existing wells. They are converted into the sequence or radius of influences, $R_{wi}$, through Equation 6.18. In the case of already drilled well fields, the distances between adjacent wells can be found directly from the well location map. Let $D_{ij}$ indicate the distance between two adjacent wells sites, $i$ and $j$. In general, if there are $n$ wells there will be $n(n-1)/2$ different distance combinations within the area. Each distance should be equal to or greater than the summation of the two adjacent well radii as the distance constraint in the management program,

$$D_{ij} > R_i + R_j \qquad (i, j = 1, 2, \ldots, n) \qquad (6.21)$$

After having obtained the radius of influences, Equation 6.21 should be checked for each pair of adjacent wells. If the constraint in Equation 6.21 is not satisfied, then the radius of influence of any one or two of the wells will be reduced conveniently until the inequality or equality in Equation 6.21 is satisfied. Of course, such reductions are possible by reductions in pump discharges. This gives new values for some of the local radii of influences, and with these values the corresponding drawdowns are calculated from Equation 6.18 as

$$s_{wi} = \frac{R_i}{3000\sqrt{K_i}} \qquad (i = 1, 2, 3, \ldots, n) \qquad (6.22)$$

In this manner all the adjacent well pairs can be checked, and final sequences of drawdown and radii of influence can be obtained. With these values at hand, the optimum discharge for each well can be obtained from Equation 6.17. Subsequently, optimum groundwater velocities, $v_{oi}$, can be calculated from Equation 6.13 at the well surface as

$$v_{oi} = \frac{Q_{oi}}{2\pi m_i r_{wi}} \quad (i = 1, 2, 3, \ldots, n) \tag{6.23}$$

The following points should be taken into consideration in a simple, but effective groundwater management program.

1. The regional optimum water levels can be estimated by the simple management rule developed here.
2. Interference between wells must be avoided during any groundwater management program.
3. The groundwater velocity must not be allowed to be more than an optimum value, so that the damages on the screen and the gravel pack on the aquifer can be avoided.
4. Actual discharges should be adjusted such that they are equal to the optimum discharges as proposed here.
5. Additional demands must be met by making better use of the aquifer storage capacity.

## 6.8   WADI GROUNDWATER MANAGEMENT SYSTEM PRINCIPLES

Groundwater resources problems are unique in arid regions not only from the physical aspects of intermittent streams or no surface runoff, depletion of existing groundwater storages, overexploitation and consumption, salt water intrusion, and pollution of shallow aquifers, but also from the managerial aspects due to lack of trained personnel, deficient institutional arrangements, and poor or nonexistent resource management rules, regulations, programs, and software. In order to cope with future rational groundwater distribution in arid countries, there is an urgent need for efficient planning and management schemes to be put into effect, taking into consideration the sustainability of the resources and the fragile nature of the arid environment.

   *Wadi groundwater management* (WGM) in arid regions includes a lexicon of several commonly used words and one of the most frequent is the *demand* for water. This single word can have at least four quite distinctive implications, as in the use of water, the consumption of water, the need for water, or the economic demand for water. Each one of these four separate terms can be defined carefully in the context of the hydrosocial balance of a region. There is a more general point that WGM is now widely seen as principally a form of *demand management*, which is not correct, in many arid or humid regions.

   In some potential wadis of an arid region, there is a real problem in groundwater resources exploitation including development and WGM, which are almost nonexistent. The groundwater is not extracted by considering optimum benefit for the present and future generations and especially for the critical periods. Simple and effective WGM rules and regulations must be put forward prior to the preparation of dynamic and adaptive management software. WGM must be programmed in such a way that abstraction must be adjusted in accordance with groundwater resources support expectations within stipulated time periods in the future, in a sustainable and *perennial supply* manner. The renewable (recharge) character of potential wadi aquifers must be considered with suitable aquifer pump rates (discharges), well field locations, and especially priority of subunits (wadis) within a single wadi system, as well as within integrated and joint wadi systems.

Balkhair (2002) has already suggested the selection of five potential wadis from arid regions of western Arabian Peninsula for the most optimum groundwater reservoir management through multicriteria, decision-making problem solving. He used 12 hydrologic and hydrogeologic variables on a lump basis for each wadi. Although his approach can be used as a preliminary step for solution, it includes many restrictive assumptions and subjective indexing of some variables.

In this section, more refined assessment for two hypothetical wadis (WA and WB) is presented with consideration of 20 different decision variables concerning subwadi drainage system hydrology, hydrogeology, and demand. The demand management is considered on the basis of recharge facilities from these potential wadi aquifers. This section concentrates on the conjunctive use of two wadis under the name of WAWB system, which is a virtual management unit that combines the properties from these two wadis so as to obtain the best solution for water supply. Different scenarios are sought and the most suitable ones are presented for the necessary basic decision plans, which may lead to the final decision making by the local and central administrators.

WGM is composed of complex interactions between a number of natural and artificial factors. It is necessary to divide the wadi area into sub-areas so as to identify the effect of each factor. Herein, each wadi is considered first as the groundwater potential area, which lies at the upper stream locations with high elevations, whereas the lower wadi portions are not considered. Additionally, upstream portions are divided into a set of wadi branches so as to facilitate both the complexity of the system and to manage them in an integrated manner in the overall management system. In each sub-area the effecting factors are characterized from different views such as climate, hydrology, hydrogeology, and hydrochemistry and groundwater environment.

## 6.8.1 BASIC DATA

The joint and integrated groundwater management of resources in WA and WB provide an opportunity to enhance the water supply. The general configuration of joint management program for WAWB system can be considered initially as in Figure 6.12.

The following points are important in the conceptualization of this system, and they must be imbedded into the final W1W2 WGM program.

1. WA and WB do not have any direct interchange but their management will depend on what goes on inside a WAWB block diagram.

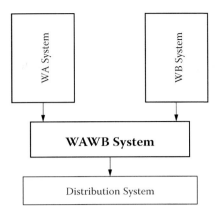

FIGURE 6.12 WAWB overall groundwater management systems.

2. WAWB block receives groundwater in different ratios from each wadi but the total water amount remains equal to what monthly water demand is.
3. The WAWB block may be thought similar to surface water reservoirs where joint water withdrawals contribute in different proportions to overall demand.
4. In deciding on the water withdrawal ratios from each wadi to WAWB system, one of the following three alternatives can be considered.

   a. Quantity-based ratio determination.
   b. Quality-based ratio determination.
   c. Quantity–quality based determination. In particular, for the initial development of the WGM program, herein the quantity based ratio determination is used.

5. The groundwater supply from the WAWB goes directly into the water distribution system.

The basic decision variables, which are taken into consideration for the WGM study, are shown in Tables 6.2 and 6.3 for WA and WB, respectively. It is obvious that there are 20 different management variables contributing to the overall multivariable decision criteria for an integrated WGM system. Numerical data are presented for 18 variables, whereas the last two variables are expressed linguistically. Although, it is possible to give certain index numbers to these linguistic variables, this is not a consideration here.

Table 6.2 has 4 columns, whereas Table 6.3 has 5 columns each for different subwadis within respective major wadis. Accordingly, in Table 6.2 there are $20 \times 4 = 80$ pieces of information and in Table 6.3, $20 \times 5 = 100$ inputs are available.

## 6.8.2 LOGICAL-CONCEPTUAL MANAGEMENT PRINCIPLES AND MODELS

Generally, only haphazard and heuristic techniques are employed without logical, rational, or expert views in any study area. Prior to a WGM system, it is necessary to

**TABLE 6.2**
**Wadi A Management Variables**

| Features of the Reach | Strategic Groundwater Planning Subwadi | | | |
|---|---|---|---|---|
| | WA1 | WA2 | WA3 | WA4 |
| Annual rainfall (mm/year) | 189 | 192 | 173 | 182 |
| Alluvium surface area ($\times 10^3$ m$^2$) | 6,612.1 | 5,816.9 | 15,494.1 | 1,652 |
| Direct rain water volume ($\times 10^6$ m$^3$/year) | 1.25 | 1.12 | 2.68 | 0.3 |
| Possible direct annual recharge ($\times 10^6$ m$^3$/year) | 0.25 | 0.17 | 0.27 | 0.075 |
| Water table area ($\times 10^3$ m$^2$) | 5,687.6 | 4,449.1 | 12,354.7 | 1,236 |
| Average saturated thickness (m) | 31.6 | 57.2 | 36.6 | 18.7 |
| Average unsaturated thickness (m) | 10.3 | 19.6 | 27.3 | 14.8 |
| Average transmissivity (m$^2$/day) | 659.7 | 1,510 | 282 | 557 |
| Average hydraulic conductivity (m/day) | 20.9 | 26.4 | 7.7 | 29.78 |
| Hydraulic gradient | 0.0126 | 0.008 | 0.005 | 0.00614 |
| Subsurface flow rate (m$^3$/m/day) | 8.31 | 12.08 | 1.41 | 3.42 |
| Average storativity | 0.09 | 0.128 | 0.045 | 0.068 |
| Average effective porosity | 0.25 | 0.25 | 0.2 | 0.3 |
| Abstractable groundwater volume ($\times 10^6$ m$^3$) | 16.17 | 32.57 | 20.35 | 1.57 |
| Rechargeable direct groundwater volume ($\times 10^6$ m$^3$) | 17.03 | 28.5 | 84.6 | 7.33 |
| Abstractable fullness ratio | 0.48 | 0.53 | 0.18 | 0.17 |
| Present groundwater consumption ($\times 10^6$ m$^3$/year) | 3.11 | 6.17 | 16.50 | 5.05 |
| Average water quality (EC, micro-mhos) | 1,150 | 2,245 | 2,293 | 1,020 |
| Recharge possibility | High | Poor | Medium | Medium |
| Geological environment | Alluvium grandiorite | Alluvium volcano-clastic | Alluvium metabasalt | Alluvium feldspar, marble |

obtain the conceptual models first of individual and then of integrated wadis. Conceptual management models describe the essential features of groundwater phenomena and identify the principal processes that take place during a management program. A complete *conceptual model* provides information about the following points.

1. Definition of the phenomenon in terms of features recognizable by observations, analysis, or validated simulations. Here, the phenomenon includes the groundwater recharge to and exploitation of two wadis, and their joint integrated operation and management for the best service during a strategic emergency situation. For this purpose, various linguistic and simple conceptual simulation alternatives are identified.

**TABLE 6.3**
**Wadi B Management Variables**

| | Strategic Groundwater Planning Subwadis | | | | |
|---|---|---|---|---|---|
| **Features of the Reach** | **WB1** | **WB2** | **WB3** | **WB4** | **WB5** |
| Annual rainfall (mm/year) | 357 | 357 | 166 | 166 | 166 |
| Alluvium surface area ($\times 10^3$ m$^2$) | 31,895 | 8,989 | 1,514 | 3,027 | 35,789 |
| Direct rain water volume ($\times 10^6$ m$^3$/year) | 11.39 | 3.21 | 0.25 | 0.50 | 5.94 |
| Possible direct annual recharge ($\times 10^6$ m$^3$/year) | 1.71 | 0.48 | 0.05 | 0.1 | 1.19 |
| Water table area ($\times 10^3$ m$^2$) | 29,907 | 7,678 | 9,367 | 1,873 | 22,997 |
| Average saturated thickness (m) | 7.27 | 8.06 | 4.36 | 5.00 | 14.59 |
| Average unsaturated thickness (m) | 31.73 | 18.94 | 37.52 | 41.69 | 43.63 |
| Average transmissivity (m$^2$/day) | 187 | 14 | 63.8 | 659.7 | 431.8 |
| Average hydraulic conductivity (m/day) | 25.7 | 1.7 | 14.6 | 139.1 | 29.6 |
| Hydraulic gradient | 0.026 | 0.017 | 0.005 | 0.0086 | 0.011 |
| Subsurface flow rate (m$^3$/m/day) | 9.88 | 0.58 | 0.16 | 7.18 | 4.90 |
| Average storativity | 0.2476 | 0.1099 | 0.0694 | 0.1836 | 0.1391 |
| Average effective porosity | 0.25 | 0.25 | 0.20 | 0.30 | 0.25 |
| Abstractable groundwater volume ($\times 10^6$ m$^3$) | 53.83 | 6.80 | 2.83 | 1.72 | 56.83 |
| Rechargeable direct groundwater volume ($\times 10^6$ m$^3$) | 253 | 42.5 | 11.36 | 37.86 | 390.4 |
| Abstractable fullness ratio | 0.17 | 0.14 | 0.2 | 0.04 | 0.12 |
| Present groundwater consumption ($\times 10^6$ m$^3$/year) | 1.2 | 1.4 | 0.14 | 2.6 | 10.6 |
| Average water quality (EC, micro-mhos) | 977 | 838 | 1,346 | 1,063 | 1,393 |
| Recharge possibility | High | Moderate | Poor | Moderate | High |
| Geological environment | Alluvium biotite | Alluvium diorite | Alluvium diorite | Alluvium diorite | Alluvium monzogranite |

2. Description of the WGM practices in terms of appearance (various alternatives), size, intensity, and accompanying groundwater conditions.
3. Logical statements about the controlling physical processes, which enable the understanding of the factors that determine the mode and rate of

groundwater recharge and consumption with time. This corresponds to the derivation of basic and simple logical rules and regulations about the overall system performance.

4. Specification of the key hydrogeological fields, demonstrating the main processes such as the recharge potential, and saturated and unsaturated zone potentials for abstraction and additional storage of groundwater.
5. Guidance for predicted hydrogeological conditions or situations using the diagnostic and prognostic fields that best discriminate between development and nondevelopment guiding to the WGM.

Conceptual models provide decision makers with the following knowledge, which can then be employed in any effective WGM study:

1. Diagnosis: This is the preliminary step that helps in understanding the internal and external activities within the whole system under consideration.
2. Synthesis: Available verbal and numerical data and information must be synthesized for arriving at preliminary conclusions that help to determine the optimum pattern of WGM.
3. Logic: This provides a "mental picture" of the WGM aspects within and between wadi systems.
4. Isolation: The basic ingredients such as subwadis and WGM variables are isolated from each other so as to assess the individual effects.
5. Main pattern: This includes the identification of main and distinctive patterns within all the complex system.
6. Numerical evaluation: Mental and logical aspects are assessed numerically by considering individual and joint WGM operations.
7. Numerical projections: Based on the previous steps, numerical predictions are produced for future situations under different scenarios.
8. Numerical products: Different tools are used for modification of numerical products.
9. Simple calculations: Rather than involved mathematical procedures, it is preferred in this paper to construct preliminary steps for a simple WGM model.
10. Data gaps: Suggested scenarios provide the possibility of filling in gaps in the data.

### 6.8.3 Conceptual Model of Wadi (WAWB) Systems

The conceptual model of subsystems in each wadi is presented in Figure 6.13 for the joint WAWB management program. This is a block diagram that shows the natural groundwater flow from main wadi branches and along the main channel in each wadi. It is to be noticed that each subunit within a wadi is connected either in a parallel or serial manner. Parallel connection implies that the groundwater withdrawal from one of the branches does not affect the other, and hence their contribution to the WAWB system is independent from each other. On the other hand, the serial connection implies that the groundwater withdrawal is interconnected between the components. In such a situation, the groundwater withdrawal should be started from

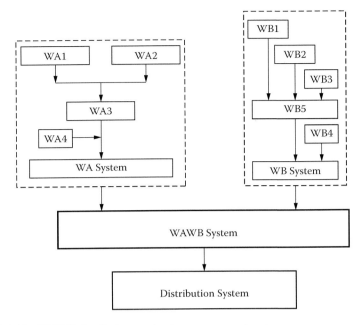

**FIGURE 6.13** WAWB detailed groundwater management system.

the lower end (downstream) component. This may delay the groundwater abstraction, because in such a case the groundwater transmission from upper branches to lower ones takes time. Depending on the aquifer transmissivity value, the time varies, and the greater the transmissivity the smaller is the required time for groundwater transportation. In the case of a serial connection, the aquifer plays a role similar to pipe flow.

As a whole, there are nine components that may contribute to the overall WAWB system in the WGM stages.

### 6.8.4　Logical Management Principles

The connections between the subwadi units are achieved by two logical connections, *AND* or *OR*. If the subwadis are in serial connection then all the serial components must function for the overall performance of the system and, therefore, the units are connected by AND, whereas parallel units contribute to the overall performance of the system by an OR connective. The statements including ANDs and ORs are referred to as a *rule base*. In this manner, it is possible to extract the relevant rules for each conceptual model under consideration. Each rule indicates the independent response of the system to the demand. Let us consider WA configuration only as in Figure 6.12. This leads to the rule bases shown in the following frame, which presents an opportunity to assess the whole WA basin subwadi units in an exhaustive management program.

1. (WA1)AND(WA3)
OR
2. (WA2)AND(WA3)
OR
3. (WA1)AND(WA2)AND(WA3)
OR
4. (WA1)AND(WA2)OR(WA3)
OR
5. (WA1)AND(WA2)AND(WA3)AND(WA4)
OR
6. (WA1)AND(WA2)AND(WA3)OR(WA4)
OR
7. (WA1)OR(WA2)AND(WA3)OR(WA4)

Similarly, consideration of the subunit configuration from Figure 6.13, the logical management rules for WB can be deduced as in the following frame:

1. (WB1)AND(WB5)
OR
2. (WB2)AND(WB5)
OR
3. (WB3)AND(WB5)
OR
4. (WB4)AND(WB5)
OR
5. (WB1)OR(WB2)AND(WB5)
OR
6. (WB1)OR(WB3)AND(WB5)
OR
7. (WB2)OR(WB3)AND(WB5)
OR
8. (WB1)AND(WB2)AND(WB3)AND(WB5)
OR
9. (WB1)AND(WB2)AND(WB3)AND(WB5)AND(WB4)
OR
10. (WB1)AND(WB2)AND(WB3)AND(WB5)OR(WB4)

Hence, for WB within-basin management, there are ten independent logical management rules.

In order to supply groundwater from WAWB system, there are many independent rules, and in total there are $7 \times 10 = 70$ alternatives. It is not necessary to write down all these rules, but the most logical ones are listed in the following frame.

1. (WA1)AND(WA2)AND(WA3)AND(WA4)AND(WB1)AND(WB2)AND
(WB3)AND(WB5)AND(WB4)
OR

2. (WA1)AND(WA2)AND(WA3)AND(WB1)AND(WB2)AND(WB3)AND(WB5)

OR

3. (WA1)AND(WA2)AND(WA3)AND (WB1)AND(WB2)AND(WB5)

OR

4. (WA1)AND(WA3)AND(WB4)AND(WB5)

5. (WA1)AND(WA2)AND(WA3)AND(WA4)AND(WB4)AND(WB5)

OR

6. (WA1)AND(WA2)AND(WA3)AND(WB1)AND(WB2)AND(WB3)AND(WB5)AND(WB4)

OR

7. (WA1)AND(WA3)AND(WB1)AND(WB5)

OR

8. (WA1)AND(WA2)AND(WA3)AND(WB1)AND(WB2)AND(WB3)AND(WB5)

OR

9. (WA1)AND(WA2)AND(WA3)AND(WA4)AND(WB1)AND(WB2)AND(WB3)AND(WB5)AND(WB4)

It is a difficult task to identify the best and the most strategic joint WGM rule among these logical rules. However, as the specific aspects are taken into consideration the picture becomes more obvious. In the following subsections, different aspects are considered, and they all help to make a final decision about the strategic WGM planning.

### 6.8.5   HYDROLOGICAL MANAGEMENT PRINCIPLES

The hydrological features of subwadis in WA and WB are given in Tables 6.5 and 6.6, respectively.

The overall comparison of these two tables indicates that, in general, WB is hydrologically richer than WA, in that WB receives more rainfall, especially at the upstream reaches of WB1 and WB2. Similar comparisons indicate that from direct rain water and possibly the direct recharge possibility point-of-view, WB is preferable over WA. It is possible to construct a decision matrix for WAWB system on the basis of annual rainfall amounts as in Table 6.7.

In this matrix, each cell includes the total annual rainfall amount that falls over concerned subwadi pairs. This decision matrix has five rows and four columns, and

**TABLE 6.5**
**Wadi A Hydrological Features**

| Features of the Reach | Strategic Groundwater Planning Reach | | | |
| --- | --- | --- | --- | --- |
| | WA1 | WA2 | WA3 | WA4 |
| Annual rainfall (mm/year) | 189 | 192 | 173 | 182 |
| Direct rain water volume ($\times 10^6$ m³/year) | 1.25 | 1.12 | 2.68 | 0.3 |
| Possible direct annual recharge ($\times 10^6$ m³/year) | 0.25 | 0.17 | 0.27 | 0.075 |

### TABLE 6.6
### Wadi B Hydrological Features

| Features of the Reach | Strategic Groundwater Planning Reach | | | | |
|---|---|---|---|---|---|
| | WB1 | WB2 | WB3 | WB4 | WB5 |
| Annual rainfall (mm/year) | 357 | 357 | 166 | 166 | 166 |
| Direct rain water volume ($\times 10^6$ m³/year) | 11.39 | 3.21 | 0.25 | 0.50 | 5.94 |
| Possible direct annual recharge ($\times 10^6$ m³/year) | 1.71 | 0.48 | 0.05 | 0.1 | 1.19 |

### TABLE 6.7
### Annual Rainfall (mm) Decision Matrix

| WB | WA | | | |
|---|---|---|---|---|
| | WA1 | WA2 | WA3 | WA4 |
| WB1 | *564* | *449* | *530* | *539* |
| WB2 | **564** | 449 | *530* | **539** |
| WB3 | 355 | 358 | 183 | 348 |
| WB4 | 355 | 358 | 183 | 348 |
| WB5 | 355 | 358 | 183 | 348 |

therefore its dimension is (5 × 4). Any cell in this decision matrix can be represented by mathematical expression as follows.

$$C_{i,j} = \sum_{k=1}^{N} R_k \quad (i=1,2,\ldots L)$$

$$(j=1,2,\ldots M)$$ 

$$(k=1,2,\ldots N)$$

(6.24)

where $R_k$ is the annual rainfall amount in wadi $k$; $L$, $M$, and $N$ are the number of sub-units in wadi $i$ and wadi $j$, and the number of wadis (here, $N = 2$). In the same table bold numbers indicate the maximum possible value, which can be expressed as

$$M_1 = Maximum\ (C_{i,j})$$

$$i=1,2,\ldots L$$

$$j=1,2,\ldots M$$

(6.25)

$$C_{i,j} = \sum_{k=1}^{N} r_k \quad (i=1,2,\dots L)$$

$$(j=1,2,\dots M)$$

$$(k=1,2,\dots N)$$

(6.26)

Herein, $r_k$ indicates the direct recharge availability from wadi $k$ annually. This decision matrix indicates to priorities for the integrated groundwater resources in sequence of significance as bold, bold italic, and bold underlined $f$ features. These three different feature priorities yield the following alternatives.

First alternative: (WA1)AND(WB1)OR(WA1)AND(WB2)
Second alternative: (WA3)AND(WB1)OR(WA1)AND(WB2)
Third alternative: (WA4)AND(WB1)OR(WA4)AND(WB2)

It is also possible to prepare a joint decision matrix for direct recharge to various combinations among subwadis in WAWB system which is shown in Table 6.8. This annual possible direct recharge decision matrix indicates the three preferable alternatives:

First alternative: (WA3)AND(WB1)
Second alternative: (WA1)AND(WB1)
Third alternative: (WA2)AND(WB1)

On the other hand, the annual rainfall volume decision matrix is presented in Table 6.9. This decision matrix leads to the following preferences:

First alternative: (WA3)AND(WB1)
Second alternative: (WA1)AND(WB1)
Third alternative: (WA2)AND(WB1)

On the basis of the combined assessment of these last three decision matrices, the preferred subwadis from the hydrological points of view for WAWB are from WB in the sequence of significance as WB1 and WB2 and from WA as WA1, WA2, WA3,

**TABLE 6.8**
**Annual Direct Recharge (x10⁶ m³) Decision Matrix**

| WB | WA | | | |
|---|---|---|---|---|
| | WA1 | WA2 | WA3 | WA4 |
| WB1 | *1.96* | <u>1.88</u> | 1.98 | 1.785 |
| WB2 | 0.73 | 0.65 | 0.75 | 0.555 |
| WB3 | 0.3 | 0.22 | 0.32 | 0.125 |
| WB4 | 0.35 | 0.27 | 0.37 | 0.175 |
| WB5 | 1.44 | 1.36 | 1.46 | 1.265 |

**TABLE 6.9**
**Annual Rainfall Volume (x10⁶ m³) Decision Matrix**

|        | WA      |       |       |       |
|--------|---------|-------|-------|-------|
| **WB** | **WA1** | **WA2** | **WA3** | **WA4** |
| **WB1** | *12.64* | 12.51 | **14.07** | 11.69 |
| **WB2** | 4.46 | 4.33 | 5.89 | 3.51 |
| **WB3** | 1.5 | 1.37 | 2.93 | 0.55 |
| **WB4** | 1.75 | 1.62 | 2.93 | 0.8 |
| **WB5** | 7.19 | 7.06 | 8.62 | 6.24 |

and WA4. It is, therefore, advised not to include from the 80 different logical rules the ones that do not include especially wadis WB1, WA1, and WA2.

### 6.8.6  Hydrogeological Management Principles

Hydrogeological characteristics of each subunit (basic wadis) are specified by the quantities as indicated in Tables 6.10 and 6.11 for WA and WB, respectively.

Integrated joint hydrogeological WGM decisions can be taken on the basis of the following decision matrices for WAWB system.

The decision matrix of wadi couples is presented in Table 6.12 for the average saturated thickness. The groundwater storage is available in the saturated zone of the aquifer. The total available groundwater storage can be calculated as the multiplication of the saturated thickness with the water table area and the storage coefficient.

*Saturation thickness* provides the possibility of groundwater presence in the wadi alluvium, and therefore the decision matrix is calculated by taking the average saturated thicknesses in both wadi subunits. This decision matrix leads to the following three consecutive integrated decisions:

First alternative: (WA2)AND(WB5)
Second alternative: (WA2)AND(WB2)
Third alternative: (WA2)AND(WB1)

**TABLE 6.10**
**Wadi A Hydrogeological Features**

|                                           | Strategic Groundwater Planning Reach |       |       |       |
|-------------------------------------------|---------|---------|---------|---------|
| **Features of the Reach**                 | **WA1** | **WA2** | **WA3** | **WA4** |
| Average saturated thickness (m)           | 31.6 | 57.2 | 36.6 | 18.7 |
| Average unsaturated thickness (m)         | 10.3 | 19.6 | 27.3 | 14.8 |
| Average transmissivity (m²/day)           | 659.7 | 1,510 | 282 | 557 |
| Average hydraulic conductivity (m/day)    | 20.9 | 26.4 | 7.7 | 29.78 |
| Subsurface flow rate (m³/m/day)           | 8.31 | 12.08 | 1.41 | 3.42 |
| Abstractable groundwater volume (×10⁶ m³) | 16.17 | 32.57 | 20.35 | 1.57 |

**TABLE 6.11**
**Wadi B Hydrogeological Features**

| Features of the Reach | Strategic Groundwater Planning Reach | | | | |
|---|---|---|---|---|---|
| | WB1 | WB2 | WB3 | WB4 | WB5 |
| Average saturated thickness (m) | 7.27 | 8.06 | 4.36 | 5.00 | 14.59 |
| Average unsaturated thickness (m) | 31.73 | 18.94 | 37.52 | 41.69 | 43.63 |
| Average transmissivity (m²/day) | 187.00 | 14.00 | 63.80 | 659.70 | 431.80 |
| Average hydraulic conductivity (m/day) | 25.70 | 1.70 | 14.60 | 139.10 | 29.60 |
| Subsurface flow rate (m³/m/day) | 9.88 | 0.58 | 0.16 | 7.18 | 4.90 |
| Abstractable groundwater volume (×10⁶ m³) | 53.83 | 6.80 | 2.83 | 1.72 | 56.83 |

**TABLE 6.12**
**Average Saturated Thickness (m) Decision Matrix**

| WB | WA1 | WA2 | WA3 | WA4 |
|---|---|---|---|---|
| WB1 | 38.87 | <u>64.47</u> | 43.87 | 25.97 |
| WB2 | 39.66 | *65.26* | 44.66 | 26.76 |
| WB3 | 35.96 | 61.56 | 40.96 | 23.06 |
| WB4 | 36.60 | 62.20 | 41.60 | 23.70 |
| WB5 | 46.19 | **71.79** | 51.19 | 33.29 |

**TABLE 6.13**
**Average Unsaturated Thickness (m) Decision Matrix**

| WB | WA | | | |
|---|---|---|---|---|
| | WA1 | WA2 | WA3 | WA4 |
| WB1 | 42.03 | 51.33 | 59.03 | 46.53 |
| WB2 | 29.24 | 38.54 | 46.24 | 33.74 |
| WB3 | 47.82 | 57.12 | <u>64.82</u> | 52.32 |
| WB4 | 51.99 | 61.29 | *68.99* | 56.49 |
| WB5 | 53.93 | 63.23 | **70.93** | 58.43 |

The decision matrix for the average *unsaturated thickness* is presented in Table 6.13. This thickness indicates the potential of the wadi to store groundwater after storm rainfall and flood by direct or indirect recharge. The total volume of possible groundwater storage capability of the wadi can be calculated by multiplying the water table area by this thickness and the aquifer material porosity.

**TABLE 6.14**

**Average Transmissivity (m²/day) Decision Matrix**

| WB | WA | | | |
|----|-----|-----|-----|-----|
|    | WA1 | WA2 | WA3 | WA4 |
| WB1 | 423.35 | <u>848.50</u> | 234.50 | 372.00 |
| WB2 | 336.85 | 762.00 | 148.00 | 285.50 |
| WB3 | 361.75 | 786.90 | 172.80 | 310.40 |
| WB4 | 659.70 | **1084.85** | 470.85 | 608.35 |
| WB5 | 545.75 | *970.90* | 356.90 | 494.40 |

The unsaturated thickness decision matrix yields the following three integrated decision results:

First alternative: (WA3)AND(WB5)
Second alternative: (WA3)AND(WB4)
Third alternative: (WA3)AND(WB3)

Table 6.14 includes the decision matrix for average transmissivity, which indicates the abstractability of groundwater from the aquifer with ease. The greater the transmissivity value, the easier the water abstraction from the aquifer.

The following management alternatives emerge from average transmissivity decision matrix:

First alternative: (WA2)AND(WB5)
Second alternative: (WA2)AND(WB4)
Third alternative: (WA2)AND(WB1)

The subsurface flow rate is the volume of water that can be transmitted naturally from the upstream portions toward the downstream within the aquifer. It is expressed as the volume of groundwater discharge from a 1 m depth in the saturation thickness. In the calculation, the hydraulic gradient values are used as presented in Tables 6.2 and 6.3. The subsurface flow rate decision matrix for this variable is presented in Table 6.15.

**TABLE 6.15**

**Average Subsurface Flow Rate (m³/m/day) Decision Matrix**

| WB | WA | | | |
|----|-----|-----|-----|-----|
|    | WA1 | WA2 | WA3 | WA4 |
| WB1 | <u>9.10</u> | **10.98** | 5.65 | 6.65 |
| WB2 | 4.46 | 6.33 | 1.00 | 2.00 |
| WB3 | 4.24 | 6.12 | 0.79 | 1.79 |
| WB4 | 7.75 | *9.63* | 4.30 | 5.30 |
| WB5 | 6.60 | 8.49 | 3.15 | 4.16 |

**TABLE 6.16**
**Abstractable Groundwater Volume (x10⁶ m³) Decision Matrix**

| | WA | | | |
|---|---|---|---|---|
| WB | WA1 | WA2 | WA3 | WA4 |
| WB1 | 70.00 | 86.4 | <u>74.18</u> | 55.40 |
| WB2 | 22.97 | 39.37 | 27.15 | 8.37 |
| WB3 | 19.00 | 35.40 | 23.18 | 4.40 |
| WB4 | 17.89 | 34.29 | 22.07 | 3.29 |
| WB5 | 73.00 | **89.40** | *77.18* | 58.4 |

Based on the *average subsurface flow* rate decision matrix, one can obtain the following three subsequent alternatives:

First alternative: (WA2)AND(WB1)
Second alternative: (WA1)AND(WB1)
Third alternative: (WB2)AND(WB4)

The amount of groundwater that can be withdrawn from each wadi is calculated in million cubes of water and presented in the following table for the joint management of the two wadis.

For this variable the meaningful alternatives that can be used in the WGM system are given as follows:

First alternative: (WA2)AND(WB5)
Second alternative: (WA3)AND(WB1)
Third alternative: (WA3)AND(WB5)

### 6.8.7 WATER DEMAND MANAGEMENT PRINCIPLES

Another significant point is the exploitation of aquifers in wadis according to demand levels. For this purpose, three variables—namely, annual *abstractable groundwater volume, fullness ratio,* and consumption volumes—are considered as given in Tables 6.17 and 6.18 for WA and WB, respectively.

The system components, their possible management strategies, and the strategic WGM can be decided by consideration of available information sources.

The amount of groundwater recharge possibilities in each wadi plays a significant role in the overall WGM system, and their joint composition is presented in Table 6.19.

The groundwater *recharge possibility* is one of the most important factors in any WGM operation. The previous decision matrix indicates the following management integrations.

First alternative: (WA3)AND(WB5)
Second alternative: (WA2)AND(WB5)
Third alternative: (WA1)AND(WB5)

TABLE 6.17
**Wadi A Water Demand Features**

| Features of the Reach | Strategic Groundwater Planning Reach | | | |
|---|---|---|---|---|
| | WA1 | WA2 | WA3 | WA4 |
| Rechargeable direct groundwater volume ($\times 10^6$ m$^3$) | 17.03 | 28.50 | 84.60 | 7.33 |
| Abstractable fullness ratio | 0.48 | 0.53 | 0.18 | 0.17 |
| Present groundwater consumption ($\times 10^6$ m$^3$/year) | 3.11 | 6.17 | 16.50 | 5.05 |
| Recharge possibility | High | Poor | Medium | Medium |

TABLE 6.18
**Wadi B Water Demand Features**

| Features of the Reach | Strategic Groundwater Planning Reach | | | | |
|---|---|---|---|---|---|
| | WB1 | WB2 | WB3 | WB4 | WB5 |
| Rechargeable direct groundwater volume ($\times 10^5$ m$^3$) | 253.00 | 42.50 | 11.36 | 37.86 | 390.40 |
| Abstractable fullness ratio | 0.17 | 0.14 | 0.20 | 0.04 | 0.12 |
| Present groundwater consumption ($\times 10^6$ m$^3$/year) | 1.20 | 1.40 | 0.14 | 2.60 | 10.60 |
| Recharge possibility | High | Poor | Poor | Poor | High |

On the demand side of the WGM program is the groundwater consumption amounts as presented in Tables 6. and 6.3 for each wadi; Table 6.20 present the joint groundwater demand in the area.

The three preferred integrated water consumption rates from WA and WB basin sub-units are as follows:

First alternative: (WA3)AND(WB5)
Second alternative: (WA3)AND(WB2)
Third alternative: (WA3)AND(WB1)

TABLE 6.19
**Rechargeable Direct Groundwater Volume (x10$^6$ m$^3$) Decision Matrix**

| WB | WA | | | |
|---|---|---|---|---|
| | WA1 | WA2 | WA3 | WA4 |
| WB1 | 270.03 | 231.50 | 337.60 | 260.33 |
| WB2 | 59.53 | 71.00 | 127.10 | 49.83 |
| WB3 | 28.39 | 39.86 | 95.96 | 18.69 |
| WB4 | 54.89 | 66.36 | 122.46 | 45.19 |
| WB5 | 407.43 | *418.90* | **475.00** | 397.73 |

**TABLE 6.20**
**Present Groundwater Consumption (x10$^6$ m$^3$/year)**

|  | WA | | | |
| --- | --- | --- | --- | --- |
| WB | WA1 | WA2 | WA3 | WA4 |
| WB1 | 4.31 | 7.37 | <u>17.70</u> | 6.25 |
| WB2 | 4.51 | 7.57 | *17.90* | 6.45 |
| WB3 | 3.25 | 6.31 | 16.64 | 5.19 |
| WB4 | 5.71 | 8.77 | 19.10 | 7.65 |
| WB5 | 13.71 | 16.77 | **27.10** | 15.65 |

## 6.9   INTEGRATED MANAGEMENT PRINCIPLES

There are many ways of combining the aforementioned specific logical principles so as to arrive at a few sets of integrated WGM principles. Therefore, rather than standardized numerical calculations, the suitable WGM practices are chosen based on expert knowledge with the contribution of different specialists who worked on each one of the wadis. The following tables and the scores of each subunit (wadi) are considered to represent overall decisions about the three most suitable strategic plans in the WAWB system. Tables 6.21 and 6.22 show the scores in WA and WB management decisions, respectively.

**TABLE 6.21**
**Wadi A Strategic Planning Preferences**

|  | WA RULE BASIS | | | |
| --- | --- | --- | --- | --- |
|  | Strategic Groundwater Planning Units | | | |
| Preference | WA1 | WA2 | WA3 | WA4 |
| First | 2 | 5 | 5 | 0 |
| Second | 4 | 3 | 4 | 1 |
| Third | 2 | 4 | 4 | 2 |
| Total | 8 | 12 | 13 | 3 |

**TABLE 6.22**
**Wadi B Strategic Planning Preferences**

|  | WB RULE BASIS | | | |
| --- | --- | --- | --- | --- |
|  | Strategic Groundwater Planning Units | | | |
| Preference | WB1 | WB2 | WB3 | WB4 | WB5 |
| First | 4 | 1 | 0 | 0 | 5 |
| Second | 5 | 3 | 0 | 2 | 2 |
| Third | 6 | 1 | 0 | 3 | 2 |
| Total | 15 | 5 | 0 | 5 | 9 |

The combined considerations in these two tables indicate the following significant points in the strategic planning of interwadi management.

First strategic alternative: (WA3)AND(WB1)
Second strategic alternative: (WA2)AND(WB5)
Third strategic alternative: (WA1)AND(WB2)

### 6.9.1 STRATEGIC MANAGEMENT CONFIGURATION SELECTION

*Strategic management* is the process of specifying a set of objectives, developing policies and plans to achieve these objectives, and allocating resources so as to implement the plans. It provides overall direction to the whole enterprise of such a task. An organization's strategy must be appropriate for its resources, circumstances, and objectives. The process involves matching the expected strategic advantages to the environment that an organization faces. A good corporate strategy should integrate an organization's goals, policies, and action sequences (tactics) into a cohesive whole. Improvements in the operation of integrated WGM in arid zones can be brought by introducing additional storage into the local water supply system, which may include the following points:

1. Detention storages for flood control to damp out short duration of discharge peaks, and hence increase the groundwater recharge.
2. Seasonal carry-over storage to hold water that is in excess of immediate requirements until it is necessary to make up for deficits in future runoffs.

In temperate climates, both these storage functions are fulfilled simultaneously in nature. In general, a rational approach to water resources development based on a major *ephemeral* wadi channel system in arid zones is the adaptation of an integrated conjunctive WGM strategy, which utilizes both sporadic flood runoff and shallow groundwater. The basic elements of such an approach should include the following points.

1. The level of the flood peaks must be controlled by spate breakers so that the kinetic energy of the runoff is reduced.
2. In order to regulate runoff in the wadi channel both for direct diversion to cultivated areas and for recharge management, weir structures must be used.
3. Wells and ancillary equipments must be used in order to withdraw from groundwater storage during periods when there is no flow in the ephemeral stream channels.

In arid regions, groundwater is the only natural water supply, and therefore it must be carefully exploited in a sustainable and strategic manner. If groundwater abstraction approaches or exceeds recharge, then the precision of the water balance becomes more important for future planning. When groundwater abstraction is less than mean recharge, it is said that the *safe yield* is not exceeded. However, in many arid zones

**TABLE 6.23**
**Exploitable Water (x10$^6$ m$^3$)**

| WB | WA | | | |
|---|---|---|---|---|
| | WA1 | WA2 | WA3 | WA4 |
| WB1 | 64.95 | *76.64* | 68.740 | 51.64 |
| WB2 | 21.41 | 34.04 | 23.040 | 5.94 |
| WB3 | 17.41 | 32.08 | 21.120 | 4.08 |
| WB4 | 16.45 | 31.11 | 20.268 | 3.17 |
| WB5 | 67.15 | **80.63** | <u>69.730</u> | 52.62 |

only limited abstraction will result in overdraft. It is a matter of decision of how rapidly to use groundwater in storage.

The level of WGM must remain within certain limits so as not to exceed safe discharge levels or the high water table, and so as not to fall so low that poor quality of water will intrude, that pumping levels will become uneconomical, or even worse, that the aquifer becomes dewatered and overlying owners are deprived of a sustainable water supply without recourse to an alternative.

In the following scenarios, integrated joint operation and management of different alternatives are suggested for strategic WGM by making use of joint groundwater resources in a WAWB system. The entries in scenario matrices are based on the possible direct annual recharge and abstractable groundwater resources in the respective units.

In the first stages of WGM for the WAWB system, it is necessary to assess presently available and exploitable groundwater resources that can provide an immediate response to any emergency situation. For this purpose, the subwadis are analyzed on the basis of their joint groundwater availability volumes, which are shown in Table 6.23. Pairwise, the three available groundwater-rich combinations in this table are (WB5–WA2), (WB5–WA3), and (WB1–WA2). On the other hand, the most critical (minimum) three alternatives are (WB4–WA4), (WB3–WA4), and (WB2–WA4) combinations. Since the last three cases are only between minor subwadis, their exploitations are preserved for the last and most emergency case, and they will be regarded as standby groundwater supply alternatives in the WGM program. Each alternative in Table 6.23 is processed for groundwater supply purposes by considering only 50 l/day/person, in which case this scenario will be able to support the number of people as shown in Table 6.24.

It is possible to interpret the entries in this table pairwise and collectively as follows:

1. The joint exploitation of subwadis, WA2 and WB5 can support 4,418,082 persons per day with 50 l of water.
2. The population numbers in the last column (row) of Table 6.24 indicate the joint exploitation of the WA (WB) branch in the second column (row) with all the subwadis of WB (WA) per day.

**TABLE 6.24**
**Number of People Needing 50 l/d of Water**

|       |            |            | WA         |           |            |
| ----- | ---------- | ---------- | ---------- | --------- | ---------- |
| WB    | WA1        | WA2        | WA3        | WA4       | Total      |
| WB1   | 3,558,904  | 4,199,452  | 3,766,575  | 2,829,589 | 14,354,521 |
| WB2   | 1,173,150  | 1,865,205  | 1,262,466  | 325,480   | 4,626,301  |
| WB3   | 953,973    | 1,757,808  | 1,157,260  | 223,562   | 4,092,603  |
| WB4   | 901,370    | 1,704,658  | 1,110,575  | 173,699   | 3,890,301  |
| WB5   | 3,679,452  | 4,418,082  | 3,820,822  | 2,883,288 | 14,801,644 |
| Total | 10,266,849 | 13,945,205 | 11,117,699 | 6,435,616 | 41,765,369 |

3. If all the subwadis are exploited jointly the maximum population that can be supported for one day is about 42 million people (41,765,369).

Table 6.24 can be converted to number of days that a certain population can be supported, where the question is how many days, say, 5 million people can be supported without any other alternative water supply such as a desalination plant or treated water. Here, a set of the population is considered as 5, 15, and 25 million, and the results are shown in Table 6.25. Again, the basic critical 50 l/d/person are adopted for the calculations.

The values in this table can be presented in graphical form for visual interpretation, and Figure 6.14 shows the decrease in the days with increase in the population.

The values in this figure are based on the absolute exploitation of groundwater resources in the WAWB system. It is practical and preferable to consider exploitation on the basis of a certain risk factor. Therefore, a 10% risk level has been adopted, and the resulting populations in Table 6.25 are recalculated with this risk level. The results are presented in Table 6.26.

The application of WGM principles provides an example for other arid regions of the world for simple but effective management programs. The following overall conclusions can be drawn from this study.

1. Under ever-increasing pressure for groundwater use, human activities have important implications for WGM programs in the formulation of simple rules and regulations that are expected to lead to wiser development and sustainable productivity for the improvement of overall welfare in the arid region, especially during emergency periods.
2. The WGM program is presented for the integrated management of two wadis, which are jointly referred to as the WAWB system.
3. The most important basis in any WGM is the soundness of data, which is presented here for the WAWB system in terms of 20 different hydrology, hydrogeology, and demand variables.
4. Preliminary steps for an effective WGM are the deduction of a set of rational, logical, and simple rules and regulations.

**TABLE 6.25**

**Number of Critical Days for Different Populations in WAWB System**

| WB | WA | | | | |
|---|---|---|---|---|---|
| | WA1 | WA2 | WA3 | WA4 | Total |
| | | N = 5 million persons | | | |
| WB1 | 260 | 306 | 275 | 206 | 1,047 |
| WB2 | 86 | 136 | 92 | 24 | 338 |
| WB3 | 70 | 128 | 84 | 16 | 299 |
| WB4 | 66 | 124 | 81 | 13 | 284 |
| WB5 | 269 | 322 | 279 | 210 | 1,080 |
| Total | 749 | 1,018 | 812 | 470 | 3,049 |
| | | N = 15 million persons | | | |
| WB1 | 86 | 102 | 92 | 68 | 349 |
| WB1 | 28 | 45 | 31 | 7 | 112 |
| WB3 | 23 | 42 | 28 | 5 | 99 |
| WB4 | 22 | 41 | 27 | 4 | 95 |
| WB5 | 89 | 107 | 93 | 70 | 360 |
| Total | 250 | 339 | 271 | 156 | 1,016 |
| | | N = 25 million persons | | | |
| WB1 | 52 | 61 | 55 | 41 | 210 |
| WB2 | 17 | 27 | 18 | 5 | 67 |
| WB3 | 14 | 26 | 17 | 3 | 60 |
| WB4 | 13 | 25 | 16 | 2 | 57 |
| WB5 | 54 | 64 | 56 | 42 | 216 |
| Total | 150 | 203 | 162 | 94 | 610 |

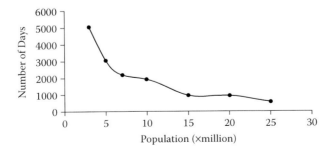

**FIGURE 6.14** Number of days from complete exhaustion of WAWB system.

**TABLE 6.26**
**Critical Day Number in WAWB System at 10% Risk Level**

| WB | WA | | | | |
|---|---|---|---|---|---|
| | WA1 | WA2 | WA3 | WA4 | Total |
| N = 5 million persons | | | | | |
| WB1 | 234 | 275 | 248 | 185 | 942 |
| WB2 | 77 | 122 | 83 | 22 | 304 |
| WB3 | 63 | 115 | 76 | 14 | 269 |
| WB4 | 59 | 112 | 73 | 12 | 256 |
| WB5 | 242 | 290 | 251 | 189 | 972 |
| Total | 674 | 916 | 731 | 423 | 2,744 |
| N = 15 million persons | | | | | |
| WB1 | 77 | 92 | 83 | 61 | 314 |
| WB2 | 25 | 41 | 28 | 6 | 101 |
| WB3 | 21 | 38 | 25 | 5 | 89 |
| WB4 | 20 | 37 | 24 | 4 | 86 |
| WB5 | 80 | 96 | 84 | 63 | 324 |
| Total | 225 | 305 | 244 | 140 | 914 |
| N = 25 million persons | | | | | |
| WB1 | 47 | 55 | 50 | 37 | 189 |
| WB2 | 15 | 24 | 16 | 5 | 60 |
| WB3 | 13 | 23 | 15 | 3 | 54 |
| WB4 | 12 | 23 | 14 | 2 | 51 |
| WB5 | 49 | 58 | 50 | 38 | 194 |
| Total | 135 | 183 | 146 | 85 | 549 |

5. In the WA basin, subwadis WA1 and WA2 couple indicates the best scenario, predicting that they can be developed and managed jointly for better benefits.
6. WB basin has the most preferable joint management strategy in terms of subwadis has wadi WB1 from the upper part of the whole drainage basin with theWB5. Especially, WB4 can be exploited by itself as a parallel support to the previous integrated combination.
7. At the first stages of any WGM in WAWB system, it is not possible to consider major subwadis, namely WA3 and WB5, as the first integrated management units, but many tributary wadis such as (WA4, WB3, and WB4 must be exploited at early stages in a general WGM system).
8. Necessary tables for the joint exploitation of wadis in WAWB system are presented on the population basis with 10% risk level.

## 6.10   ARTIFICIAL MIXING OF POTABLE AND SALINE GROUNDWATERS

In arid regions of the world, groundwater quality may vary within short distances. There are many wells abandoned due to increasing groundwater salinity. New wells are drilled at nearby locations where better quality groundwater is available. In order to make the best use of the groundwater in an aquifer, it is suggested that rather than abandoning the wells completely, their waters be mixed with better quality waters so as to obtain a usable water quality. Desalinized water with very low mineral content is mixed with groundwater that is relatively richer in minerals. To establish standard mixture water qualities, it is necessary to have a technique whereby a decision can be made as to at what proportions to mix the rich or potable water with poor or saline water. It is possible that by mixing different quality waters, the availability of potable groundwater in an aquifer may be increased at least for some short or moderate periods of time.

The conceptual model of groundwater withdrawal and artificial mixture can be idealized as in Figure 6.15 (Şen et al., 2005). Two wells with different groundwater quality pump water from an aquifer (or different aquifers) with respective discharges $Q_S$ and $Q_P$. During a specific time period, t, the pumped waters are placed in a common reservoir prior to release to the consumer so that an artificial mixture takes place.

An economical, fast, and convenient way to approximate the saline content and total dissolved solids is to measure the electrical conductivity (or specific conductance, EC), of a water sample. The more ions in a groundwater sample, the higher is its electric conductivity. Individual EC values of groundwater at each well location can be measured readily and easily in the field. In any groundwater supply problem, the mixture water quality should be predetermined based on the desired use. For instance, a simple water quality criterion is presented by Wilcox (1955) as shown in Table 6.27.

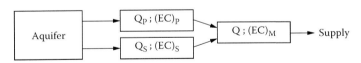

**FIGURE 6.15**   Two-well artificial mixture model description of variables.

**TABLE 6.27**
**Water Quality**

| Water Type | Electric Conductivity (μ mhos/cm) |
|---|---|
| Excellent | <250 |
| Good | 250–750 |
| Moderate | 750–2,000 |
| Permissible | 2,000–3,000 |
| Doubtful | >3,000 |

For the sake of simplicity, a two-well approach was used in this study. In Figure 6.15 the water quality variables in the wells, namely the electrical conductivity of saline $(EC)_S$, and potable $(EC)_P$ waters, and $(EC)_M$ of the artificially mixed water are known. In fact, $(EC)_S$ and $(EC)_P$ are determined by measurements; whereas, $(EC)_M$ is taken from standard tables such as Table 6.15. The question is then in what volume ratios should one mix water from each well so as to achieve the predetermined $(EC)_M$ value for the water supply.

### 6.10.1 EXPERIMENTAL MODEL

Although there are many studies that have been conducted on salt water intrusions, water supply problems of islands, and poor groundwater quality from the use of connate water, one is unable to find any published literature discussing the potential for artificial mixture. Water samples were collected from two different sources with great difference in their EC values. One of the sources was from the Red Sea, which has the greatest salinity among the oceans of the world. The other source was potable groundwater from a nearby wadi (Wadi Fatimah) in the vicinity of Jeddah city (see Figure 6.16). The electrical conductivity values for these two sources are 59,000 µmhos/cm and 1,300 µmhos/cm, respectively.

Two sets of laboratory experiments were carried out; namely, addition of saline water to fixed volume of potable water and vice versa. After each addition, the mixture volume and the $(EC)_M$ value of the artificially mixed water are recorded. The experiments were carried out under the same laboratory temperature of 20°C. Representative results are given in Tables 6.28 and 6.29.

The graphical representations of the complete results are exhibited in Figures 6.17 and 6.18. Each curve in Figure 6.17 represents the change of electrical conductivity of the mixture with the increase of potable water volume addition to 50, 100, 200,

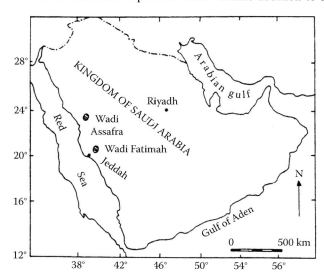

**FIGURE 6.16** Location of the study areas.

## TABLE 6.28
## Potable Water Additions to 50 ml Saline Water

| Incremental Addition of Potable Water (ml) | Total Volume of Potable Water (ml) | $(EC)_M$ $(x10^6 \text{ } \mu\text{mhos/cm})$ |
|---|---|---|
| 50 | 50 | 31.8 |
| 50 | 100 | 22.5 |
| 50 | 150 | 17.7 |
| 50 | 200 | 14.7 |
| 50 | 250 | 12.5 |
| 50 | 300 | 11.1 |
| 50 | 350 | 10.1 |
| 50 | 400 | 9.0 |
| 50 | 450 | 8.2 |
| 50 | 500 | 7.8 |
| 50 | 550 | 7.2 |
| 50 | 600 | 6.8 |
| 50 | 650 | 6.4 |
| 50 | 700 | 6.0 |
| 50 | 750 | 5.8 |
| 50 | 800 | 5.5 |
| 50 | 850 | 5.3 |
| 50 | 900 | 5.1 |
| 50 | 950 | 4.9 |
| 50 | 1,000 | 4.8 |
| 50 | 1,050 | 4.6 |
| 50 | 1,100 | 4.5 |
| 50 | 1,150 | 4.4 |
| 50 | 1,200 | 4.3 |
| 50 | 1,250 | 4.2 |
| 50 | 1,300 | 4.1 |
| 200 | 1,500 | 3.7 |
| 500 | 2,000 | 3.1 |
| 1,000 | 3,000 | 2.5 |
| 1,000 | 4,000 | 2.2 |
| 1,000 | 5,000 | 2.1 |
| 2,000 | 7,000 | 1.9 |
| 3,000 | 10,000 | 1.7 |
| 10,000 | 20,000 | 1.5 |
| 20,000 | 40,000 | 1.4 |

**TABLE 6.29**
**Saline Water Additions to 50 ml Potable Water**

| Incremental Addition of Potable Water (ml) | Total Volume of Potable Water (ml) | $(EC)_M$ $(x10^6\ \mu mhos/cm)$ |
|---|---|---|
| 2 | 2 | 4.1 |
| 4 | 6 | 8.9 |
| 7 | 13 | 15.3 |
| 10 | 23 | 21.8 |
| 27 | 50 | 31.6 |
| 50 | 100 | 40.7 |
| 50 | 150 | 45.2 |
| 50 | 200 | 47.8 |
| 50 | 250 | 49.6 |
| 50 | 300 | 50.9 |
| 50 | 350 | 51.9 |
| 50 | 400 | 52.6 |
| 50 | 450 | 53.2 |
| 60 | 500 | 53.7 |
| 50 | 550 | 54.0 |
| 50 | 600 | 54.5 |
| 50 | 650 | 54.7 |
| 50 | 700 | 55.0 |
| 50 | 750 | 55.2 |
| 50 | 800 | 55.4 |
| 500 | 1,300 | 56.4 |
| 500 | 1,800 | 56.8 |
| 500 | 2,300 | 57.1 |
| 1,000 | 3,300 | 57.5 |
| 1,000 | 4,300 | 57.7 |
| 1,000 | 5,300 | 57.8 |

30, and 500 ml of fixed saline water volumes. All of the curves have the same original EC value, which is 59,000 μmhos/cm, and they converge asymptotically toward the potable water EC, which is 1,300 μmhos/cm. Figure 6.18 shows the relationship between EC of the mixed water and the volume of saline water for various fixed potable water volumes. The curves start from a common initial value corresponding to the potable water EC and increase asymptotically toward the saline water EC.

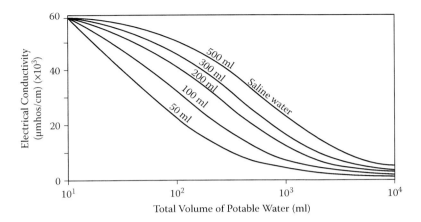

**FIGURE 6.17** Relationship between mixture electrical conductivity, $(EC)_M$, and volume of potable water added to given volumes of saline water.

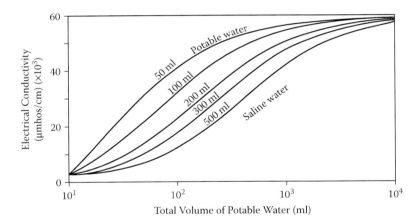

**FIGURE 6.18** Relationship between mixture electrical conductivity, $(EC)_M$, and volume of saline water added to given volumes of potable water.

## 6.10.2 Standard Curves

In order to arrive at a universally usable curve between the volumes and the EC values, all of the curves in Figures 6.17 and 6.18 are rendered into a dimensionless form by defining two sets of ratios. The first one is referred to as the volume ratio, $V_r$, which is defined as

$$V_r = \frac{V_S}{V_P} \qquad (6.27)$$

where $V_S$ and $V_P$ are the saline and potable water volumes, respectively. The second one is the electrical conductivity ratio, $(EC)_R$, which is a dimensionless value and defined as:

$$(EC)_R = \frac{(EC)_M - (EC)_P}{(EC)_S - (EC)_P} \times 100 \qquad (6.28)$$

It is obvious that the $(EC)_R$ assumes any value between 0 and 100. Furthermore, it is interesting to note that when each one of the curves in Figures 6.17 and 6.18 is transformed into a dimensionless form, i.e., $V_r$ versus $(EC)_R$, there appears a dimensionless and standard curve as shown in Figure 6.19. When equal amounts of potable and saline water are mixed, $(V_S = V_P)$, $V_r = 1.0$, then correspondingly, $(EC)_R = 50$. However, excess of potable water, $V_P > V_S$, i.e., $(V_r < 1.0)$ results in $(EC)_R$ value between 0 and 50. Otherwise, if $V_S > V_P$, then $V_r > 1.0$, and consequently, $50 < (EC)_R < 100$. After different sets of experiments in the laboratory, it was observed that whatever the electrical conductivity values of the two sources are, the same dimensionless standard curve is obtained. Hence, it is universal and ready for use in the case where the water quality from the two sources differs.

The standard curve volume ratio 1 in Figure 6.19 corresponds to an electrical conductivity ratio value of 50. When the two sources contribute the same amount of volume to the artificial mixture, then this mixture's electrical conductivity is equal to the arithmetic mean of the original electrical conductivities.

### 6.10.3  THEORETICAL MODEL

In order to find an analytical expression for the standard curve, a mass balance equation was used to describe the mixture of different water qualities. The solution is assumed as an ideal case with each ion behaving independently. The magnitude of

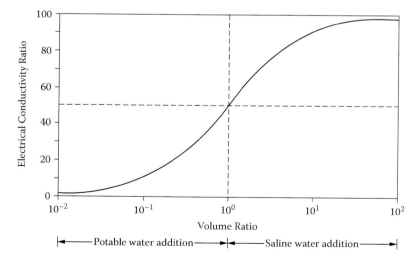

**FIGURE 6.19**  Standard curve for electrical conductivity ratio–volume ratio relationship.

interactions depends, among other factors, on the ionic concentration and on ion electric charge. The effects of these two major factors can be combined through the *ionic strength*, I. The relationship between I and the total dissolved solids, TDS, was estimated by Langelier (1936) as

$$I = 2.5 \times 10^{-5}\,TDS. \qquad (6.29)$$

Ionic strength can be estimated with a reasonable degree of accuracy by including the concentration of only the major ions and cations (Şen and Dakheel, 1985). When two different quality waters are mixed, the ionic strength of each individual water changes and attains a new value for the mixture. Let the volumes of potable, saline, and mixture waters be denoted by $V_P$, $V_S$, and $V_M$, respectively. The mass balance equation by considering TDS values of constituents and the mixture can be written as

$$V_S\,(TDS)_S + V_P\,(TDS)_P = V_M\,(TDS)_M = (V_S + V_P)\,(TDS)_M$$

Substitution of Equation 6.29 for TDS values leads to

$$V_S\,I_S + V_P\,I_P = (V_S + V_P)\,I_M \qquad (6.30)$$

In order to render this expression into a more practical form, the ionic strength can be expressed in terms of the electrical conductivity as given by Şen and Dakheel (1985)

$$I = C \times EC \qquad (6.31)$$

in which C is a proportionality factor depending on the type of water quality. For groundwaters in the Kingdom of Saudi Arabia, Şen and Dakheel (1985) gave the following relationships:

$$I = 1.47 \times 10^{-5}\,EC \ (NaCI\ type\ and\ EC < 2{,}000\ \mu mhos/cm)$$

$$I = 1.87 \times 10^{-5}\,EC\ (mixed\ type)$$

$$I = 2.25 \times 10^{-5}\,EC\ (CaSO_4\ type\ and\ EC > 2{,}000\ \mu mhos/cm)$$

The substitution of Equation 6.31 into Equation 6.30 gives

$$V_S\,C_S\,(EC)_S + V_P\,C_P\,(EC)_P = (V_S + V_P)\,C_M\,(EC)_M \qquad (6.32)$$

in which Cs are the water quality factors. If the source waters are of the same type—say, NaCI type—then $C_S = C_P = C_M$. Equation 6.32 becomes

$$V_S\,(EC)_S + V_P\,(EC)_P = (V_S + V_P)\,(EC)_M \qquad (6.33)$$

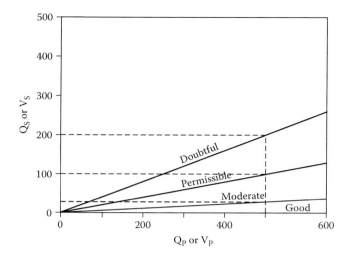

**FIGURE 6.20** Quality and discharge relationship.

Division of both sides by $V_P$ gives

$$V_r(EC)_S + (EC)_P = (1 + V_r)(EC)_M \qquad (6.34)$$

Solution of $(EC)_P$ from this expression and its substitution into Equation 6.28 leads to

$$(EC)_R = 100V_r/(1 + V_r) \qquad (6.35)$$

In fact, this expression gives the experimentally obtained standard curve in Figure 6.20.

### 6.10.4  APPLICATION

An application of the standard curve will be shown for Wadi As-Safra in the northeastern part of the Kingdom of Saudi Arabia (see Figure 6.16). More groundwater is needed to irrigate additional agricultural areas. The most distinctive characteristic of this wadi is that within the same unconfined aquifer there are areas of extremely different water qualities. This is due to the hydraulic connection of the aquifer with adjacent aquifers through several lateral faults along the wadi course. The groundwater from adjacent aquifers is comparatively far better than the Wadi As-Safra waters. Two of the wells in the adjacent aquifers have EC values as $(EC)_S = 9,000$ μmhos/ cm and $(EC)_P = 300$ μmhos/cm. They are about 1,000 m apart from each other. From the classification in Table 6.27, one of the wells in Wadi As-Safra is close to an excellent type as it has an EC of 300 μmhos/cm; whereas, the other one is not usable at all. By means of artificial mixture it is possible to obtain any EC value between 300 to 9,000 μmhos/cm, depending on the volume mixture ratio. A solution will be given, herein, for all the possible water quality levels given in Table 6.27.

A good groundwater quality of $(EC)_M = 750$ μmhos/cm was chosen as the representative artificially mixed water. In order to attain this value, saline water (well within 9,000 μmhos/cm) must be mixed with fresh water. If they are mixed at equal volumes, then the mixture EC would be $(9,000 + 300)/2 = 4,650$ μmhos/cm. This is less than the volume ratio that should be considered as $V_r = V_S/V_P$. The application of the aforementioned electrical conductivity ratio in Equation 6.28 yields $(EC)_R = 5.17$. The corresponding volume ratio from Figure 6.19 is $V_r = 4.7 \times 10^{-2}$ which implies that $V_S = 4.7 \times 10^{-2}V_P$. Thus, the volume of water available for use in the good category can be increased by 4.7% by mixing saline and potable water from different sources.

In order to obtain permissible water quality for agricultural uses from the artificial mixture, it is sufficient to substitute $(EC)_M = 2,000$ μmhos/cm in the $(EC)_R$ equation (Equation 6.28), which leads to the dimensionless value 19.54. With this value at hand, Figure 6.20 gives the volume ratio value as $2 \times 10^{-1}$ which is greater than the previous ratio value for good water quality mixture. This volume ratio implies that $V_S = 2 \times 10^{-1}V_P$. This statement is tantamount to saying that water available for agricultural uses would be increased by 20% by mixing potable and saline water to yield a mixture having $(EC)_M = 2.000$ μmhos/cm.

Finally, the mixture of the same groundwater sources to obtain $(EC)_M = 3,000$ μmhos/cm after similar calculations one sees that $V_S = 3.9 \times 10^{-1}V_P$. Thus, the volume of water available for use in the permissible category can be increased by 39% by mixing saline and potable water having the characteristics used in this example. The complete relationships between the saline and fresh water on the basis of groundwater quality have been shown in Figure 6.20. The discharges or volumes are without units in this figure, which means that any unit can be employed. Any point within the same quality region gives the volume ratio as well as the quality. According to this ratio, the two volume amounts from each well can be determined rather arbitrarily. For instance, if the potable water well is pumped at 500 l/min, then the saline water well must pump between 0 to 23 l/min for good quality after artificial mixture; between 23 to 100 l/min for mode rate water quality; and finally, between 100 to 195 l/min for permissible water quality for the artificial mixture.

The laboratory investigation in this study has led to understanding an important relationship between the electrical conductivity and the volume or discharge amounts of the saline and potable water in the groundwater studies. The standard curve which emerged as a result of this study provides the relationship between the electrical conductivity and volume ratios. It can be helpful to many ground water users throughout the world. The use of such a curve is shown for mixing the two different water quality sources. The purpose was to obtain a predetermined electrical conductivity for the artificial mixture depending on the desired water quality for a particular use such as agriculture. The adoption of electrical conductivity in the experimental setup is due to the fact that it determines adequate water quality on a large scale and is an inexpensive method for monitoring water supply systems.

## REFERENCES

Aitcheson, J. and Brown, J. A. C., 1957. *The Log-Normal distribution with Special Reference to Its Uses in Economics.* Cambridge University Press, p. 176.

Al-Hajeri, F.Y., 1977. *Groundwater studies of wadi Qudaid.* Institute of Applied Geology, King Abdulaziz University, Research Series, No. 2, p. 222.

Balkhair, K. S., 2002. Outranking strategic groundwater basins in western Saudi Arabia using multi-criterion decision making techniques. In *Groundwater Hydrology*, Sherif, Sing, and Al-Rashed (Eds.), Swets and Zeitlinger, Lisse, pp. 325–342.

Davis, S. N., 1969. Porosity and permeability of natural materials. In *Flow Through Porous Media*. DeWeist, R. J. H. (Ed.) Academic Press, New York: 54–89.

Davis, G., 1982. Prospect risk analysis applied to groundwater reservoir evaluation. *Ground Water*, Vol. 20: 657–662.

Delhomme, J. P., 1978. Kriging in the hydrosciences. *Adv. Water Resour. Res.*, Vol.1: 251–266.

Gheorghe, A., 1978. *Processing and Synthesis of Hydrogeological Data*. Abacus Press, Kent, p. 390.

Holting, B., 1980. Hydrogeologie. Einfuhrung in die Allgemeine und Angewandte Hydrogeologie.

Huisman, L., 1972. *Groundwater Recovery*. Winchester Press, New York.

Journel, A. G., 1985. The deterministic side of geostatistics. *J. Int. Math. Geol.*, Vol. 7: 1–15.

Koch, G. S. and Link, R. F., 1971. *Statistical analysis of geological data*. Dover Publications, New York, Vol. 1, p. 375.

Krumbein, W. C. and Graybill, F. A., 1965. *An Introduction to Statistical Models in Geology*. McGraw-Hill Book Co., New York, p. 475.

Langelier, W. F., 1936. The Analytical Control of Anti-corrosion Water Treatment. *J. Am. Water Works Assoc.* 28, p. 1500.

Margat, J. and Saad, K. F., 1983. Concepts for the utilization of non-renewable groundwater resources in regional development. *Natural Resources Forum*. Vol. 7, 4; U.N., New York.

Meinzer, O. E., 1920. Quantitative methods of estimating groundwater supplies. *Bull. Geological Society of America*. 31: 329–338.

Papadopulos, I. S. and Cooper, H. H., 1967. Drawdown in a well of large diameter. *Water Resour. Res.*, Vol. 3, pp. 241–244.

Salas, J. D., Delleur, J. W., Yevjevich, V. and Lane, W. L., 1985. *Applied Modeling of Hydrologic Time Series*. Water Resources Publications, p. 484.

Schultz, E. F., 1973. *Problems in Applied Hydrology*. Water Resources Publications, Colorado State University, p. 510.

Seabear, P. R. and Hollyday, E. F., 1966. Statistical analysis of regional aquifers. *Am. Geophys.* Union Meeting, San Francisco.

Sichard, W., 1927. Das Fassungsvermogen von Bohrbrunnen und seine Bedeutung fur die grundwassesabsenkung inbesondere fur grossere Absenktiefen. Diss. Technische Hochshule, Berlin.

Şen, Z., 1978. Risk and reliability analysis in hydrologic design. *Int. Sym. on Risk and Reliability in Water Resources*, University of Waterloo, Vol. 2: 364–375.

Şen, Z., 1986. Determination of aquifer parameters by the slope matching method. *Ground Water*, Vol. 24: 217–223.

Şen, Z., 1995. *Applied Hydrogeology for Scientists and Engineers*. CRC, Lewis Publishers, Boca Raston, p. 444.

Şen, Z., 1999. Simple probabilistic and statistical risk calculations in an aquifer. *Ground Water*, Vol. 37, No. 5, 748–754.

Şen, Z. and Dakheel, A., 1985. Hydrochemical facies evaluation in Umm Er Radhuma limestone eastern Saudi Arabia. *Ground Water,* 24(5): 626–635.

Şen, Z., Saud, A., Altunkaynak, A., and Özger, M., 2005. Increasing water supply by mixing of fresh and saline ground waters. *J. Am. Water Resour. Assoc.*

Şen, Z. and Somayien, M. S., 1991. Some simple management criteria for confined aquifer in Tabuk Region, Saudi Arabia. *Water Resour. Manag.,* Vol. 5, 161–17.

Theim, G., 1906. *Hydrologische Methoden,* Gebhart, J. M., Leipzig, 56.

Todd, D. K. (1976). *Ground Water Hydrology.* (2nd ed.). John Wiley & Sons Inc., New York.

Way, J. H., 1968. Bed thickness analysis of some carboniferous fluvial sedimentary rocks near Joggin, Nova Scotia. *J. Sedimentary Petrology,* Vol. 83: 424–435.

Wilcox, L. V., 1955. *Classification and Use of Irrigation Waters.* U.S. Dept. Agric. Circ. 969, Washington, D.C., p. 19.

# 7 Sediment Transport in Arid Regions

## 7.1 GENERAL OVERVIEW

Climate, geology, and the age of the ground surface define the features of an arid environment. In general, arid regions are characterized by two kinds of morphology, namely, shields (platforms) and shelves (basins) (see Figure 1.3). The former is located in extremely stable seismic zones of tectonic origin such as the western Arabian Peninsula, the Sahara in Africa and southern Africa, parts of Asia, India, and Australia. Arid region shields are often dominated by eroded surfaces on volcanic rocks, which constitute the base of stratigraphic sequences. There are also platforms developed on horizontally layered sedimentary rocks like the Nubian sandstone in North Africa. The origin of these plains is not connected to their current aridity. Intermountain basin deserts are dominated by a succession of mountains and troughs often characterized by closed drainage basins.

Arid and semi-arid surfaces are subject to weathering and mass wasting processes. Most of the weathering processes are atmospheric, mechanical (physical), and chemical in origin. Similarity in the morphology among different regions does not mean that they are generated by the same geological mechanics. In fact, different processes can give rise to comparable forms in different places, starting from different beginnings.

Sediment yield at a catchment outlet, or at any section downstream, is the integrated results of upland, gully, and channel erosion, transportation, and depositional processes. The external dynamic agents of sediment yield are water, wind, gravity, temperature change, ice, and biological activities. Although each may be important locally, water is the most widespread agent of erosion and accounts for the bulk of sediments transport.

Although there have been extensive research efforts in the past, unfortunately even today there is not a unified approach that is generally accepted among researchers and engineers alike for sediment yield formulations. This chapter describes erosion, sedimentation, and dispersion phenomena, their significance in engineering applications, and the derivation of some formulations from wadi geomorphology features. It will help to establish philosophical, logical, and rational foundations for future research possibilities and application opportunities.

## 7.2 SURFACE COVER FEATURES

The essence of a warm arid environment is its sparse vegetation cover resulting from aridity. Vegetation cover shows arid region adaptations in terms of dry lands when

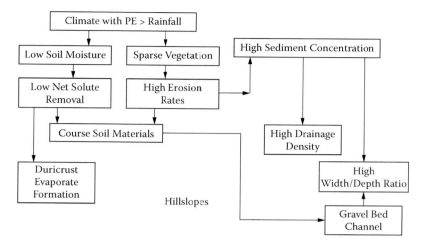

**FIGURE 7.1** Dominant geomorphologic processes in arid environments. (From Bull, L. J. and Kirby, M. J., 2000. *Dryland Rivers: Hydrology and Geomorphology of Semi-arid Channels*. Kirby and Bull, 2000.)

the rainfall is less than the potential evapotranspiration (PE) rate for all or part of the year, creating a permanent or seasonal soil moisture deficit. Many arid lands show a strong seasonal variation in *moisture deficit*, or a seasonal alteration between deficit and surplus. Arid and semi-arid climates produce a characteristic balance of hill slope and channel processes that gives dry land rivers (wadis) their special features. Arid climates are associated with intense rainstorms, which over sparsely vegetated surfaces, generate locally high rates of overland flow that lead to hill slope erosion by wash processes. Little subsurface flow is available for solute removal, so that soil tends to weather only slowly, and younger dry land areas are characteristically coarse grained with little information of clay mineral. The products of weathering tend to remain *in situ*, leading to the gradual accumulation of more soluble components such as duri-crusts and evaporation in suitable locations (Figure 7.1). Wadis occur in a range of tectonic settings. Most of the arid regions are associated with active tectonics, so that the area shows widespread quaternary uplift, with more erodible late tertiary deposits such as basalts.

The most obvious and important property of arid lands is that flow is ephemeral, occurring only for a short period during and after rainstorms, unless they have their source outside the arid area similar to the Nile in Africa and the Colorado River in North America. Hence, the relative importance of many fluvial processes, especially the magnitude and frequency of their operation, differs considerably from more humid regions. At the start of a storm, the advance of flood waves is limited by channel infiltration except where there is sealing by fines. The frequency distribution of flood discharges is therefore very different from humid regions.

High erosion rates and limited runoff give high concentrations in rivers leading to closely spaced channels and a high drainage density. In headwater areas this may lead to gullying and badland development. In the downstream, the high sediment concentrations and coarse grain sizes give rise to channels with high width–depth ratios. Although intense storms in all climates tend to have a limited areal extends, average storm intensity tends to fall off more rapidly with area for dry land climates than for temperate climates.

*Erosion* is the displacement of solids (soil, mud, rock, and so forth) by the agents of wind, water, ice, and movement in response to gravity or living organisms (in the case of bioerosion). Although the processes may be simultaneous, erosion is to be distinguished from weathering, which is the decomposition of rock *in situ*. Erosion is an important natural process, but in many places it is increased by human activities. Some of those activities include deforestation, overgrazing, and road or trail building. Likewise, humans have sought to limit erosion by terrace-building and tree planting.

Stream erosion occurs with continued water flow along a linear feature. The erosion deepens the valley and, headward, extends the valley into the hillside. In the earliest stage of stream erosion, the erosive activity is dominantly vertical, the valleys have a typical V cross-section, and the stream gradient is relatively steep. When some base level is reached, the erosive activity switches to lateral erosion, which widens the valley floor and creates a narrow floodplain. The stream gradient becomes nearly flat and lateral deposition of sediments becomes important as the stream meanders across the valley floor. In all stages of stream erosion by far the most erosion occurs during times of flood when more and faster moving water is available to carry a larger sediment load.

Despite extensive research efforts, knowledge of erosion and sediment transport still remains incomplete, and there is no generally accepted formula to be used for an accurate solution of the sediment transport rate and watershed sediment yield (Maidment, 1993). Especially, in arid regions erosion and sediment yields are closely related to surface geology, directly through deterioration of the bedrock exposed to the flow of water and indirectly through the character of the parent material whose properties are similar to the bedrock. The geological conditions of the area and soil properties affect *rock weatherability* and *soil erodibility*, respectively, and thereby also intensity of erosion processes. As a result of erosion and sedimentation, geologic, topographic, and hydrologic characteristics of an area may vary in horizontal, lateral, and vertical directions (Al-Suba'i, 1991, Şen and Al-Suba'i, 2002).

Erosion and sediment yield phenomena are studied by a wide range of methods, but there is no clearly defined and accepted approach to the problem. According to Lal (1985) either our conceptual understanding of the erosion–sedimentation problem on the earth's surface is far from being complete, especially in arid and semi-arid regions, or erosion–sedimentation research techniques are still more of an art than a science.

Soil type, topography, and soil cover are among the passive forces affecting erosion and transport of sediment from land surface. Soil mass and soil constituents as soil characteristics are the main factors for the sedimentation. Soil mass properties include permeability, volume change and dispersion properties, moisture content, and frost susceptibility. Permeability determines percolation rate and affects infiltration and runoff rates. Volume change and dispersion properties cause soil swelling losses and dispersion of soil thereby reducing the cohesion and facilitating dislodgement and transport. Moisture content reduces cohesion and lengthens erosion period by increasing the period of precipitation excess.

Topography is mainly concerned with the slope (orientation, degree, and length). Climatic force effectiveness is determined by orientation, the energy of flow by degree of slope, and length of the slope affects quantity or depth of flow. Depth and velocity affect turbulence both of which markedly affect erosion, transportation, and consequent sedimentation rates.

The external dynamic agents of sediment are water, wind, gravity, temperature (climate) change, ice, and biological activities. Although each may be important locally, water is the most widespread agent of erosion and accounts for the bulk of sediment transportation. In hydrological analysis of sediment climate, catchment morphology, soil, vegetation, and human activities are considered. Overland flow is an important process in arid lands; considerable quantities of sediment are moved by runoff. Indeed, drylands —particularly semideserts with about 250 to 300 mm of rain per year—have long been known to produce record levels of sediment yield (Langbein and Schumm, 1958). The exact nature of the generalized relation between sediment yield and rainfall is complex and will continue to evolve as more and more data become available. However, the transfer of such large amounts of material provides considerable problems for dryland water resource management in that impoundment structures are rapidly compromised by the reduction of reservoir volume as sediment accumulates. In order to minimize the nuisance caused by these high rates of sedimentation and to optimize the utilization of water resources, it is essential for managers and water engineers to understand the physical processes that govern the sedimentary behavior of wadis, whether these are in a state which is largely natural or affected in various ways by human agency.

## 7.3   SEDIMENTATION IN ARID ENVIRONMENTS

Arid and semi-arid regions are prone to sedimentation due to their weak vegetation cover and direct soil and rock outcrop exposition to rainfall, runoff, and wind phenomena. Although there are many geomorphologic and hydrologic factors that affect sedimentation, the two most significant sediment yield agents from geological point of view are *rock weatherability* and *soil erodibility*, which are demanding in arid environments. It is necessary to represent critically the rainfall, runoff, wind, and dry climate conditions in relation to sedimentation processes in arid regions. Rock weatherability and soil erodibility maps should be prepared for arid and semi-arid regions. The weatherability classifications are rather fuzzy and have "very low," "low," "medium," and "high" classes. Additionally, soil erodibility classifications can be based on the dispersion ratio and have "negligible," "low," "medium," "high," and "very high" fuzzy classes.

Due to low vegetation and strong physical rock weathering in the arid regions, the sediment yield is very large and results in environmental, economic, and social risks. The factors that affect the sedimentation process in arid region include not only rainfall, runoff, and wind but additionally other relevant variables such as the drainage area, total stream length, drainage density, mean bifurcation ratio, transport efficiency, mean slope, sedimentation index, sediment movement index, elongation ratio, mean channel slope, acceleration of gravity, rainfall erosivity, peak discharge, wind velocity, density of water, viscosity of water, rock weatherability, soil erodibility, vegetation cover, land use, and management indices. Erosion and sediment yield are also closely related to soil properties through the susceptibility of soil to both detachment and transport. This susceptibility defines soil erodibility, which varies with soil texture, aggregate stability, shear strength, infiltration capacity, and organic as well as chemical contents.

The earliest work on the erodibility indices of soil is presented by Middleton (1930). He analyzed samples of several erodible and nonerodible soils and found that some physical properties of soil such as dispersion and *erosion ratios* could be used to differentiate between *erodible* and *nonerodible* soils. These two indices are defined as such that the dispersion ratio is equal to *suspension percentage* divided by ultimate silt plus clay, and the erosion ratio is equal to *dispersion ratio* divided by ratio of colloid percentage to moisture equivalent. From a study of erosion in Southern California, Anderson (1951) supported the use of Middleton's dispersion ratio and later he introduced a new index of soil erodibility. It is the surface–aggregation ratio, which is defined as the surface area requiring binding, divided by aggregated silt and clay content. He observed that this index is well correlated with suspended-sediment discharge from 33 watersheds in western Oregon. Andre and Anderson (1961) and Scott and Williams (1978) concluded that both dispersion and surface–aggregation ratios are significantly related to soil-geologic rock type and such indices provide practical tools in evaluating erosion and sediment yields. Semi-arid and arid environments are distinctive in terms of factors affecting erosion and sediment yield (Jansson, 1982; Hadley, 1986; Pearce, 1986; Walling, 1986; Walling and Webb, 1986). Al-Suba'i (1991) presented a detailed evaluation of erosion and sedimentation features in arid regions.

The quantitative prediction of sediment transport across areas of net deposition is an essential part of assessing the offsite effects of soil erosion on hillslopes. An understanding of sediment delivery from the hillslopes to the stream requires information on rock weatherability, soil erosion, and sediment deposition. Although a vast number of studies concerning different aspects of soil erosion processes at various spatial and temporal scales exist, there are very few detailed studies on sediment transport by overland flow through net deposition zones.

### 7.3.1 RAINFALL AND SEDIMENTATION

Sediment is closely related to rainfall partly through the detaching and splash power of raindrops striking the surface materials and partly through the contribution of rain to runoff. The rainfall potential to cause erosion is dependent upon its kinetic energy, which is determined by fall velocity and raindrop diameter, and the total number of raindrops per unit time, which is represented by rainfall intensity (Onstad, 1986). Because rainfall consists of a series of different intensity events, it is a major factor that explains temporal variations in erosion and sediment yield. Generally, the most important rainfall features of the arid and semi-arid environments can be summarized as follows, (Al-Suba'i, 1991).

1. Rainfall can be very varied and erratic, spatially as well as temporally.

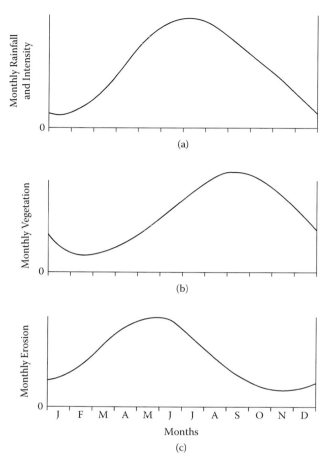

**FIGURE 7.2**   Seasonal cycles of rainfall, vegetation cover, and erosion.

2. Individual storm-total can be very high, where in many cases the single storm rainfall exceeds the mean annual rainfall.
3. Rainfall intensities can be very high. Sediment yield can increase greatly due to the high rainfall intensity.
4. The amount of runoff is increased by the scaling effects of *rainfall impact*. Rainfall impact on runoff increases its *transport capacity*, and due to the schematically represented monthly patterns of rainfall, vegetation, and erosion shown in Figure 7.2, the most valuable period for erosion and sediment yield is the early part of the wet season when the rainfall is high, but the vegetation has not grown sufficiently to protect the surface. Nevertheless, questions arise of how much rain is required to induce significant erosion and sediment yield, which events are to be employed (moderate or extreme), and what is the most suitable expression of the rainfall erosivity.

**TABLE 7.1**
**Coefficient of Determination for Erosivity Expressions at Various Sites**

| Erosivity Factor | Watkinsville, Georgia | Morris, Minnesota | Presque Isle, Maine | Campinas, Brazil | Katumani, Kenya |
|---|---|---|---|---|---|
| $A_r$ | 0.46 | 0.11 | 0.01 | 0.45 | 0.66 |
| R0 | 0.64 | 0.71 | 0.18 | 0.62 | 0.71 |
| E | — | — | — | — | 0.64 |
| $EI_{30}$ | 0.70 | 0.67 | 0.35 | 0.58 | 0.69 |
| $AI_{30}$ | 0.62 | 0.56 | 0.23 | — | 0.72 |
| $EI_{30} + R0$ | 0.74 | 0.84 | 0.38 | 0.70 | 0.74 |

*Note:* $A_r$ = rainfall depth, R0 = runoff depth, E = total storm kinetic energy, $I_{30}$ = maximum 30 min intensity.

Rainfall threshold values may vary, however, with the variation of other factors controlling erosion and sediment yield. For instance, intensity of 25 mm/h is discussed in the literature as a threshold of erosive rains (Hudson, 1981). Quoting from research in Zimbabwe, he stated that 50% of the annual soil loss occurred in only two storms in one year, and 75% of the erosion took place in 10 min. The opinion that most erosion and sediment takes place during events of moderate frequency and magnitude exemplified by rainstorms yielding 30 to 60 mm with frequency of 1 to 10 times per year is generally accepted by geomorphologists (Morgan, 1986). Roose (1977) stated that rainfall of extreme intensity (e.g., storms with a 10- to 100-year recurrence interval) may accomplish a large part of erosion, transportation, and mass-movement work in some climatic settings, as in the arid and semi-arid regions of the Sahara and Mediterranean zones. Wischmeier and Smith (1978) concluded that the best representation is the product of the total storm kinetic energy, E, and the maximum 30-min intensity, $I_{30}$. They further stated that the expression $EI_{30}$ is widely used in the United States as the basic erosivity expression for the universal soil loss equation. Lal (1976) in Nigeria reported a better correlation between the product of total rainfall, $A_r$, and the maximum 30-min intensity, $I_{30}$, than with $EI_{30}$. Foster and Meyer (1985) indicated that rainfall amounts and maximum 30-min intensity are the two most important general measures of rainfall erosivity. Table 7.1 shows a coefficient of determination for some erosivity expressions applied to three locations in the United States, and one each in Brazil and Kenya (Onstad, 1986).

Values in this table show wide variations in coefficients of determination, which indicate that expressions for erosivity need to be independently examined for various regions. As can be seen from this table, the $A_r$-$I_{30}$ index shows about the same amount of variability in the data as the $EI_{30}$ index, but the former can be computed more easily.

### 7.3.2 RUNOFF FEATURES

Running water is the most important agent of erosion in humid areas, but arid lands, in which water is in short supply, have their own distinctive landforms. Surface features in arid areas are stark and sharp. Much bare rock is exposed. The alternation of blistering daytime heat and chilly nights shatters the rocks, and the lack of moisture limits the softening effects of chemical weathering. Few plants cloak the contours of the land. The wind carries away the finer particles of weathered material and leaves the surface littered with a desert pavement of stones too large for it to move.

Rain is rare in arid areas, but once or twice a year violent thunderstorms pummel the earth with great torrents of water which cannot soak into the sun-baked ground, and it is carried away by sudden flash floods. The floodwater is so choked with sediment that the streams cannot erode their valleys deeper, but they undercut the valley sides to create box canyons with steep sides and flat bottoms. The floodwater pours into shallow lakes and soon evaporates. The soils of arid areas contain large amounts of soluble alkaline salts that would be removed by solution in a more humid area. Some of these salts are dissolved by the water that runs off in *flash floods*. When the water evaporates, the salts are deposited on the dry lake beds as glistening salt flats caked with white alkaline salts.

Desert pavement, box canyons, and salt flats all are distinctive features of arid lands. The stream dumps its load of sediment and dams its own valley when its speed is reduced at the change in gradient. It repeatedly seeks and dams new channels, and gradually builds up a cone-shaped deposit of alluvium with its apex at the point where the stream leaves the mountains.

Runoff indicates sediment yield by detachment of the sediment from the surface and by the subsequent accumulation and transportation of eroded sediment. It occurs when the rainfall intensity is higher than the infiltration capacity and when surface depression storage has been filled and the soil is saturated. Osborn and Lane (1969) have established the minimum amount of rainfall required to initiate runoff in semi-arid environment small catchments as 0.81 cm. High intensity rain, as in semi-arid zones, seals the surface of bare soil very quickly and, consequently, only a shallow depth of soil moisture may be achieved before pounding and surface runoff start. When runoff is concentrated in rills or when there is a deep sheet of flow on a slope, erosion and transportation by flow are important. The water exerts a drag force on the surface particles and with increasing depth and gradient, the velocity and, hence, the shear stress increase and the water have a greater ability both to erode and transport materials. Quite a lot of evidence suggests that the rate of detachment by flowing water decreases as the sediment load increases, (Willis, 1971; Meyer and Monke, 1965) or that the rate of deposition is directly proportional to the difference between the sediment concentration in the flow and the equilibrium concentration for that flow condition (Einstein, 1968). On the other hand, Foster and Meyer (1975) expressed this concept by the following relation:

$$\frac{D_R}{D_C} + \frac{S_L}{T_C} = 1 \tag{7.1}$$

where $D_R$, $D_C$, $S_L$ and $T_C$ are the *detachment rate, detachment capacity, sediment load,* and *transport capacity,* respectively. Equation 7.1 indicates that riling will be greatest when the sediment load is low, but that little or no riling will occur when

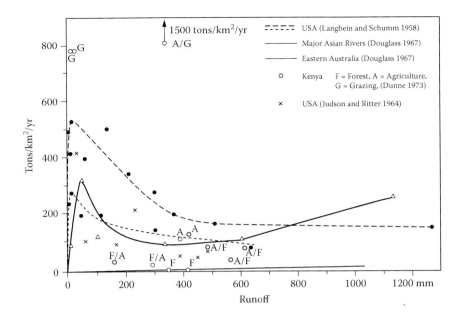

**FIGURE 7.3**   Sediment yield runoff relationship.

the transport capacity of flow is satisfied with the sediment produced by raindrop-impact erosion. It also shows that deposition may occur when transport capacity of sediment-load flow is decreased due to reduced flow gradient or increased hydraulic roughness. Many researchers have correlated runoff with sediment yield in several regions. For instance, in the United States, Langbein and Schumm (1958) showed that in very large basins with considerable human activities, the sediment yield per square kilometer is maximum at about 10 to 20 mm of runoff, decreasing sharply on both sides of the maximum (see Figure 7.3). A curve by Douglass (1967) based on data from main Asian rivers exhibits two maximums: one at about 50 mm runoff and another at about 1,145 mm runoff. Under natural vegetation in east Australia, he indicated that sediment yield per square kilometer is very low even at high runoff.

Due to the technical difficulties in measuring bed load and the relative paucity of observations on solute concentrations, most sediment yield estimation studies tend to relate water discharge to suspended sediments. This simple relationship has been questioned by Walling (1977) who stated that the error of ±50% or more may be associated with many *rating curve* estimates of sediment load.

In his study on discharge frequency compared to long-term sediment yield Neff (1967) pointed out that there is a direct relationship between the variability of annual peak discharge and the amount of sediment moved by less frequent flows. He stated that in arid and semi-arid areas, which have the greatest variability in annual peak discharge, only 40% of the long-term sediment load is moved by flows having a frequency of less than 10 years. There is a direct relationship between the variability of annual peak discharge, which may also be implied by climate change, and the amount of sediment moved by less frequent flows. Arid and semi-arid areas have the greatest variability in annual peak discharge and only 40% of the long-term sediment load is moved by flows having a frequency of less than 10 years. Obviously, generalizations about the influence of arid and semi-arid runoff on sediment yield are difficult to make but the following features are important (Al-Suba'i, 1991).

1. There is a great difference between the theoretical limits of the basin, the divide or watershed, and the effective area that yields runoff and sediment.
2. Irregular patterns of flow events make collection and interpretation of discharge data difficult or rather impossible in case of very extreme conditions. This is not peculiar to events only, but within the event itself there is irregularity where the flash flood events vary in time and space with very large variation in discharge over a short period of time during the rising stage (Campbell, 1977; Al-Khafif, 1986).
3. Extreme events of high magnitude and low frequency may be associated with sediment yield exceeding the mean annual value by 50 times or more.
4. There are several types of sediment yield pattern associated with hydrograph rise (Schick and Sharon, 1974). Rising stage is normally associated with a turbulent front wave of high energy carrying a large amount of eroded materials. It would be practically impossible to carry out accurate sampling of sediment concentration or precise measurement of flow parameters during this stage. Hence, it will be equally impossible to correlate the sediment discharge to the flow. This sharp rise is followed by a relatively sharp crest before the relatively smooth recession occurs. The sediment discharge rate varies markedly between these stages. Hence, computation of erosion or sediment yield based on such data becomes highly speculative.
5. In rock gullies, torrential runoff can cause movement of all size of material so that tributary channels are at times swept clean of debris. At gully mouths, there is a hazard of small fan and delta development in the main channel, which may lead to bank erosion, the formation of gravel sheets, lobs, bars, and alterations of bed elevation.
6. Deposits from high-magnitude tributary flows may produce temporary terraces that are steadily removed by smaller magnitude flows or even swept clean by a furthest major spate.
7. As with rainfall, the most vulnerable time for runoff to contribute to sediment yield is the early part of the wet season when the rainfall is high but vegetation has not grown sufficiently to protect the surface (Figure 7.2).

Hill slopes are the primary source of sediment transported by rivers with channel bank and flood plains as a secondary source. The sediment yield phenomenon is, therefore, divided into two broad categories:

1. The upland phase (off-stream)
2. The lowland stream or in-channel (in-stream) phase

The upland phase emphasizes the erosion process of detachment and transport in rills and inter-rill areas, where the mechanics of the precipitation event and surface flow are the major agents. Major variables influencing the yield in this phase are rock units, soil type, condition, and moisture content at the start of the event; slope and slope length; vegetation and litter cover; and rainfall amount, intensity, and duration. In the in-channel phase, sediment transport and deposition processes predominate and, consequently, channel transport capacity becomes more important. Pertinent variables in this phase are the velocity, depth of flow, channel slope, wash load, water temperature, and median size and grain size distribution of bed material. Although not all of the subprocesses occur on all source areas, each has its part in the total sediment yield process.

Temporal and spatial streamflow and flood frequency variations largely affect the erosion and sedimentation. The drainage density is closely related to the flood. Clearly, factors other than average rainfall intensity can affect the PDF of flood peaks, and hence the critical assumption of the design storm method is often non-valid. The weaknesses of the design storm method are especially critical when it is used to evaluate complex strategies for flood mitigation (engineering).

*Flash floods*, which are short-lived extreme events, are exceptional. They usually occur under slowly moving or stationary thunderstorms, which last for less than 24 hours. The resulting rainfall intensity exceeds infiltration capacity, so runoff takes place very rapidly. Flash floods are frequently very destructive as the energy flow can carry much sedimentary materials.

Natural forces will continue to alter the face of the earth, but their effects are being accentuated by human actions. The removal of vegetation, climate change effects, and cutting roads in steep hillsides encourages soil erosion; overgrazing or intensive cultivation of crops is transforming vast areas into deserts; pollution is creating acid rain, which hastens chemical weathering. There is now an urgent need for improving the understanding of how natural forces, and in this context, how climate change shapes the landscape and how to use this vital knowledge to limit environmental degradation.

### 7.3.3 WIND AND SEDIMENTATION

The action of wind on exposed sediments and friable rock formations causes erosion (abrasion) and entrainment of sediment and soil particles. Eolian action also forms and shapes sand dunes, streamlined bedrock hills, and other landforms. Subsurface deposits and roots are commonly exposed by wind erosion. Wind can also reduce vegetation cover in wadis and depressions, scattering the remains of vegetation. Stone pavements may result from the removal of fine material from the surface, leaving a residue of course particles. Blowouts in the forms of erosion troughs and depressions in coastal dune complexes are important indicators of changes in wind erosion. The potential for removal is generally increased by shoreline erosion or washovers, vegetation die-back due to soil nutrient deficiency or to animal activity, and by human actions such as recreation and construction.

Changes in the wind-shaped surface morphology and vegetation cover that accompany *desertification*, *drought*, and *aridification* are important gauges of environmental change in arid lands. Wind erosion also affects large areas of croplands in arid and semi-arid regions, removing top soil, seeds, and nutrients.

Wind erosion occurs when soil bore of vegetation are exposed to high-velocity wind. When its velocity outcomes the gravitational and cohesive forces of the soil

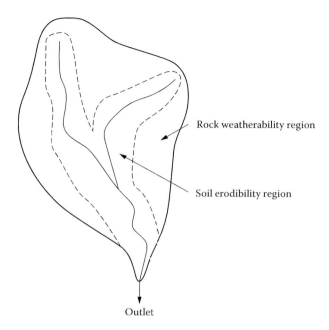

Outlet

**FIGURE 7.4**   Weathering and erosion regions.

particles, wind will move soil and carry it away in suspension. Wind moves soil particles of 0.1 to 0.5 mm in size in hopping or bouncing fashion, which is known as saltation, and those greater than 0.5 mm by rolling what are known as *soil creeps*. The finest particles (less than 0.1 mm) detach into suspension. Wind erosion is most visible during the suspension stage, as dust storms.

In arid regions, apart from the sediment-generation agents such as rainfall, runoff, and wind phenomena and in addition to morphological features including slope, rock weatherability and soil erosivity play significant roles. Especially in arid regions due to very weak vegetative cover, rock weatherability from necked rock surfaces at high elevations and soil erodibility, which takes place at lower elevations around the wadi channel network (see Figure 7.4), become very important basic factors for sediment generation.

## 7.4   ROCK WEATHERABILITY

It is possible to view weatherability in two categories, namely, meteorological and mechanical. Surficial and selective features are generated by the meteorological weathering processes, which affect both rock outcropping and the detritus made up of sand and cobbles. Such a weathering process does not allow high infiltration rates and temperature variations in the layer close to the earth's surface. Rocks in arid and semi-arid regions are severely exposed to atmospheric weathering because these regions are not covered by soil or vegetation, and consequently lithologic and structural characteristics of the rocks control weathering. The degradation process is tended, concentrated, and controlled due to the existence of fractures, faults, crystal edges, and other planes of weakness exposed at the rock's surface. The alteration varies according to rock exposure that depends on local temperature and humidity variations. Surface rocks are subject to *splintering* (shattering into coarse, angular flakes), *exfoliation* (desquamation), *breakage* (pebbles and spherical blocks that look like they were split by an axe), and *granular disaggregation* phenomena resulting in a huge quantity of debris.

Mechanical weathering is more significant than chemical weathering in the presence of acutely angular broken rocks and detritus in the vast majority of deserts. In arid and semi-arid regions, mechanical weathering is more important than chemical or biological. Some mechanical operations such as salt weathering actually depend on a chemical agency. It is possible to distinguish three major mechanical weathering processes:

1. *Thermoclastism* (insolation, temperature weathering): Temperature weathering causes rock and mineral breakup as a result of wide-ranging daily and seasonal temperatures, as well as significant contrasts between heat and minerals. This causes expanding and contracting at different rates, depending upon the coefficients of thermal expansion and other properties that may trigger breakage. Weathering by dry insolation can induce microfractures, which increase rock permeability and open a way to other erosive processes. If the humidity is associated with wide-ranging daily temperatures, water trapped in the capillaries of the rock can generate internal pressure great enough to fracture the rock. This type of weathering by humid insolation depends upon the structure of the pores of the material, thermal variation, and the availability of water. Repeated heating and cooling over long periods of time will also provoke detachment into rock splinters.
2. *Cryoclastism* (ice weathering): The crumbling of rock when water freezes in its cramped internal spaces and the expanding volume of water crystals exercises pressure.
3. *Haloclastism* (salt weathering): One of the most active processes in hot and cold deserts, it is rare in the temperate zones. The most favorable climate conditions for salt weathering are those typical of hot and arid environments: low relative humidity and heavy evaporation during the day, high diurnal temperatures, and fluctuating relative humidity.

Evaporation from the top desert strata induces the continuous capillary ascent of water above the piezometric surface of groundwater. The capillary fringe, or the area immediately above the water table where capillary action occurs, becomes saturated with salt due to this continual evaporation. The thickness of fringe varies, especially as a function of the gronulometry, or particle size, of its sediments. In the case of very fine sediments and extreme desert conditions, it may be more than 3 m (10 ft) thick. The fringe may lie immediately below the desert surface where its upper edge can host a growth of crystals, which are often made of gypsum and produce the so-called desert roses. In other cases, it reaches the surface by creating an area of salt blooms and crusts, often in the form of needle-shaped crystals that spurt out of the ground (Stoppato and Bini, 2003).

Salts are concentrated in three main environments, namely, humid deserts on arid western coasts, *playas*, and *sabkhas*. Salt also accumulates within and around drainage basins including playas and salt lakes. Sabkhas may form in the areas of coastal aggradations, especially along inland seas like the Arabian Gulf and the Red Sea. The capillary ascension of sea water is the primary cause of extensive accumulations of salt on these wide tidal plains.

For sediment yield investigations, Al-Suba'i (1991) considers only the catchments of the central part of Tihamat Asir region, extending over the Red Sea littoral zone between latitudes 16° 52' N and 18° 05' N and stretching eastward from the coast at longitude 42° 35' E to the summit of the Red Sea escarpment at longitude 43° 30' E, as shown in Figure 4.33. It thus embraces mainly the western slopes of the central part of Asir highlands with an area of about 15,000 km$^2$ in the Asir and Jizan provinces of Saudi Arabia and in Sa'da province of Yemen.

For rock weatherability evaluations, 50 samples of slightly weathered rocks were collected from rock surfaces of the major rock groups. The number of samples from each rock type was proportional to the variability in lithology of each geological unit. The rock weatherability index, $W_i$, is estimated on the basis of *slake durability index*, $I_d$ determined test, following the procedure suggested by the International Society for Rock Mechanics (ISRM) (Brown, 1981). The results obtained from the tested samples and the geological map is used to delineate the areas that are almost similar in their degree of weatherability as shown in Figure 7.5.

The weatherability classification is given by utilizing the durability classification produced by Gamble (Brown, 1981) with the necessary modification of terms. Table 7.2 shows the values of rock weatherability index $W_i$ of the 50 tested rock samples, which are in the range of 60 to 99%.

These results are used to classify the surface rocks in the study region into four groups (as A, B, C, and D) with weatherability varying from "very low" to "high" (Table 7.3) following the classification of Gamble quoted by Brown (1981) with the necessary modification in terms. The main rock units ranked as "high," "medium," "low," and "very low" in terms of their weatherabilities, together with their areal extent expressed as percentages of the study area, are shown in Table 7.3 which, together with the geologic map, is used to draw the boundaries of these classes shown in Figure 7.5.

From this map, it can be seen that the areas of high weatherability with $W_i$ between 60 to 85% are found mainly in the central part as well as in the center of eastern and southern sides of the study area. Rocks of medium weatherability occur mainly in the north, north central, east central, southeast central, and southeast parts. On the other hand, areas of low and very low weatherability are found in the northern, eastern, and southern parts, in addition to the central part and a narrow strip along the western part of the study area. The weighted average of the weatherability indices for different wadis is also shown in Table 7.4.

## 7.5  SOIL ERODIBILITY

In order to evaluate soil erodibility, soil samples from 70 locations throughout the study area are collected for laboratory testing (Al-Suba'i, 1991). The samples are

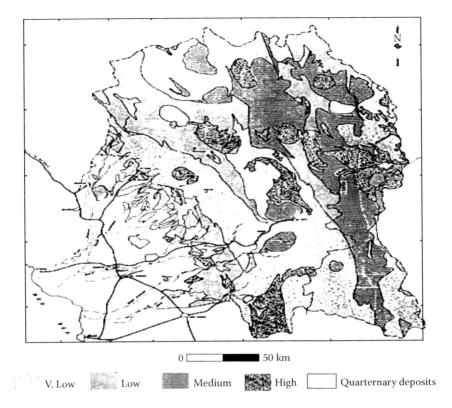

0 [▭▬▬▬▬] 50 km

V. Low      Low      Medium      High      Quarternary deposits

**FIGURE 7.5**    Weatherability map of the Tihamat Asir region.

selected from the major soil geological units under conditions as nearly uniform as possible with respect to slope (8–30%), annual rainfall (300–500 mm), and natural vegetation cover. Attempts are made to obtain a representative distribution among geological parent materials and physiographic zones. *Dispersion ratio* as defined by Middleton (1930) is determined by using wet sieving through 0.05 mm for estimation of silt, plus the use of a clay content and hydrometer method for suspension percentage. A sample of undisturbed air-dry soil equivalent to 30 gr of oven-dry soil and the ASTM H152-type hydrometer are used in analyzing particle size distribution. Based on the dispersion ratio and the geological map, the study area is divided into subareas, almost, similar in their degree of soil erodibility as shown in Figure 7.6.

The ranges and the mean values of dispersion ratio of all samples from the different parent rock groups are shown in Table 7.5. It is observed that the lowest value of dispersion ratio is 27%, which is higher than the 15% that is considered by Middleton (1930) as the upper boundary for nonerodible soil. This means that all the soils in the mountainous part of the study area are erodible.

It is interesting to notice that these results indicate some trends, which enables a classification of soil erodibility into five groups as "negligible," "low," "medium," "high," and "very high." The proposed classification, as well as the main parent

**TABLE 7.2**
**Summary of Test Results for Rock Weatherability Index**

|  | $W_1$ (%) | |
|---|---|---|
| **Geological Units** | **Mean** | **Range** |
| Ba | 98.50 | 98.0–99.00 |
| gb, gd, dp | 98.11 | 97.9–98.30 |
| Tb, Tgb, Tgl | 98.30 | 98.0–99.00 |
| Qb | 98.60 | — |
| Hv | 98.50 | — |
| Sa | 97.30 | 95.0–98.00 |
| Gdn | 94.85 | 94.8–94.90 |
| gdh | 96.30 | 96.0–97.00 |
| gs, al | 97.14 | 96.0–98.00 |
| Sy | 94.90 | 94.6–95.20 |
| Am | 91.60 | 91.0–92.00 |
| ma | 95.60 | 95.0–96.00 |
| hs, h | 91.50 | 88.0–94.00 |
| di, dgb | 93.50 | 92.0–95.00 |
| thn | 95.00 | 94.5–95.50 |
| gn | 74.90 | 66.0–85.00 |
| gm | 72.50 | 60.0–85.00 |
| OCW | 71.70 | 59.5–84.00 |

**TABLE 7.3**
**Weatherability Classification of the Rocks of the Study Area**

| Group | Geological Units | No. of Tested Samples | Rock Weatherability Index $W_1$(%) | Class | Group Area (km²) | Areas as Percentage of the Study Area |
|---|---|---|---|---|---|---|
| A | ba, gb, hv, Qb, Tb, Dp, gd, Tgl, Tgb | 16 | 98–100 | Very low | 2,040 | 17.0 |
| B | Sa, gdn, gdh, gs, al | 15 | 9–98 | Low | 3,910 | 32.0 |
| C | Sy, am, ma, hs, h, dgb, thn, di | 12 | 85–95 | Medium | 1,760 | 14.6 |
| D | gn, gm, OCW | 7 | 60–85 | High | 1,070 | 8.9 |

**TABLE 7.4**
**Weighted Average Weatherability Index**

| Basin | Sub-Basin | Area (km²) | Weighted Average Weatherability Index (%) | Weatherability Class |
|-------|-----------|-----------|-------------------------------------------|----------------------|
| Baysh | — | 4,570 | 92.5 | Medium |
| Akas | — | 29 | 96.5 | Low |
| Gara | Tafshah | 46 | 96.5 | Low |
| Shadan | — | 91 | 93.5 | Medium |
| Wasia | — | 64 | 97.2 | Low |
| Sabya | Sabya | 74 | 96.8 | Low |
| Damas | — | 569 | 94.2 | Medium |

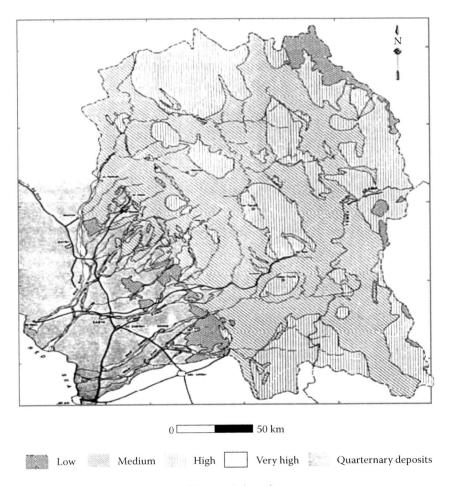

0 ⬜⬜⬜⬛⬛ 50 km

▨ Low   ▨ Medium   ▥ High   ☐ Very high   ▨ Quarternary deposits

**FIGURE 7.6**   Erodibility map of the Tihamat Asir region.

**TABLE 7.5**
**Dispersion Ratio as Related to Geological Units**

| | Dispersion Ratio (%) | |
|---|---|---|
| Parent Geological Type | Mean | Range |
| Tl | 28.0 | 27–29 |
| Tb, Tgb, Tgl | 37.5 | 37–38 |
| Qb | 41.9 | 39–45 |
| Sa | 59.6 | 45–70 |
| hv, hs, h | 51.0 | 47–57 |
| ma | 52.5 | 52–53 |
| gb, dp, gd, dgb | 59.5 | 58–60 |
| ba | 60.2 | 58–62 |
| am | 59.8 | 59–60 |
| di | 62.5 | 60–64 |
| Sy | 76.0 | 75–77 |
| OCW | 77.0 | 75–78 |
| gs | 77.6 | 76–79 |
| thn | 81.7 | 78–85 |
| gn | 80.8 | 79–83 |
| gm, al | 84.6 | 84–85 |
| gdn, gdh | 85.6 | 82–90 |
| Qal | 94.3 | 90–100 |

geological rock-soil types and the areal extent of soil of each class within the study area, are given in Table 7.6.

Based on this classification, together with the geological map, the soil erodibility map of Figure 7.6 is prepared. It divides the study region into areas of different degrees of soil erodibility. This map shows that the areas of "low" soil erodibility are, generally, small in size and scattered along the eastern boundary and the

**TABLE 7.6**
**Proposed Classification of Erodibility Based on Dispersion Ratio**

| Dispersion Ratio (%) | Degree of Erodibility | Parent Geological Units (Al Su'bai, 1992) | Class Area | Areas as Percentage of the (km²) Study Area |
|---|---|---|---|---|
| Up to 15 | Negligible | — | — | — |
| 15–45 | Low | Tl, Tb, Tgb, Tgl, Qb | 552 | 4.58 |
| 45–75 | Medium | Sa, hv, hs, h, ma, gb, dp, gd, dgb, ba, am, di | 5,052 | 41.94 |
| 75–90 | High | Sy, OCW, gs. thn, gn, gm, al, gdn, gdh | 3,177 | 26.40 |
| 90–100 | Very high | Qal | 278 | 2.30 |

**TABLE 7.7**
**Weighted Average Soil Erodibility Index**

| Basin | Sub-Basin | Area (km²) | Weighted Average Soil Erodibility Index, $e_i$ (%) | Degree of Erodibility |
|-------|-----------|------------|---------------------------------------------------|----------------------|
| Baysh | — | 4,570 | 68.70 | Medium |
| Akas | — | 29 | 60.00 | Medium |
| Gara | Tafshah | 46 | 60.00 | Medium |
| Shadan | — | 91 | 64.20 | Medium |
| Wasi | — | 64 | 62.40 | Medium |
| Sabya | Sabya | 74 | 62.20 | Medium |
| Damas | — | 569 | 66.60 | Medium |

southeastern parts. Soils of "medium" erodibility cover most of the study area while those of "high" erodibility are predominantly in the north, east, south, and southeast. Soils of "very high" erodibility generally occupy beds of the channels of the active drainage systems. The weighted averages of soil erodibility index of different wadis are presented in Table 7.7. These results indicate that the weighted *erodibility index* for all the wadis are within the "medium" range except those of Wadi Wu'al and Wadi Jizan, which are close to the upper limit of the "medium" range.

Soil erosion is one factor that hinders water resource development. Its direct impact results in the decrease of reservoir storage capacities and the efficiency of water transfer canals. The main factors that govern these phenomena are climate—mainly rain and wind—topography, soil media, land use, and human activities. In wadi hydrologic systems, characterized by arid and semi-arid climates, the soil nature and the land use density considerably increase the erosion process and limit water resources development.

## 7.6   EROSION CONDITIONS AND CONCLUDING RESULTS

Based on the physiographic, geologic, hydrologic, rock weatherability, and soil erodibility studies, as well as the review of the previous works, the following remarks on the erosion conditions can be noted.

1. Tectonic processes are generally important, and they exceed erosional process in their role in landscape (e.g., drainage-net characteristics and hill slope morphology) formation. Furthermore, in steep areas, gravitational force acts directly on sediment particles in various forms of mass movement, and it is in some local areas more important than the indirect entrainment through normal stress runoff.
2. With the exception of glacial action, all forms of erosion are almost active in arid regions, and the movement of sediment as bed load or suspended load in stream runoff is dominant and the most important.

3. Geologically young terrain (the interplay of relief, weatherability, and erodibility) and sparse vegetation cover give rise to the parent materials, which can be considered logically as the dominant soil forming factors in arid regions. This point is supported by the observation made by Andre and Anderson (1961) in the western United States region, which has tectonic and climatic conditions almost similar to the study region in the previous sections. They showed that parent material explains a large amount of the variation in erodibility between samples, and this amount of the variation in erodibility between samples was more dominant than any other soil forming factor for soil in that region.

4. It is possible to formulate a model for the sediment in the wadis (Figure 7.7), as lateral supply of sediment to stream channels is almost a continuous process achieved to a significant degree during the dry season by gravitational forces and in the wet season by overland flow and by mass movements.

## 7.7  SEDIMENT YIELD MODELING IN ARID REGIONS

Sediment yield has been studied in the past by two basic types of analysis, namely, hydrologic and hydraulic approaches. The choice largely depends on the type of data used in the analysis. In hydrological analysis, climate, catchment morphology, soil, vegetation, and human activities are taken into consideration. In the case of hydraulic analysis, factors such as properties of the fluid and entrained solids, geometric characteristics of boundaries, presence of density interfaces, and other factors are taken into consideration. Runoff erosion largely concerns the transportation of loose materials by turbulent water flowing in sheets, rills, or gullies, although some detachment of particles can occur in runoff erosion, (Cooke and Doornkamp, 1974; Throne et al., 1987).

It is possible to divide the sediment yield affecting factors into four groups as shown in Figure 7.8.

Considerable knowledge related to the nature and influence of the factors involved in sediment yield has been accumulated during the past few decades. In the next few paragraphs, the more important processes and factors related to the present study, and the type and effectiveness of the various approaches and methods of analysis are reviewed critically and evaluated to decide the approach and methodology to be followed here.

Deterministic conceptual sediment yield models have not been excessively developed and still include gross simplifications. The requirement of these models for many input parameters, the assumptions which have to be made about spatial and temporal variations, and the striking knowledge gaps in understanding of parameters controlling the sedimentation process mean that the inputs of these models are not better defined than empirical models. Furthermore, the climate and hydrologic variables in deterministic models are essentially the same as those used in empirical models. Because information on sedimentation process is generally imperfect, one may resort to empirical modeling approaches. Sediment yield system complexity, with its numerous interacting factors, creates difficulties including a high number of factors in a single predictive regression equation.

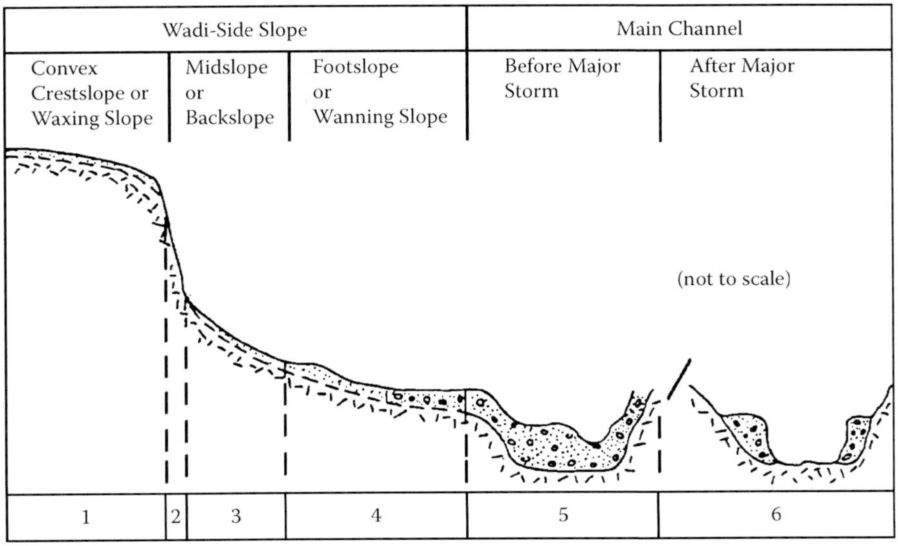

| Wadi-Side Slope | | | Main Channel | |
|---|---|---|---|---|
| Convex Crestslope or Waxing Slope | Midslope or Backslope | Footslope or Wanning Slope | Before Major Storm | After Major Storm |

(not to scale)

| 1 | 2 | 3 | 4 | 5 | 6 |

| Zone No | Name of the Zone | Surface Materials | Range of the Slope Angle | Predominant Erosion Conditions |
|---|---|---|---|---|
| 1 | Convex crestslope | Thin soil cover with bare patches of rocks | 1°–4° | Sheet erosion (surface wash) and creep |
| 2 | Free face midslope (cliffed) | Almost bare hard basement rocks | >45° normally 65°–90° | Gravity (fall and slide) as well as structural control rill and gully erosion |
| 3 | Transportational midslope | Talus and debris wash with bare patches of soil and rocks | 28°–36° | Sliding, flowing and debris wash with rills and gullies action |
| 4 | Footslope (sometimes don't occur) | Colluvial and alluvial deposits | 2°–15° | Gully, rill, and sheet erosion |
| 5 | Main channel before major storms | Channel fill of debris-flow deposits, stream alluvial, and colluvial derived directly from adjacent hill slopes | 2°–10° | Between major storm, Wadi-side slopes are undercut by water and gravity actions and the main channels are time-continuous fill. These channels may be choked by the infilling materials |
| 6 | Main channel after major storms | Bedrock and debris levees | 2°–10° | During major storms, the above mentioned main channels deposits are mostly scoured to bedrock and abnormal sediment transportation is expected. |

**FIGURE 7.7**  Typical erosion conditions in the mountainous parts.

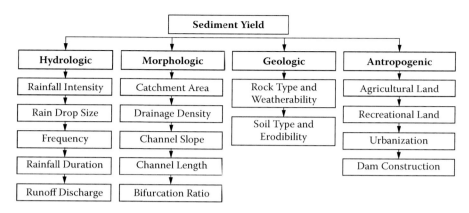

**FIGURE 7.8**   Sediment yield components.

The explanatory theories of runoff, erosion and sediment transportation in the upland phase of sediment yield are available. The first theory was formulated by Horton (1945) and its important components are the rainfall intensity, which exceeds the infiltration capacity of the soil, and water accumulation in depressions on the surface before spilling over to run down slope as an irregular sheet or turbulence flow. As the flow thickens and accelerates with increasing down-slope, the share stress imposed on the surface by sheet flow exceeds the critical attractive force of the soil and, consequently, erosion occurs. We derived the critical distance from the divide beyond which erosion starts to occur as a function of the *runoff intensity* (i.e., the difference between rainfall intensity and infiltration capacity), the hydraulic roughness and critical attractive force of the surface, and the local gradient. Beyond the belt of no erosion, sheet wash erosion is proportional to total shear stress (taken up by vegetation and large mobile grains) and, therefore, to the local product of the flow depth and water surface gradient (taken equal to surface gradient). A consequence of this is that an erosive sheet flow is fundamentally unstable and sheet erosion implies the formation of either a rilled or gullied surface. According to this model, the critical distance required for sheet flow to become erosive is the same as that required for channel incision, and it is equal to one-half the reciprocal of drainage density.

As an alternative to the Horton's overland flow and erosion theory, the saturated through flow model has been developed to explain the continuous sequence of possible conditions for gully development. In this model emphasis is placed on the movement of water down-slope through the upper soil horizons (Kirby, 1969). It takes place as concentrated flow in subsurface pipes; its erosive effects through pipe collapse and gully formation are well known (Morgan, 1986). Actual studies by Roose (1977) in Senegal revealed that throughflow contributes only about 1% of the total material eroded from a hillside and that this is mainly in the form of colloids and minerals in ionic solution. Cooke and Doornkamp (1974) stated that overland flow and throughflow are likely to be two extremes of continuous sequences of possible conditions for gully development, with overland flow more common in semi-arid areas and throughflow more common in humid areas.

Theoretical explanations of in-channel sediment transport mechanism involve complex considerations of open channel and loose boundary hydraulics and fluid mechanics, which are beyond the scope of this section. However, the subject has been reviewed exhaustively by various researchers (Bogardi, 1974; Graf, 1971; Herbertson, 1969; Morris and Wiggert, 1972; Shen, 1971; Simons and Şentürk, 1977; Vanoni, 1975, and Yalin, 1972).

Sediment yield for a given stream or reservoir site is the end result of three processes, namely, erosion, transportation and deposition. These processes are complex in nature, and they are controlled by many factors. The existing data on these processes in the study area do not enable one to make direct evaluation of sediment yield. Moreover, the facilities and time do not allow for extensive field experiments leading to measurements for obtaining reliable and additional data. Furthermore, it is economically impractical, and physically and technically difficult, to study all catchments, especially in the mountainous regions by plot method or direct measurements of erosion. Therefore, indirect method of modeling and prediction is used. The development of a model for this purpose will be based on various factors related to eroding and transport agents, basin morphology, geology and soil, vegetal cover, land use, and land management. Here, the sediment yield model has been developed by following three steps:

1. Selection and definition of the variables
2. Model formulation
3. Factor of proportionality estimation

### 7.7.1 Selection of Variables

Selection of the sediment yield variables are based on logic and a series of initial postulates, which are valid in light of the observations and experiences from the pervious studies. Based on the literature survey on erosion assessments, it is considered in general that the sediment yield variables are the total stream lengths, $L_t$; catchment area, A; drainage density, $D_d$; mean bifurcation ratio, $R_b$; transport efficiency, $T_e$; mean ground slope, $\Theta$; sediment area index, $S_a$; sediment movement index, $S_m$; elongation ratio, $R_e$; relief ratio, $E_{rr}$; mean channel slope, $\Theta_c$; acceleration due to gravity, g; rainfall erosivity, $E_{el}$; annual mean discharge with 100-year return period, $Q_{100}$; density of water, $o_{wn}$; viscosity of water, $\mu$; rock weatherability index, $W_i$; soil erodibility index, $e_i$; vegetation cover index, $C_{sv}$; land use index, $U_1$; and land management factor, $M_1$. Table 7.8 shows a summarized form of the definitions, dimensional forms, and the type of proportionality (direct, D, or inverse, I) of these variables with the sediment yield.

Accordingly, these variables are taken as a preferred set for the modeling of sediment yield.

### 7.7.2 Model Formulation

The choice of a method for model formulation should be based on present knowledge, purpose, scale, and reliability. However, conventional approaches based partly on scientific and empirical considerations are still being adopted. It is difficult to

**TABLE 7.8**

**Definition, Dimensional Formula, and Proportionality of Variables in Sediment Yield Model**

| Variable Group | Variable Name | Symbol | Definition | Dimensional Formula | Proportionality to Sediment Yield D: direct I: indirect |
|---|---|---|---|---|---|
| Variable defining the geometric and kinematic factors of the catchment | Total stream length | $L_t$ | Sum of stream length | $L$ | D |
| | Catchment area | $A$ | Planimetric measure of the area | $L^2$ | I |
| | Drainage density | $D_d$ | $L_t/A$ | $L^{-1}$ | D |
| | Mean bifurcation ratio | $R_b$ | Mean ratio of number of streams of order i to those of order (i + 1) | Dimensionless | D |
| | Transport efficiency | $T_e$ | $L_t \times R_b$ | $L$ | D |
| | Mean group slope | ⊠ | Weighted average of slope of the whole catchment area | Dimensionless | D |
| | Sediment area index | $S_a$ | $A/\cos\theta$ | $L^2$ | D |
| | Sediment movement index | $S_m$ | $S_a x \sin\theta$ | $L^2$ | D |
| | Rainfall erosivity index | $E_r$ | Max. 30-min rainfall intensity × its depth | $L^2/T^{-1}$ | D |
| | Discharge with T-year return period | $Q_T$ | Discharge volume per unit time | $L^3/T$ | D |
| | Mean channel slope | $\theta_0$ | Average slope of main channel | Dimensionless | D |
| | Elongation ratio | $R_e$ | Ratio of diameter of a circle with an area equal to that of the catchment divided by max. catchment length | Dimensionless | D |
| | Relief ratio | $R_r$ | Ratio of catchment relief to overall catchment length | Dimensionless | D |

## TABLE 7.8 (CONTINUED)
## Definition, Dimensional Formula, and Proportionality of Variables in Sediment Yield Model

| Variable Group | Variable Name | Symbol | Definition | Dimensional Formula | Proportionality to Sediment Yield D: direct I: indirect |
|---|---|---|---|---|---|
| Variable defining kinematic and field of force | Density of water | $\rho_w$ | Mass per unit volume | $ML^{-3}$ | D |
| | Viscosity of fluid | $\mu$ | Internal friction between fluid particles that resist forces tending to cause flow | $ML^{-1}T^{-1}$ | I |
| | Acceleration due to gravity | $g$ | Distance per unit time square | $LT^{-2}$ | D |
| Variable defining properties and resistance of the surface materials of the catchment | Rock weatherability index | $W_i$ | Ratio of final to initial dry weight of tested samples | Dimensionless | I |
| | Soil erodibility index | $e_i$ | Ratio of suspended silt and clay to total silt + clay | Dimensionless | D |
| | Vegetation cover index | $C_v$ | Ratio of relatively dense vegetation to catchment area | Dimensionless | I |
| | Land use index | $U_i$ | Ratio of areas affected by trampling and grazing to total catchment area | Dimensionless | D |
| | Land management index | $M_i$ | Ratio of terraced lands to total catchment area | Dimensionless | I |

include large numbers of the interacting factors noted earlier in the conventional regression analysis, in deterministic models, or in the basic universal soil loss equation (USLE) and its modified versions. Here, the general dimensional analysis, which has become of prime importance for forming physical laws in many scientific fields, is used for sediment yield modeling. This technique has surprising power and simplicity in determining physical laws, regardless of the number of input variables, and it permits quantitative sediment yield studies to be placed on a sound geometrical, mechanical, and environmental basis. Dimensional analysis has been the concern of many engineers and physicists throughout the ages. A brief statement of the development and concepts of this operational tool has been given by Staicu (1982). Among the most recent examples outside engineering practices and pure physics are applications in structural geology (Hubbert, 1937), geophysics (Ramberg, 1967), geomorphology (Strahler, 1958), and biology (Thompson, 1959). Numerous applications of its basic ideas are also evident in hydraulics, hydrology, geography, economics, business, sociology, and physiography (Backer, 1976).

Formulation of the sediment yield model based on dimensional analysis comprises the following four steps.

1. Expressing the formula of the model as a power product of nominal relationship between variables in the form

$$P^p L^1 M^m N^n = k A^a B^b C^c \qquad (7.2)$$

where $P$ is the dependent variable, $L$, $M$, $N$, ..., and $A$, $B$, $C$, ... are the independent variables, and $k$ is the factor of proportionality. The magnitude of each $L$, $M$, and $N$ is inversely proportional to the magnitude of $P$, whereas each $A$, $B$, and $C$ is directly proportional to $P$. The powers $p$, $1$, $m$, $n$, ..., $a$, $b$, $c$ are unknown exponents.

2. Describing the precise set of variables entering into the model in terms of $r$ fundamental dimensions, namely, length, mass, and time, and expressing them as dimensional matrix with $r$ ranks.

3. Expressing the condition of dimensional homogeneity in Equation 7.2 through a set of $r$ homogeneous linear equations formed from unknown exponents.

4. Solving these Diophantine equations by stages: First, the equations are considered in the increasing order of their number of terms; then positive, integral, minimum, or small values are sought for the unknown exponents. The unknown exponent values so determined provide the required solution for the model, except for the factor of proportionality.

Accordingly, the general functional relationship between sediment yield $S_y$ (m³/km²/year) and the selected variables (Table 7.8) can be expressed implicitly as

$$f(L_t, R_b, \Theta, T_c, S_e, S_m, D_d, E_r, Q_{100}, A, \mu, \rho_w, g, R_r, R_e, \Theta_c, e_i, U_1, C_v, W_i, M_1, S_y) = 0$$

However, from Table 7.8, it is observed that $S_y$ varies directly with the 16 variables, i.e., $L_1$, $D_d$, $R_1$, $S_c$, $S_m$, $E_r$, $Q_{100}$, $T_e$, $\Theta_1$, $\rho_1$, $g$, $R_r$, $R_e$, $U_1$, $\Theta_c$, and $e_i$, and inversely with the 5 variables $A$, $\mu$, $C_v$, $W_i$, and $M_1$. The 8 variables Re, $R_e$, $\Theta_c$, $e_i$, $U_1$, $C_v$, $W_i$, and $M_1$, are not taken into consideration at this step because they have no dimensional forms. Moreover, the variables $L_i$, $R_b$, and $\Theta$ are not considered because their effects are reflected implicitly through $D_d$, $T_e$, and $R_r$, respectively. Thus, the revised form of the relation involving the 11 variables is

$$f_1(S_y, A, \mu) = f_2 (D_d, S_a, S_m, E_r, Q_{100}, T_e, \rho_w, g) \tag{7.3}$$

The dimensional matrix of the 11 variables based on the three fundamental dimensions, namely, length (L), time (T), and mass (M), is

$$
\begin{array}{c c c c c c c c c c c c}
 & L_1 & A & \mu & D_d & A_a & S_m & E_r & Q_{100} & T_e & \rho_w & g \\
L & 1 & 2 & -1 & -1 & 2 & 2 & 2 & 3 & 1 & -3 & 1 \\
T & -1 & 0 & -1 & 0 & 0 & 0 & -1 & -1 & 0 & 0 & -2 \\
M & 0 & 0 & 1 & 0 & 0 & 0 & 0 & 0 & 0 & 1 & 0
\end{array}
\tag{7.4}
$$

The nominal relation with unknown exponents for these variables becomes

$$S_y^a A^b \mu^c = k D_d^e S_a^f S_m^h E_r^i Q_{100}^j T_e^k \rho_w^l g^m \tag{7.5}$$

The system of linear equations that assembles the unknown exponents and expresses the condition of dimensional heterogeneity is

$$
\begin{aligned}
c - 1 &= 0 & \text{(M)} \\
-a - c + i + j + 2m &= 0 & \text{(T)} \\
a + 2b - c + e - 2f - 2h - 2i - 3j - k + 31 - m &= 0 & \text{(L)}
\end{aligned}
\tag{7.6}
$$

Bearing in mind the rule of positive, integral, minimum, and small values that are applied for Diophantine systems of equations, the solution of these indeterminate systems leads to

$$c = 1, 1 = 1, a = 3, i = 1, j = 1, m = 1, b = 2, e = 2, f = 1, h = 1, k = 1 \tag{7.7}$$

Hence, the final relation as expressed by physical quantities and values for the exponents becomes

$$S_y^3 = k_1 \frac{D_d^2 S_a S_m T_e E_r Q_{100} \rho_w g}{A^2 \mu} \times \frac{R_r R_e U_1 \Theta_c e_i}{C_v W_i M_i}$$

or, more conveniently,

$$S_y = \left[ k_1 \frac{D_d^2 S_a S_m T_e E_r Q_{100} \rho_w g}{A^2 \mu} \times \frac{R_r R_e U_1 \Theta_c e_i}{C_v W_i M_i} \right]^{1/3} \tag{7.8}$$

**TABLE 7.9**

**Evaluation of Variables in Sediment Yield Model**

| Variable | Symbol | Estimated Value | Units |
|---|---|---|---|
| Catchment area | A | 1,430.00 | $km^2$ |
| Drainage density | $D_d$ | 0.48 | $km^{-1}$ |
| Transport efficiency | $T_e$ | 2,760.00 | km |
| Sediment area index | $S_a$ | 1,443.00 | $km^2$ |
| Sediment movement index | $S_m$ | 194.60 | $km^2$ |
| Rainfall erosivity | $E_r$ | 60.50 | mm.mm/h |
| Mean annual discharge with 100-year return period | $Q_{100}$ | 3,362.00 | $m^3/s$ |
| Density of water | $\rho_w$ | 997.00 | $kg/m^3$ |
| Viscosity of water (ab.) | $\mu_w$ | 0.0087 | gr/cm.s |
| Acceleration of gravity | g | 9.81 | $m/s^2$ |
| Mean channel slope | $\theta_c$ | 0.095 | — |
| Elongation ratio | $R_e$ | 0.66 | — |
| Relief ratio | $R_r$ | 0.039 | — |
| Weighted average rock Weatherability index | $W_i$ | 0.904 | — |
| Weighted average soil Erodibility index | $e_i$ | 0.709 | — |
| Vegetation cover index | $C_v$ | 0.091 | — |
| Land use index | $U_i$ | 0.26 | — |
| Land management index | $M_i$ | 0.30 | — |
| Sediment yield | $S_y$ | 706.30 | $m^3/km^2/year$ |

*Source:* From Al-Suba'i, K. A. M. G., 1991. Erosion-sedimentation and seismic considerations for dam siting in the central Tihamat Asir region. Unpublished Ph.D. Thesis, King Abdulaziz University, Faculty of Earth Sciences, Kingdom of Saudi Arabia, p. 343.

By fixing the numerical factor of proportionality, k, the model becomes completely determinate.

### 7.7.3 FACTOR OF PROPORTIONALITY ESTIMATION

The numerical factor of proportionality is the link between the derived model and the data found for the relevant variables by measurements carried out in laboratory conditions or in actual practice. For the application of Equation 7.8 it is necessary to express numerically the involved variables. Table 7.9 summarizes the determined numerical values of these variables for Wadi Jizan, which is located in the southwestern province of the Kingdom of Saudi Arabia (see Figure 4.33).

Substitution of the resulting values in the model gives a value of k = 0.0989141. Use of this value in the derived model in Equation 7.8 leads to the completely determinate form of the model as

$$S_y = 0.0989141 \left[ \frac{D_d^2 S_a S_m T_e E_r Q_{100} \rho_w g}{A^2 \mu} \times \frac{R_r R_e U_1 \Theta_c e_i}{C_v W_i M_i} \right]^{1/3} \tag{7.9}$$

This model provides a tool for estimating sediment yield rate for planning and design. Besides, it is calibrated with a probable mean annual discharge of return period of 100 years, which is a reasonable value (Saudiconsult, 1987).

The study of sediment yield has proven to be very complex and difficult because of the number of variables involved, the inconsistency or interrelation among those variables, and the lack of an appropriate method to quantify some of those variables. Measurements involved in evaluating the variables are straightforward, and the model uses variables representing vegetation cover, land use, and land management.

## 7.8  SIMPLE SEDIMENT YIELD MODEL FORMULATION

The simple formulations of the sediment yield model are based on dimensional analysis as explained in the previous sections. In light of the aforementioned principles of dimensional analysis and selected sediment variables, two simple models are proposed in the following sequel.

### MODEL 1

Consideration of four variables as the sediment yield, $S_y$, wadi area, $A$, discharge, $Q$, and slope, $S$, together with the fundamental expression in Equation 7.2 gives

$$S_y^a A^b = k_1 Q^c S \tag{7.10}$$

where $k_1$ is the factor of proportionality and $a$, $b$, and I are the unknown exponents. There is no exponent for the slope variable since by definition it is dimensionless. Equation 7.10 corresponds to a multiple-linear regression because, if logarithms of both sides are taken, then one can obtain

$$a \log(S_y) + b \log(A) = \log(k_1) + c \log(Q) + \log(S) \tag{7.11}$$

Many sediment yield models have logarithmic relationships, as has been already observed by various researchers. In Equation 7.11, the constants correspond to the multiple regression coefficients, and they represent the degree of correlation between the dependent variable, $S_y$, and each one of the corresponding independent variables.

There are two different approaches in the determination of model parameters, depending on the data availability. If a set of data are available for each variable (dependent and independent), then formal multiple regression procedure helps to estimate the model parameters, under a set of assumptions such as the linearity, normality, stationary, etc. On the other hand, if data is not available then logic and the dimensional analysis help to identify the model exponents of Equation 7.10, except for the factor of proportionality. In order to guarantee dimensional equality, it is possible to rewrite Equation 7.10 in its basic dimensional form as,

$$\left(\frac{L}{T}\right)^a L^{2b} = \left(\frac{L^3}{T}\right)^c$$

The dimensional matrix of the three variables based on the two fundamental dimensions, namely, length [L] and time [T], can be written as,

|   | $S_y$ | $A$ | $Q$ |
|---|---|---|---|
| $L$ | 1 | 2 | 3 |
| $T$ | -1 | 0 | -1 |

The system of linear equations, which assembles the unknown exponents and expresses the condition of dimensional heterogeneity, is

$$a + 2b = 3c$$

and

$$a = c$$

The basic rule of dimensional analysis is the selection of exponents such that positive, integral, minimum and small values should be taken and hence, the solution becomes, $a = 1; b = 1; c = 1$. With these exponent values Equation 7.10 takes the following form:

$$S_y = k_1 \frac{Q}{A} S \tag{7.12}$$

Provided that all the variables are measured for any drainage basin (wadi, in arid lands) the constant $k_1$ can be determined from Equation 7.12.

## MODEL 2

Let us consider the model with drainage density, $D_d$, wadi area, $A$, discharge, $Q$, and slope, $S$, as the independent variables in the sediment yield formulation. According to Equation 7.2 it will have the following formulation.

$$S_y^a A^b = k_2 D_d^c Q^d S \tag{7.13}$$

In this case the dimensional matrix for the variables is,

|   | $S_y$ | $A$ | $D_d$ | $Q$ |
|---|---|---|---|---|
| $L$ | 1 | 2 | $1^{-1}$ | 3 |
| $T$ | $-1$ | 0 | 0 | $-1$ |

Under the light of this matrix, the dimensional expression and the relationship between its exponents yield

$$a + 2b = -c + 3d$$

and

$$a = d$$

The substitution of the last expression into the previous one gives $2b = -c + 2d$. Hence, integer, positive and minimum exponents can be found as $a = 2$, $b = 1$, $c = 2$, and $d = 2$. Finally, the substitution of these exponent values into Equation 7.13 gives

$$S_y^2 A = k_2 D_d^2 Q^2 S$$

or the sediment yield dependent variable as a subject becomes

$$S_y = k_2 D_d Q \sqrt{\frac{S}{A}} \tag{7.14}$$

## 7.8.1 Application of the Models

The study region is characterized by a semi-arid climate with high intensity rainfall and flash floods leading to a large amount of eroded sediments. Some of the wadis in the central part of the region (see Figure 4.33) have runoff amounts in excess of 100 MCM/year. Wadi floods with a discharge of 10,000 cusec are of common occurrences in the region. These floods usually wash away temporary small earth embankments and sometimes uproot plants and seeds, carry away valuable alluvial soil, and in many cases lead to loss of livestock and destruction of houses, roads, and other utilities (Al-Suba'i, 1991).

The relevant data for the determination of model constants are taken from Table 7.9 and represented in Table 7.10.

First of all the proportionality factors, $k_1$ and $k_2$, are obtained by substitutions of the relevant quantities into Equations 7.12 and 7.14, which yield $k_1 = 10^{-4}$ and $k_2 = 1.74 \times 10^{-6}$, respectively. Substitution of these constants into respective equations gives the final practical forms of the models as

$$S_y = 0.0001 \frac{Q}{A} S \tag{7.15}$$

## TABLE 7.10
### Sediment Yield Model Variables for Wadi Baysh

| Variable | Symbol | Estimated Value | Units |
|---|---|---|---|
| Catchment area | A | 1,430.00 | km² |
| Drainage density | $D_d$ | 0.48 | km⁻¹ |
| Mean annual discharge with 100-year return period | $Q_{100}$ | 3,362.00 | m³/sec |
| Mean channel slope | S | 0.095 | — |
| Sediment yield | $S_y$ | 706.30 | m³/km²/year |
| | | $2.239 \times 10^{-11}$ | m³/m²/sec |

## TABLE 7.11
### Sediment Yield Calculation Results for Wadi Baysh (Model 1)

| Sub-Basin No. | Area (km²) | S | Q(m³/s) | $S_y$ (m³/m²/sec) | $S_y$ (m³/sec) | $S_y$ (m³/year) |
|---|---|---|---|---|---|---|
| 1 | 146.5 | 0.048 | 104.011 | 3.44E-12 | 0.0005038 | 15886.75 |
| 7 | 135.3 | 0.059 | 81.142 | 3.54E-12 | 0.0004794 | 15119.57 |
| 11 | 118.7 | 0.041 | 72.671 | 2.52E-12 | 0.0002988 | 9423.82 |
| 19 | 144.5 | 0.032 | 91.54 | 2.06E-12 | 0.0002976 | 9385.12 |
| 20 | 128.2 | 0.042 | 74.847 | 2.48E-12 | 0.0003176 | 10014.98 |
| 21 | 99.57 | 0.025 | 46.881 | 1.22E-12 | 0.0001211 | 3819.06 |
| 42 | 535.9 | 0.002 | 198.319 | 8.37E-14 | 4.49E-05 | 1415.09 |
| 44 | 112.4 | 0.022 | 63.0147 | 1.29E-12 | 0.0001446 | 4560.23 |
| 47 | 53.73 | 0.028 | 37.057 | 1.97E-12 | 0.0001061 | 3345.58 |
| 53 | 124 | 0.087 | 91.522 | 6.45E-12 | 0.0008 | 25227.87 |

and

$$S_y = 0.00000174 D_d Q \sqrt{\frac{S}{A}} \tag{7.16}$$

The application of these formulations is presented for Wadi Baysh, which is already mentioned in Chapter 4. The necessary basic data for 54 sub-basins in this wadi are already given in Table 4.9 and their substitution into Equation 7.15 yields the results in Table 7.11. Although the calculations are available for all the sub-basins, herein, only 10 of them are presented.

On the other hand, substitution of the same basic data into Equation 7.16 yields results in Table 7.12.

It is noticed that the sediment yield values with the most involved formulation in Equation 7.16 including all the factors together (A, S, $D_d$, and Q) yields smaller results.

## TABLE 7.12
## Sediment Yield Calculation Results for Wadi Baysh (Model 2)

| No. | Area (km²) | S | Q (m³/sec) | D ( km⁻¹) | $S_y$ (m³/m²/sec) | $S_y$ (m³/sec) | $S_y$ (m³/year) |
|---|---|---|---|---|---|---|---|
| 1 | 146.5 | 0.048 | 104.014 | 0.25 | 8.23E-13 | 0.0001205 | 3800.8 |
| 7 | 135.3 | 0.059 | 81.142 | 0.33 | 9.74E-13 | 0.0001317 | 4154.41 |
| 11 | 118.7 | 0.041 | 72.671 | 0.43 | 1.01E-12 | 0.0001201 | 3788.28 |
| 19 | 144.5 | 0.032 | 91.54 | 0.54 | 1.29E-12 | 0.0001864 | 5879.06 |
| 20 | 128.2 | 0.042 | 74.847 | 0.4 | 9.48E-13 | 0.0001215 | 3831.51 |
| 21 | 99.57 | 0.025 | 46.881 | 0.57 | 7.49E-13 | 7.46E-05 | 2351.62 |
| 42 | 535.9 | 0.002 | 198.319 | 0.1 | 7.09E-14 | 3.80E-05 | 1198.31 |
| 44 | 112.4 | 0.022 | 63.014 | 0.45 | 7.05E-13 | 7.92E-05 | 2498.98 |
| 47 | 53.73 | 0.028 | 37.057 | 0.56 | 8.33E-13 | 4.48E-05 | 1412.28 |
| 53 | 124 | 0.087 | 91.522 | 0.38 | 1.61E-12 | 0.0001992 | 6282.77 |

## REFERENCES

Al-Khafif, S. M., 1986. Sedimentation control of wadi Jizan reservoir. UTEN/SAU/013/SAU field document, No. 22 FAO, Rome, p. 58.

Al-Suba'i, K. A. M. G., 1991. Erosion-sedimentation and seismic considerations for dam siting in the central Tihamat Asir region. Unpublished Ph.D. Thesis, King Abdulaziz University, Faculty of Earth Sciences, Kingdom of Saudi Arabia, p. 343.

Anderson, H. W., 1951. Physical characteristics of soils related to erosion. *J. Soil Water Conserv.* Vol. 6, 129–133.

Andre, J. E. and Anderson, H. W., 1961. Variation of soil erodibility with geology, geographic zone, elevation and vegetation type in northern California wild lands. *J. Geophys. Res.*, Vol. 66, 3351–3358.

Backer, H. A., 1976. *Dimensionless Parameters: Theory and Methodology.* Applied Sciences, London, p. 128.

Bogardi, J. L., 1974. *Sediment Transport in Alluvial Streams.* Akademiai Kiado, Budapest, p. 82.

Brown, E. T., 1981. *Rock Characterization: Testing and Monitoring*, ISRM Suggested Methods, Pergamon Press, Oxford, 2110.

Bull, L. J. and Kirby, M. J., 2000. *Dryland Rivers: Hydrology and Geomorphology of Semiarid Channels.*

Campbell, I. A., 1977. Stream discharge, suspended sediment and erosion rates in the Red Deer river basin, Alberta, Canada. *Proc. Erosion and Solid Matter Transport in Inland Waters Symp.* Paris, July 1977, IAHS-AISH Publ. No. 122.

Cooke, R. V. and Doornkamp, J. C., 1974. *Geomorphology in Environmental Management: An Introduction.* Clarendon Press, Oxford, p. 413.

Douglass, I., 1967. Man, vegetation and sediment yield of rivers. *Nature*, Vol. 215, 925–928.

Einstein, H. A., 1968. Deposition and suspended particles in a gravel bed. ASCE, *J. Hydraul. Div.*, Vol. 94, 1197–1205.

Foster, G. R. and Meyer, L. D., 1985. Mathematical simulation of upland erosion by fundamental erosion mechanics in present and perspective technology for predicting sediment yields and sources. *USDA Agr. Res. Ser. Publ.*, ARS-S-40, 190–207.

Graf, W. H., 1971. *Hydraulics of Sediment Transport.* McGraw-Hill, New York, p. 544.

Herbertson, J. G., 1969. A critical review of conventional bed load formulae. *J. Hyd.*, Vol. 8, 1–26.

Hadley, R. F., 1986. Fluvial transport of sediment in arid and semi-arid regions. *Proc. Int. Symp. on Erosion and Sedimentation in Arab Countries. Iraqi J. Water Res.*, Vol. 5, 335–348.

Hubbert, M., 1937. Theory of scale models as applied to the study of geological structures. *Geol. Soc. Am. Bull.*, Vol. 48, 1459–1520.

Horton, R. E., 1945. Erosional development of stream and their drainage basins: Hydrophysical applications of quantitative morphology. *Bull. Geol. Soc. Am.*, Vol. 56, 275–370.

Hudson, N. W., 1981. *Soil Conservation.* Bratsford Ltd., London.

Jansson, M. B., 1982. Land erosion by water in different climates. Uppsala Univ., Dept. Phys. Geography, INGI Rapport, 57, p. 151.

Kirby, M. J., 1969. Infiltration, throughflow and overland flow, and erosion by water on hillslopes. In Chorley, R. J. (Ed.). *Water, Earth and Man.* Methuen, London, 215–238.

Lal, R., 1976. Soil erosion on alfisols in western Nigeria, III effects of rainfall characteristics. *Geoderma*, Vol. 16(5), 389–401.

Lal, R., 1985. Soil erosion and sediment transport research in tropical Africa. *Hydrol. Sci. J.*, Vol. 30, 239–256.

Langbein, W. B. and Schumm, S. A., 1958. Yield of sediment in relation to mean annual precipitation. *Trans. Am. Geophy. Uni.*, Vol. 39, 1076–1084.

Maidment, D. R., 1993. *Handbook of Hydrology.* McGraw-Hill Book, 12.1–12.61.

Meyer, L. D. and Monke E. J., 1965. Mechanics of soil erosion by rainfall and overland flow. *Trans. Am. Soc. Agr. Eng.*, Vol. 8, 572–577.

Middleton, H. E., 1930. Properties of soils which influence soil erosion. *U.S. Dep. Agr. Tech. Bull.*, 178, 1–16.

Morgan, R. P. C., 1986. *Soil Erosion and Conservation.* Longman, Essex, p. 298.

Morris, H. M. and Wiggert, J. M., 1972. *Applied Hydraulics in Engineering.* 2nd. ed. Ronald Press Comp., New York, p. 629.

Neff, K. L., 1967. Discharge compared to long-term sediment yield. *Proc. Cern Symp.*, IAHS-AISH Publ., No. 75.

Onstad, C. A., 1986. Current techniques for modeling and predicting erosion and sediment yield. *Proc. Int. Symp. on Ero. and Sed. in Arab Countries, Iraqi J. Water Res.* Vol. 5, 530–550.

Osborn, H. B. and Lane, L., 1969. Precipitation-runoff relations for very small semi-arid range-land watersheds. *Wat. Resour. Res.*, Vol. 5, 419–425.

Pearce, A. J., 1986. Geomorphic effectiveness of erosion and sedimentation events. *Proc. Int. Symp. Ero. Sed. in Arab Countries, Iraqi J. Water Res.*, Vol. 5, 551–569.

Roose, E., 1977. Importance relative l'erosion du drainage oblique et vertical dans la pedogenese actuelle d'un sol ferrallitique de moyenne C'ote d'Ivoire Cah. OSTROM, Ser. Pedol 8, 469–482.

Ramberg, H., 1967. *Gravity, Deformation and Earth's Crust.* Academic Press.

Saudiconsult, 1987. Sedimentation control project for Jizan dam reservoir. Draft feasibility report. Saudi Arabian Minist. Agric. Wat., Unpublished Report.

Scott, K. M. and Williams, R. P., 1978. Erosion and sediment yields in the transverse ranges. Southern California. USGS Prof. Paper 1030, p. 38.

Shen, H. W., (Ed.). 1971. *River mechanics*, Vol. 1 and 2, Colorado State University, Fort Collins, Colorado.

Schick, A. P. and Sharon, D., 1974. Geomorphology and climatology of arid watersheds. *Mimeogr. Rep. Dept. Geogr. Hebrew Univ.*, Jerusalem.

Simons, D. B. and Şentürk, F., 1977. *Sediment Transport Technology.* Water Resources Publication, Colorado State University, Fort Collins, Colorado, p. 807.

Staicu, G. I., 1982. *Restricted and General Dimensional Analysis: Treatment of Experimental Data*. Abacus Press, Kent, p. 303.

Strahler, A. N., 1958. Dimensional analysis applied to fluvially eroded landforms. *Bull. Geol. Soc. Am.*, Vol. 69, 279–300.

Stoppato, M. C. and Bini, A., 2003. *Deserts*. A Firefly Book.

Şen, Z. and Al-Suba'i, K., 2002. Hydrologic considerations for dam siting in arid regions: A Saudi Arabian study. *Hydrol. Sci. J.*, 47 (1), 1–19.

Thompson, D. W., 1959. *On Growth of Form*. Vol. I and II, Cambridge University Press.

Throne, C. R., Bathurst, J. C., and Hey, H. D., 1987. *Sediment transport in gravel-bed rivers*. John Wiley & Sons, p. 995.

Vanoni, V. A., (Ed.), 1975. *Sediment Engineering*, ASCE, New York, p. 745.

Walling, D. E., 1986. Sediment yields and sediment delivery dynamics in Arab Countries. *Proc. Int. Symp. on Erosion and Sedimentation in Arab Countries. Iraqi J. Water Res.*, Vol. 5, 775–799.

Walling, D. E. and Webb, B. W., 1986. Solute transport by rivers in arid environments: An overview. *Proc. Int. Symp. on Erosion and Sedimentation in Arab Countries. Iraqi J. Water Res.*, Vol. 5, 800–822.

Walling, D. E., 1977. Limitations of the rating curve technique for estimating suspended sediment loads with particular reference to British rivers. *Proc. Paris Symp.*

Willis, J. C., 1971. Erosion by concentrated flow. ARS41–179.

Wischmeier, W. H. and Smith D. B., 1978. Predicting rainfall erosion losses. *U.S. Department of Agriculture Handbook No. 537*, p. 58.

Yalin, M. S., 1972. *Mechanics of Sediment Transport*. Pergamon Press, Oxford, p. 290.

# Index

## A

Abstractable groundwater volume, 272, 280, 281, 283
Actual retention, 111
Agricultural drought, 6, 15, 91
Alluvial
   fan, 21, 22, 167, 180, 181, 343
   fill, 10, 20, 23–24, 216
   plain, 20, 21, 29
   quaternary (*See* Quaternary alluvium)
Annual rainfall volume, 146, 147, 279
Annual runoff volume, 133, 147
Aquifer(s), 175
   alluvial, 169, 185
   bounded, 202
   climate change and, 9, 202
   coastal, 246
   confined, 201, 202, 218, 220, 221, 240, 247, 253, 267, 301
   defined, 243
   discontinuous, 185
   fractured, 205, 233, 235
   generalized, 185
   gneissic, 184
   groundwater replenishment to, 169, 232
   heterogeneous, 239, 240
   homogenous and isotropic, 215, 218
   karstic, 233
   layered, 202
   leaky, 205, 206, 207, 218, 219, 220, 225, 237, 239, 240
   management, 249, 253–265
   monitoring, 234
   multiple systems, 238, 239, 240
   overexploitation, 244
   parameter determination, 175
   parameter estimates, 216
   perennial yield, 244
   probabilistic risk management, 253–265
   safe yield, 244
   in sedimentary basin, 247
   semiconfined, 247
   shallow, 233
   storage capacity, 269
   test, 13, 203–207, 210, 212, 213, 216, 234, 238, 251
   transmissivity, 202, 219, 225, 275
   unconfined, 201, 202, 233, 245, 247, 248, 253, 260, 298
   volcanic rocks and, 27
   vulnerability to contamination, 233
   water acceptance capacity, 181

Aquitards, 247
Arabian Peninsula, 2, 4, 11, 24, 26, 28, 46, 50, 62, 129, 135, 145, 147, 148, 183, 193, 195, 343
Arid zone, 14, 24, 181
   climatic constraints, 21
   geographic distribution, 2
   geomorphology, 17
   hydrological measurements in, 77
   hydrological processes in, 12, 36, 236
   modeling, 1
   operation of integrated WGM in, 286
   primary damage channels of, 29
   rainfall, 40, 42, 196
   recharge estimation in, 239
   soil types, 21
Aridification, 313
Aridity, 27
   defined, 7
   geographical distribution, 3
   humidity and, 2
   index, 42–43, 44
   North African, 303
   recharge and, 108
   semi, 42, 62
Artificial mixture, 249, 291, 292, 296, 298, 299
Average slope, 31, 34, 326
Average subsurface flow, 83

## B

Basalt, 165, 182, 184, 250, 304
   flow, 185
Base flow, 104, 105, 164
Basin(s)
   drainage, 12, 17, 21, 105, 134, 169
      average slope of, 34
      centroid point of, 126
      channel network of, 121
      closed, 303
      composition of, 253
      deltas at downstream ends of, 17
      depression over the area of, 180
      determining AAR over, 75, 77
      divisions of, 72, 75, 252
      for engineering structures, 108
      features of, 17, 28–33, 126
      geomorphologic features of, 126
      in humid environments, 11
      low-lying, 27
      maximum initial retention characteristics of, 112